Communications
in Computer and Information Science   2515

Series Editors

Gang Li, *School of Information Technology, Deakin University, Burwood, VIC, Australia*

Joaquim Filipe, *Polytechnic Institute of Setúbal, Setúbal, Portugal*

Zhiwei Xu, *Chinese Academy of Sciences, Beijing, China*

Rationale

The CCIS series is devoted to the publication of proceedings of computer science conferences. Its aim is to efficiently disseminate original research results in informatics in printed and electronic form. While the focus is on publication of peer-reviewed full papers presenting mature work, inclusion of reviewed short papers reporting on work in progress is welcome, too. Besides globally relevant meetings with internationally representative program committees guaranteeing a strict peer-reviewing and paper selection process, conferences run by societies or of high regional or national relevance are also considered for publication.

Topics

The topical scope of CCIS spans the entire spectrum of informatics ranging from foundational topics in the theory of computing to information and communications science and technology and a broad variety of interdisciplinary application fields.

Information for Volume Editors and Authors

Publication in CCIS is free of charge. No royalties are paid, however, we offer registered conference participants temporary free access to the online version of the conference proceedings on SpringerLink (http://link.springer.com) by means of an http referrer from the conference website and/or a number of complimentary printed copies, as specified in the official acceptance email of the event.

CCIS proceedings can be published in time for distribution at conferences or as post-proceedings, and delivered in the form of printed books and/or electronically as USBs and/or e-content licenses for accessing proceedings at SpringerLink. Furthermore, CCIS proceedings are included in the CCIS electronic book series hosted in the SpringerLink digital library at http://link.springer.com/bookseries/7899. Conferences publishing in CCIS are allowed to use Online Conference Service (OCS) for managing the whole proceedings lifecycle (from submission and reviewing to preparing for publication) free of charge.

Publication process

The language of publication is exclusively English. Authors publishing in CCIS have to sign the Springer CCIS copyright transfer form, however, they are free to use their material published in CCIS for substantially changed, more elaborate subsequent publications elsewhere. For the preparation of the camera-ready papers/files, authors have to strictly adhere to the Springer CCIS Authors' Instructions and are strongly encouraged to use the CCIS LaTeX style files or templates.

Abstracting/Indexing

CCIS is abstracted/indexed in DBLP, Google Scholar, EI-Compendex, Mathematical Reviews, SCImago, Scopus. CCIS volumes are also submitted for the inclusion in ISI Proceedings.

How to start

To start the evaluation of your proposal for inclusion in the CCIS series, please send an e-mail to ccis@springer.com

Bin Xu · Jianlong Qiu
Editors

Cognitive Computation and Systems

Third International Conference, ICCCS 2024
Linyi, China, December 20–22, 2024
Revised Selected Papers, Part I

Springer

Editors
Bin Xu
Northwestern Polytechnical University
Xi'an, China

Jianlong Qiu
Linyi University
Linyi, China

ISSN 1865-0929 ISSN 1865-0937 (electronic)
Communications in Computer and Information Science
ISBN 978-981-96-7351-3 ISBN 978-981-96-7352-0 (eBook)
https://doi.org/10.1007/978-981-96-7352-0

© The Editor(s) (if applicable) and The Author(s), under exclusive license
to Springer Nature Singapore Pte Ltd. 2025

This work is subject to copyright. All rights are solely and exclusively licensed by the Publisher, whether the whole or part of the material is concerned, specifically the rights of translation, reprinting, reuse of illustrations, recitation, broadcasting, reproduction on microfilms or in any other physical way, and transmission or information storage and retrieval, electronic adaptation, computer software, or by similar or dissimilar methodology now known or hereafter developed.
The use of general descriptive names, registered names, trademarks, service marks, etc. in this publication does not imply, even in the absence of a specific statement, that such names are exempt from the relevant protective laws and regulations and therefore free for general use.
The publisher, the authors and the editors are safe to assume that the advice and information in this book are believed to be true and accurate at the date of publication. Neither the publisher nor the authors or the editors give a warranty, expressed or implied, with respect to the material contained herein or for any errors or omissions that may have been made. The publisher remains neutral with regard to jurisdictional claims in published maps and institutional affiliations.

This Springer imprint is published by the registered company Springer Nature Singapore Pte Ltd.
The registered company address is: 152 Beach Road, #21-01/04 Gateway East, Singapore 189721, Singapore

If disposing of this product, please recycle the paper.

Preface

This volume contains the papers from the Third International Conference on Cognitive Computation and Systems (ICCCS 2024), which was held in Linyi, Shandong, China, on December 20–22, 2024. ICCCS is an international conference on the related fields of cognitive computing and systems which was initiated by the Technical Committee on Cognitive Computing and Systems of the Chinese Association of Automation in 2022. ICCCS 2024 was hosted by the Technical Committee on Cognitive Computing and Systems of the Chinese Association of Automation, and co-organized by Linyi University, Henan Polytechnic University, the Intelligent Robotics Research Center of the Artificial Intelligence Institute of Tsinghua University, and the School of Automation of Northwestern Polytechnical University.

Cognitive computing is an interdisciplinary field of research that focuses on developing intelligent systems that can process and analyze large amounts of complex data in a way that mimics human cognition. It is based on the idea that intelligent systems can be designed to learn, reason, and interact with humans in a more natural way, and that these abilities can be used to solve complex problems that would otherwise be difficult or impossible for humans to solve alone. By enabling agents to understand and process data in a more human-like way, cognitive computing can help organizations make better decisions, improve customer service, develop more effective treatments for diseases, and even predict and prevent natural disasters.

ICCCS aims to bring together experts from different areas of expertise to discuss the state of the art in cognitive computing and intelligent systems, and to present new research results and perspectives on future development. It is an opportunity to promote the research and development of cognitive computing and systems, and provides a face-to face communication platform for scholars, engineers, teachers, and students engaged in cognitive computing and systems or related fields to promote the application of artificial intelligence technology in industry.

ICCCS 2024 received 155 submissions, all of which were written in English. After a thorough single-blind peer review process, in which each paper received at least 4 reviews, 54 papers were selected for presentation as full papers, resulting in an approximate acceptance rate of 34%. The accepted papers addressed challenging issues in various aspects of cognitive computation and systems, including not only theories and algorithms in cognition, machine learning, computer vision, decision making, etc., but also systems and applications in autonomous vehicles, computer games, intelligent robots, etc.

We would like to thank the members of the Advisory Committee for their guidance, and the members of the Program Committee and additional reviewers for reviewing the papers. We would also like to thank all the speakers and authors, as well as the participants for their great contributions that made ICCCS 2024 successful. Finally, we thank Springer for their trust and for publishing the proceedings of ICCCS 2024.

December 2024

Bin Xu
Jianlong Qiu

Organization

Conference Committee

Honorary Chairs

Bo Zhang	Tsinghua University, China
Nanning Zheng	Xi'an Jiaotong University, China
Deyi Li	CAAI, China

Advisory Committee Chairs

Qionghai Dai	Tsinghua University, China
Lin Chen	Institute of Biophysics, CAS, China
Fuchun Sun	Tsinghua University, China
Zengguang Hou	Institute of Automation, CAS, China
Shusheng Zhang	Linyi University, China

General Chairs

Bin Xu	Northwestern Polytechnical University, China
Jianlong Qiu	Linyi University, China

Organizing Committee Chairs

Ancai Zhang	Linyi University, China
Jiuru Wang	Linyi University, China
Huaping Liu	Tsinghua University, China
Zhongyi Chu	Beihang University, China
Quanyong Fan	Northwestern Polytechnical University, China

Program Committee Chairs

Dewen Hu	National University of Defense Technology, China
Cong Wang	Shandong University, China
Deqian Fu	Linyi University, China
Xiangyong Chen	Linyi University, China
Bo Zhao	Beijing Normal University, China

Program Committee

Janmin Li	Tsinghua University, China
Xiaolin Hu	Tsinghua University, China
Wei Qian	Henan Polytechnic University, China
Xiaohui Liang	Northwestern Polytechnical University, China
Weixin Han	Northwestern Polytechnical University, China

Publications Chairs

Quanbo Ge	Nanjing University of Information Science & Technology, China
Janmin Li	Tsinghua University, China
Rui Zhang	Northwestern Polytechnical University, China

Finance Chairs

Feng Zhao	Linyi University, China
Ming Guo	Linyi University, China

Registration Chairs

Xing Wang Linyi University, China
Kun Zhou Linyi University, China

Local Arrangements Chair

Guochen Pang Linyi University, China

Contents – Part I

Cognitive Computing And Information Processing

Iris Recognition Based on Asynchronous Genetic Particle Swarm Optimization 3
   Du Yining and Li Aiju

Joint Torque Sensor-Based Distributed Coordinated Control of Dual-Arm Reconfigurable Manipulators 15
   Bing Ma, Hangwei Zhang, Ximing Yao, Yuan Yao, Tianjiao An, and Qiang Pan

MAMF-ResNet: Multi-Attention Mechanisms Fusion Network for Brain Tumor Classification 27
   Qingyun Huo, Meng Yang, Yuxuan Li, Yunyu Wang, Xin Zhang, and Mingtao Liu

Noise Sensitive Relation Aware Cross-Lingual Entity Alignment 39
   Yunlong Tian, Shuanglong Yao, Liqiu Shan, and Xing Wang

Origin Classification of Rice Wine Based on an Electronic Nose and Convolutional Neural Network 54
   Wenqi Sun, Ancai Zhang, Guangyuan Pan, and Wenbo Zheng

Parkinglot Obstacle Detection System Using Infrastrue-Based Cameras 63
   Yuesheng He, Fei Wang, Zihan Zong, and Ming Yang

Self-tuning Control of Manipulator Based on Fuzzy Linear Active Disturbance Rejection Control and Particle Swarm Optimization Algorithm 75
   Bo Tao, Zhuxiang Chen, Du Jiang, and Juntong Yun

The Neural Network of College Students Research on the Intelligent Scoring System 88
   Yang Ji and Huang Shen

Uniform Light Field Control System Based on IoT Technology - Solving Eye Discomfort Caused by Changes in Light Intensity 96
   Haobin Yang, Haiying Zhang, and Xunyu Qiao

Intelligent Cooperative Control

Attention-Fused Vision Mamba Model for Remote Sensing Image
Detection and Segmentation 117
    Zhao Yusen, Wu Shan, and Tian Liang

Control Allocation Based Fault-Tolerant Attitude Control of Hypersonic
Reentry Vehicle 131
    Yuan Zhang and Weixin Han

Cooperative Multi-target Enclosing Control of Multi-robots Based
on Virtual Target Points 144
    Fang Fang Zhang, Zhen Yu Zhu, and Ji Xian Gao

Event-Triggered Secure Consensus Control of Nonlinear Multiagent
Systems with hybrid DoS attacks 152
    Yingxin Zhang, Huixia Cui, and Senping Jia

FastLSLO: An Efficient LiDAR Odometry Based on Improved Lie Group
B-Splines 165
    *Xinyang Tang, Wei Yuan, Chenxi Yang, Chunxiang Wang, Bing Wang,
    and Ming Yang*

Improved BP Neural Network Based Deck Motion Prediction and Landing
Control for Carrier-Based Aircrafts 178
    Zhaoxing Li and Xingzhao Zhang

Joint Sequencing and Merging Optimization for Airplanes and Helicopters 186
    Dawei Wang, Yi Lyu, Ken Chen, Yiman Zhang, and Chengcheng Wu

Research on UAV SINS/GNSS Integrated Navigation Error Model
for Transpolar Flight 198
    Guoqiang Zhang, Qi Zhou, and Jinjiang Wang

Technical Research on Helicopter Blind Landing Under Degraded Visual
Environments (DVE) 214
    Yanwei Du, Mingdong Qi, and Heng Zhang

Learning and Systems

A Contactless Demonstration Learning Method for Robotic Systems
Based on Dynamic Parametric Regression Modeling 231
    *Meng Li, Jinzhu Peng, Jixian Gao, Nan Zhao, Yaonan Wang,
    and Mingkuo Wu*

A Multi-scale Fusion and Dynamic Upsampling Model for Road Surface
Snow Detection 243
   Lipeng Du, Xinhao Zhou, Qili Chen, Tong Wang, Lin Zhao,
   and Guangyuan Pan

A Scoring System for Single and Parallel Bars Actions Based
on Multi-view 3D Pose Estimation 254
   Yuntong Kang, Mingwei Cao, Haoran Yao, Sen Qiu, and Zhelong Wang

Adaptive PID Controller for Industrial Process Based on Reinforcement
Learning 265
   Yicong Yang and Qili Chen

Computational Analysis of Synaptic Plasticity in Echo State Network 272
   Xinyu Shen, Shaoqi Cheng, Fanjun Li, and Jiayue Feng

LEHR: LLM-Driven Evolutionary Hybrid Rewards for Multi-agent
Reinforcement Learning 283
   Yuan Wei, Xiaohan Shan, and Jianmin Li

Magnetic Core Loss Prediction: A Data-Driven NGO-GRU Model 296
   Ningning Hu, Yongqiang Mao, Lanmei Cong, Zhaohui Zhang,
   and Ziyue Han

Research on Method of Control Surface Jamming Fault Injection in Fly
Test of Fly-by-Wire Aircraft Based on Multiple Control 308
   Lei Ming, Xie Qingping, and Li Yajing

Tactile-Based Manipulation for Wire Following 319
   Jiazhen Cai, Jing Cui, and Zhongyi Chu

Author Index 333

Contents – Part II

Cognitive Computing And Information Processing

A Colour Secret Image Sharing Scheme Based on the Theory of Prime Number Distribution 3
   Jinqiu Xue and Wenyin Zhang

A Novel Design of Radial Basis Function Neural Network Integrating Markov Chain Monte Carlo Clustering Algorithm 20
   Yunlong Zhu, Zunwei Fu, and Zheng Wang

A Synthesis of Techniques for Feature Downgrading Processing in IoT Security 27
   Yifang Wang

Anomaly Data Identification and Reconstruction in Wind Power Forecasting Based on MLOF-iForest and RF 37
   Pengcheng Du and Yu Du

Broad Learning System-Enhanced Fuzzy Inference Framework for High-Efficiency Classification 47
   Yizhen Wang, Zunwei Fu, and Congcong Zhang

Design of Radial Basis Function Neural Network with Neural Gas Network Clustering for Classification of Logistics Black Plastics 55
   Hongliang Yu, Kun Zhou, Ming Guo, Xiangyong Chen, and Jianlong Qiu

Enhanced Group Method of Data Handling with Roulette Neuron Selection for Time Series Forecasting 66
   Xixin Wang, Shaoguang Shi, and Congcong Zhang

FE-ResNet50: Frequency Enhanced Attention Network for sEMG Gesture Recognition 74
   Haozhu Wang, Du Jiang, Juntong Yun, Ying Liu, and Boao Li

Huber Correntropy Kalman Filter 89
   Shuo Wang and Xiaoliang Feng

Intelligent Cooperative Control

A Framework for Visual Target Navigation for Quadcopter Based on Large Language Models in Unknown Environment 99
   Yunzhuo Liu, Zhaowei Ma, Yuqi Yang, Mengyun Wang, and Yifeng Niu

Adaptive Output Feedback Control for Nonlinear Time-Delay Systems with Neural Network and Static Gains 111
   Wenjie Li and Zhengqiang Zhang

Automatic Ground Collision Avoidance for Aircraft in Uncertain Environments 120
   Rui Li and Rui Zhang

Investigations into the Existing State of Distribution Network Security Using Dispersed Generators 134
   Jiage Zhang, Chen Lei, and Liping He

Parkinglot Obstacle Detection System Using Infrastructure-Based Cameras 149
   Yuesheng He, Fei Wang, Zihan Zong, and Ming Yang

Predefined-Time Consensus for First-Order Heterogeneous Nonlinear Multi-agent Systems 161
   Shufen Liu, Jianhua Liu, Feng Zhao, and Xiangyong Chen

Research on Automatic Throttle Speed Stability System Based on Magic Carpet Landing Technology 170
   Letian Zhao, Hang Chen, Yangyang Zhou, Xiaolei Ma, and Yidi Lei

The Decentralized Trusted Identity Management in the Collaboration of Supply Chain and Logistics Operations 182
   Zhangliang Li, Zhengyuan Yue, Zhongli Qiao, Keqing Wang, Ziqi Liu, Jianlong Qiu, and Deqian Fu

Unify the Research and Design of Power Quality Regulators 198
   Yaoyao Qin, Chen Lei, Liping He, and Longjing Liu

Learning and Systems

A Zero Trust Continuous Authentication Scheme Based on Keystroke Dynamics 215
   Ping Gong

Cognitive-Based Autonomous Orbit of Non-Cooperative Targets 222
   Jianliang Ma, Lele Zhang, Jian Yang, and Fang Deng

Finite-Time Stabilizing Control of Furuta Pendulum System Based
on Disturbance Observer and Integral Sliding Mode Method 235
   Zhujun Wang, Shuli Gong, Ancai Zhang, and Junyao Yu

Method of Heading Estimation Based on Underwater Acoustic Sensors 246
   Qinglin Wen, Chun Jia, and Jiachang Zhang

Optimal Cooperative Control for Constrained Multi-agent Systems Using
Adaptive Dynamic Programming Method 255
   Zijie Guo, Wenshuai Lin, Xiaohong Zheng, and Zilong Zhang

Pareto-Wise Ranking Generator for Multi-objective Coevolutionary
Generative Adversarial Networks 268
   Weifeng Guo, Yijun Fu, Ying Bi, and Jing Liang

Research Progress on Control Technology of Tiltrotor Aircraft 280
   Zhang Junhong, Li Hao, Zhao Letian, Liu Long, and Zhao Hai

Robust and Non-fragile Control for Non-uniform Sampled-Data Systems
Under Try-Once-Discard Protocol 292
   Jiyuan Zhang, Shipei Cai, and Jinling Liang

Tracking Control of 2-DOF Continuum Manipulator Using Fuzzy PID
Method 304
   *Youchuan Wang, Junfeng Sun, Ancai Zhang, Wenbo Zheng,
   and Haichuan Yang*

Author Index 313

Cognitive Computing And Information Processing

Iris Recognition Based on Asynchronous Genetic Particle Swarm Optimization

Du Yining and Li Aiju(✉)

Shandong Transport Vocational College, Shandong, China
2537393747@qq.com

Abstract. With the advent of the digital age, information has become more and more important, and people's identity information has also received more and more attention. Iris recognition is an efficient and convenient identification technology, which has been applied to mobile phone payment, public security, personnel attendance and other scenarios. However, as iris recognition has become more popular, a large number of small-scale iris recognition have emerged. Faced with this situation, the training samples provided to the neural network class method are not enough, and the accuracy of the simple range-class method is too low. In this regard, this paper proposes an iris recognition method based on asynchronous genetic particle swarm algorithm (GAHPSO). The algorithm uses the embedded hybrid algorithm framework for reference, and selects appropriate crossover operator, mutation operator and possible occurrence probabilities according to small-scale iris recognition problems. In addition, it selects inverted S-type inertia weight and asynchronous learning operators to replace constant values. The optimization ability of the algorithm is improved. On the basis of distance class method, Gabor filter is optimized by asynchronous genetic particle swarm algorithm, which enhances its ability to extract iris features without changing the structure. The experimental results show that the proposed method can accurately and quickly carry out small-scale iris recognition, which is better than other existing schemes.

Keywords: Iris recognition · Particle swarm optimization · Genetic algorithm · Biometric recognition · Gabor filter

1 Introduction

Iris recognition is a technology to identify identity through the radial sulcus line, color spots, recesses, centripetal sulcus and other features of the iris. It has the advantages of stability, uniqueness and non-intrusion [1], and can be applied to various identity authentication fields such as attendance, access control [2], electronic payment, and food traceability [3].

Nowadays, iris recognition has gradually become commonplace, and researchers have designed iris recognition lenses that can be applied to mobile phones [4, 5]. Iris recognition may replace mobile phone fingerprint recognition and become a new generation of identity authentication tools for daily communication devices. In addition, further recognition-multi-modal recognition has also been derived, such as: face-iris recognition [6, 7], eye-iris recognition [8, 9], iris-ear-finger recognition, It can be seen that iris recognition has broad prospects for development [10, 11].

With the increasing popularity of iris recognition, a large number of small-scale iris recognition have appeared. However, small-scale iris recognition scenarios can only provide few training samples, which is difficult to meet the needs of neural network methods, and such methods take a long time. The accuracy of conventional range-class methods is not high enough. Therefore, people urgently need a fast and accurate small-scale iris recognition method. This paper proposes a recognition method based on GAHPSO algorithm, which draws on the framework structure of embedded hybrid algorithm, and selects appropriate crossover and mutation operator and appropriate crossover and mutation probability according to small-scale iris recognition problem. Crossover operation is associated with the optimal position of another parent, and mutation operation makes full use of population information. The inverse S-type inertia weight and asynchronous learning factors are selected to replace the fixed values, which balance the global and local search ability of the algorithm, and have a greater possibility to produce excellent individuals. The Gabor filter is optimized by GAHPSO algorithm to improve its ability to extract iris features. Filter parameter optimization can make it more suitable for different kinds of iris libraries without changing the basic structure, improve the pertinence and accuracy of filtering behavior as well as the versatility of Gabor filter.

Iris recognition includes four steps: image quality assessment, preprocessing, feature extraction and matching. Quality assessment is to assess image quality through certain criteria to avoid image quality problems affecting the recognition results, and sometimes includes live detection [12]. In particular, the quality of images captured by the iris recognition system under complex imaging conditions can also be effectively discriminated [13]. Preprocessing is a pre-processing process such as localization, normalization, noise reduction and enhancement. It can determine the location of iris in the image, map iris of different sizes to a fixed model, eliminate the interference of part of image noise [14], and then enhance the contrast of image texture information. Feature

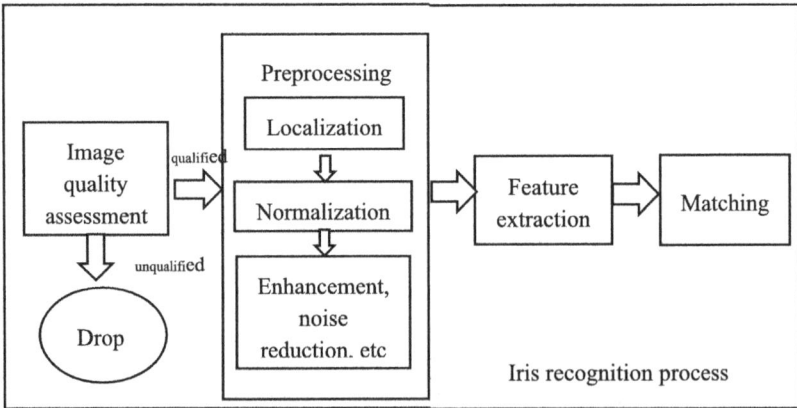

Fig. 1. Iris recognition process.

extraction is to extract the information contained in images [15] and transform it into a form that is easy to compare, which is convenient for comparison. Usually, Gabor filter method and wavelet transform method are commonly used. Feature matching is to compare and match the features extracted in the previous step with the template information and obtain the final recognition result. Usually, distance matching method and support vector machine method are commonly used [16]. The iris recognition process is shown in Fig. 1.

2 Adaptive Basis: GAHPSO Algorithm

2.1 Genetic Algorithm

It is a kind of simulated evolutionary algorithm with solid biological basis. The algorithm initializes a population, and each individual in the population is a solution to the problem. These individuals evolve through the process of selection, crossover, mutation, etc., and select the best solution according to the fitness as the next generation population and continue to evolve. The algorithm is self-organizing, intelligent, easy to parallelize, and has strong global search ability, but the convergence speed is slow, and the use of system information is insufficient.

2.2 Particle Swarm Optimization

Its basic idea is derived from the process of birds searching for food, and it is an algorithm that iterates to find the optimal solution. Each particle has its own position and speed, the former corresponds to the solution of the problem, the latter represents the direction and distance of the particle movement, and the group has a way of calculating the degree of adaptation until the end conditions are met.

The formula of the algorithm is:

$$v_i(t+1) = \omega v_i(t) + c_1 r_1 [p_i(t) - x_i(t)] + c_2 r_2 [p_g(t) - x_i(t)] \tag{1}$$

$$x_i(t+1) = x_i(t) + v_i(t+1) \tag{2}$$

where: v represents position, x represents speed, and p represents the optimal value.

2.3 GAHPSO Algorithm

Based on the framework of embedded hybrid algorithm [17], this paper adopts inverted S-type inertia factor [18] and asynchronous learning factors [19] to better adjust the global and local optimization capabilities. In addition, based on small-scale iris recognition, appropriate crossover operator is selected for particles with average adaptability, and appropriate mutation operator is selected for particles with poor adaptability. so that the particles have the ability to jump out of the local optimal value and possibly get a better solution.

Inverted S-type Inertia Weight. The performance of PSO is closely related to the inertia weight ω. Nowadays, decreasing inertia weight often replace constant inertia weight, so that global and local optimization capabilities can be adjusted more reasonably. Formula (3) is a commonly used decreasing value strategy.

$$\omega = \omega_{start} - (\omega_{start} - \omega_{end})(t/t_{max}) \tag{3}$$

where, t represents the number of iterations.

By combining formula (3) with inverse S-type function, the nonlinear value strategy of formula (4) is obtained [21]. It not only satisfies the ω decline in a suitable interval, but also combines the characteristics of slow decline in the early and late stages of the inverted S-shaped curve. In this way, the particles can fully search for the global in the early stage, and can focus more on the local in the later stage.

$$\omega = \omega_{start} - (\omega_{start} - \omega_{end})\left[1/\left(1 + e^{a-bt}\right)\right] \tag{4}$$

where, $a = 3.4$ and $b = 0.07$.

When ω_{start} is 0.9 and ω_{end} is 0.4, its image is shown in Fig. 2.

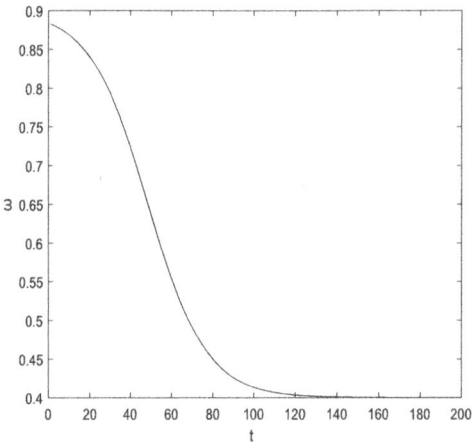

Fig. 2. Inverse S-type inertia weight trend curve.

Asynchronous Learning Factors. "$c_1 r_1 [p_i(t) - x_i(t)]$" in formula (1) is the "self" term that the particle learns from itself and is influenced by its own experience. "$c_2 r_2 [p_g(t) - x_i(t)]$" is the "social" term that the particle learns from the population and is affected by the overall situation of the particles in the population. The values of c_1 and c_2 affect the sizes of these two items. Here, asynchronous learning factors [19] are chosen to replace the conventional constant value.

$$c_1 = 2 + 0.5\cos(\pi * (t/T)) \tag{5}$$

$$c_2 = 2 - 0.5\cos(\pi * (t/T)) \tag{6}$$

where, T represents the maximum number of iterations.

Let $2 + 0.5cosx = 2 - 0.5cosx$, x is in the interval $[0, \pi]$, the solution is $x = 2/\pi$. Therefore, when $1 \leq t < 0.5T$, $c_1 > c_2$; when $t = 0.5T$, $c_1 = c_2$; when $0.5T < t \leq T$, $c_1 < c_2$. In the early stage, the larger value of c_1 and the smaller value of c_2 make the particles learn more from the individual optimal location p_i, focus more on themselves, and have stronger large-range search ability. Then, the learning of the group optimal position p_g is strengthened. The value of c_1 is smaller and the value of c_2 is larger in the later stage.

Crossover operator and Mutation Operator. Cross-operated new individuals are generated by the historically optimal location of the two parent particles. The crossover operator is:

$$x'_i(t) = \alpha x_i(t) + (1 - \alpha)p_{ij}(t) \tag{7}$$

$$x'_j(t) = (1 - \alpha)p_{ii}(t) + \alpha x_j(t) \tag{8}$$

where, α is the random number in the interval $(0,1)$, p_{ii} is the historical optimal position of parent particle i, and p_{ij} is the historical optimal position of parent particle j.

With the help of Gaussian distribution, the mutation operator is (See Fig. 3):

$$X_i^*(t) = X_i(t) + Gaussian1 * (p_i(t-1) - x_i(t)) + Gaussian2 * (p_g(t-1) - x_i(t)) \tag{9}$$

where, $Gaussian1$ and $Gaussian2$ are random numbers based on Gaussian distribution in the interval $(0,1)$.

Algorithm Steps and Flow Chart

1) Initialize particle population and set parameters such as T, ω_{start} and ω_{end}.
2) Update according to the formula (1, 2, 4–6).
3) Calculate the fitness degree of the particles.
4) Sort the fitness value in ascending/descending order.
5) The particles corresponding to the first 1/2 of the sequence remain unchanged; The particles corresponding to 1/2 - 11/12 of the sequence are crossed according to Eq. (7,8) (if a single particle cannot be paired, it is directly retained). If the generated particle has a higher degree of adaptation, the original particle is replaced; otherwise, it remains unchanged; The particles corresponding to the latter 1/12 of the sequence are mutated according to Eq. (9). The principle of particle retention is the same as above.
6) Update p_i and p_g.
7) Update the number of iterations.
8) Determine whether the conditions for termination are met. If yes, the iteration ends and the return value is output. If not, skip to step 2).

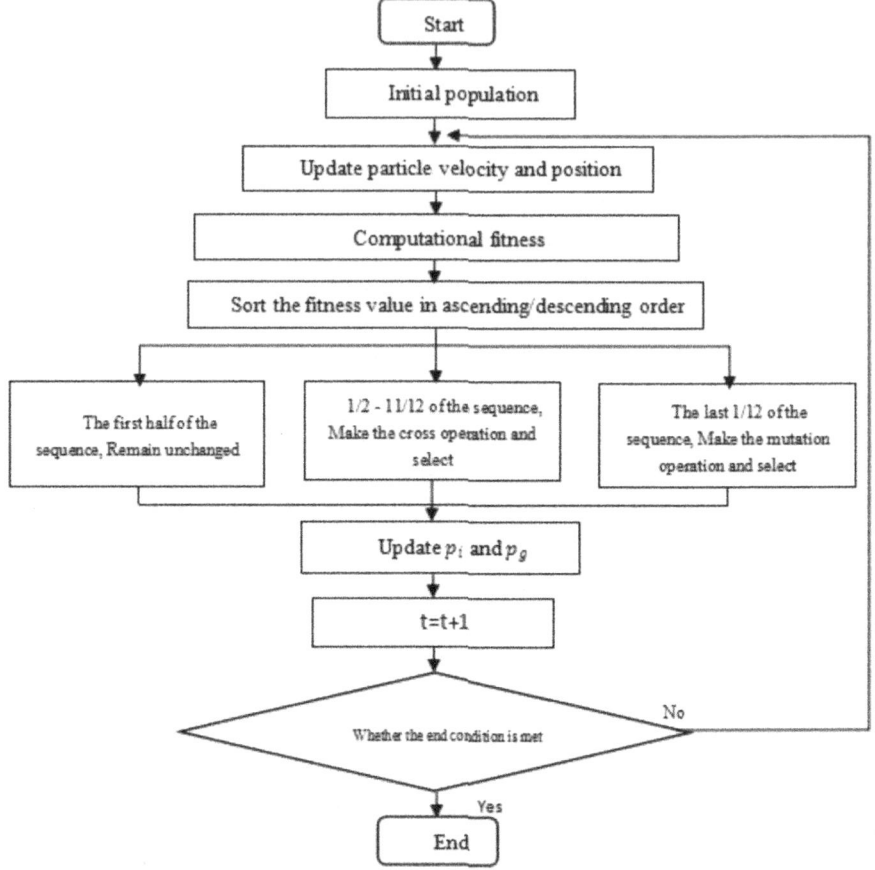

Fig. 3. Flowchart of GAHPSO algorithm.

3 Iris Recognition Process

The iris is roughly located in the image that passes the quality detection, and the ring detection operator is used to correct it. After preprocessing (image enhancement, noise reduction, etc.), the iris features were extracted by Gabor filter optimized by the algorithm in Sect. 2.3 and the features were represented by binary coding. Finally, the Hamming distance is used for rapid recognition. The process diagram of iris recognition method is shown in Fig. 4.

3.1 Adaptive Gabor Filter

Gabor filter can describe iris texture features well and has been widely used in iris recognition [20, 21], its feature extraction effect is closely related to parameters K_{max}, f_v, m and n. In the past, people often used empirical parameters, but different parameters for different problems are not the same performance, this approach often can not play the

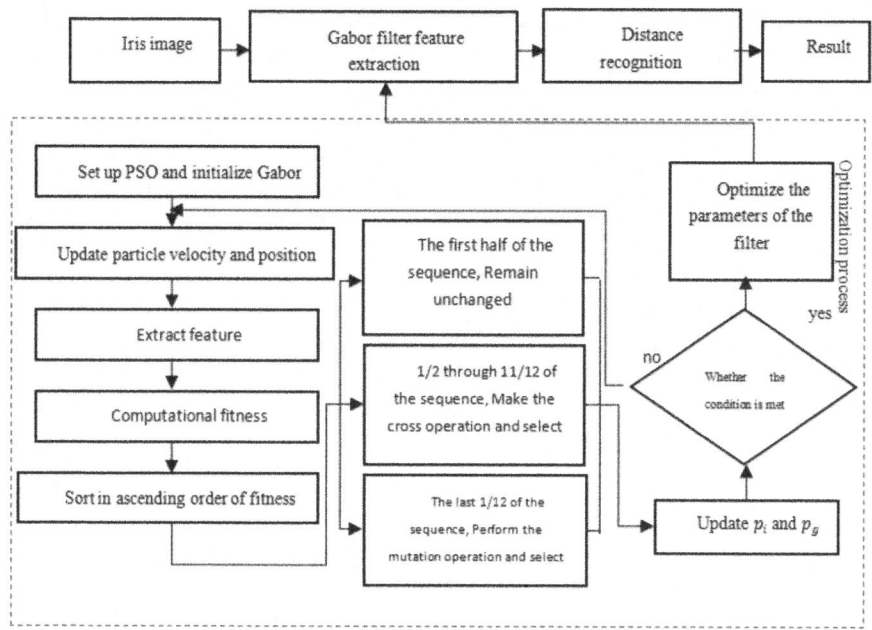

Fig. 4. Iris recognition method diagram.

maximum performance of the filter. In this paper, GAHPSO algorithm is used to optimize its maximum frequency K_{max} and frequency difference f_v, the parameter values take into account the iris texture features and the method of calculating the matching degree, so that the image information can be efficiently extracted.

The texture information in iris image is rich, and there are multi-direction and frequency texture features, so it is necessary to use multiple Gabor filters for feature extraction. The texture image has an equal probability distribution in $[0,2\pi]$, and the two-dimensional Gabor filter has angular symmetry, so only $[0,\pi]$ is considered. In addition, because it is an equal probability distribution, it is designed as a fixed Angle increment. In order to make the process as simple as possible on the premise of ensuring the function, 5×8 Gabor filter banks were selected, which has shown satisfactory performance in previous experiments [25].

The parameter optimization process is:

1) Initialize the population with 36 particles and set the parameters, where T is 300, ω_{start} is 0.9, ω_{end} is 0.4. The velocity range of the particle is $[-40, 40]$. Each particle can be thought of as a Gabor filter containing a set of K_{max} and f_v.
2) Update according to the Eqs. (1, 2, 4–6).
3) Train by a_1 test iris images extracted from the iris library and its contrast iris images (a_2 of the same type and a_2 of different types). The degree of adaptation is calculated according to formula (10). The smaller T_1 is, the greater the degree of adaptation is.

$$HD = \frac{1}{n}\sum_{i=1}^{n} C_1 \oplus C_2 \qquad (10)$$

where, HD_x is the Hamming distance between irises of the same class, and HD_y is the Hamming distance between irises of different classes.

4) Sort T_1 in ascending order.
5) The particles corresponding to the first 1/2 of the sequence do not change; For the particles corresponding to 1/2 to 11/12 of the sequence, a direct reservation is randomly selected, and the remaining random pairs are crossed by formula (7, 8). The offspring produced from crossover of particles i and j are denoted as i' and j'. If T_1 $(i') \leq T_1 (i)$, then $i = i'$; Otherwise, $i = i$. If $T_1 (j') \leq T_1 (j)$, then $j = 'j'$; Otherwise, $j = j$; For the particles corresponding to the last 1/12 of the sequence, the formula (9) is selected for mutation operation, and the offspring produced from mutation of particle i is denoted as i'. If $T_1 (i*) \leq T_1 (i)$, then $i = i'$; Otherwise, $i = i$.
6) Update p_i and p_j.
7) Determine whether the end conditions are met. If so, the iteration ends and the return value is output; If not, skip to step 2).

3.2 Gabor Filter Feature Extraction

Each filter of the Gabor filter bank has its own orientation and frequency information, and there are differences in the ability of these filters to describe texture features. The 40 optimized filters are numbered in the order of increasing frequency and direction, numbered as $Gabor_1$~ $Gabor_{40}$. And the captured 256 × 32 images evenly divided into 8 horizontal, vertical evenly divided into 4, get 32 size of 32 × 8 images, numbered I_1~I_{32}, with 40 filters directly processing I_1~I_{32} will produce huge information, Therefore, PCA is used to compress the image to 32 × 1, and $Gabor_1$~$Gabor_{40}$ is used to filter the compressed I, and then binary coding is obtained, and the binary coding of the image is obtained after sequentially splicing.

3.3 Hamming Distance Feature Matching

Hamming distance is used to construct the classifier. Its idea is to compare two sets of codes, count the number of different codes, and express its ratio to the total number of image codes by Hamming distance.

The binary code obtained in Sect. 3.2 and the feature code of the template are calculated according to formula (11). The smaller the value of HD, the more similar the two irises are.

$$T_1 = \frac{\sum_{x=1}^{a_2} |HD_x - 1|}{1/\sum_{y=1}^{a_2} |HD_y - 0|} \quad (21)$$

where, C_1 and C_2 are the feature codes of the test iris and the template iris, and n is the number of bits of feature code of the iris.

Compared with the classification threshold, if it is less than the classification threshold, it is determined to be the same iris; If it is greater than the classification threshold, it is judged to be a different iris.

4 Experimental Results and Summary

The experimental environment is Windows10 operating system, CPU: 2.5 GHz.

In this paper, iris library CASIA V4.0 provided by Chinese Academy of Sciences and JIU series iris library published by Jilin University are used for experiments. CASIA V4.0 adds three sublibraries of Thousand, Distance and Syn on the basis of CASIA V3.0's Interval, Twins and Lamp, which makes image types more abundant. The acquisition equipment of the latter is avant-garde, and the images meet the requirements of further experiments on iris. The results of the experiment were expressed by three indexes: correct recognition rate (CRR), equal error rate (EER) and time (T).

The experiment used 100 classes in CASIA V4.0, with a total of 615 images. The experiment adopts four methods: the improved filter method proposed in this paper (GAHPSO-Gabor filter + Hamming), classical improved filter method (PSO-Gabor filter + Hamming), other improved filter method (SAPSO-Gabor filter + Hamming) and neural network method (Harr + BP neural network). The results of the experiment are shown in Fig. 5.

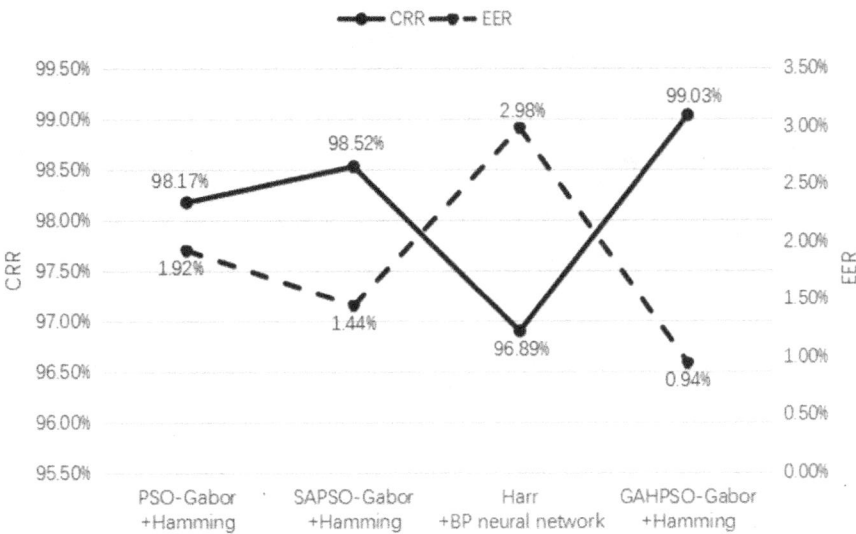

Fig. 5. CRR and EER values of different methods in CASIA V4.0 experiment.

The experiment used 80 classes in JIU, with a total of 402 images. The experiment adopts four methods: the improved filter method proposed in this paper (GAHPSO-Gabor filter + Hamming), filter method (Gabor filter + Hamming), improved filter method (HMPSO-Gabor filter + Hamming) and neural network method (Harr + BP neural network). The results of the experiment are shown in Fig. 6.

Examples of correctly identified and incorrectly identified images are shown in Fig. 7.

The recognition time is also the main factor that people consider when selecting the recognition method. The results are shown in Fig. 8.

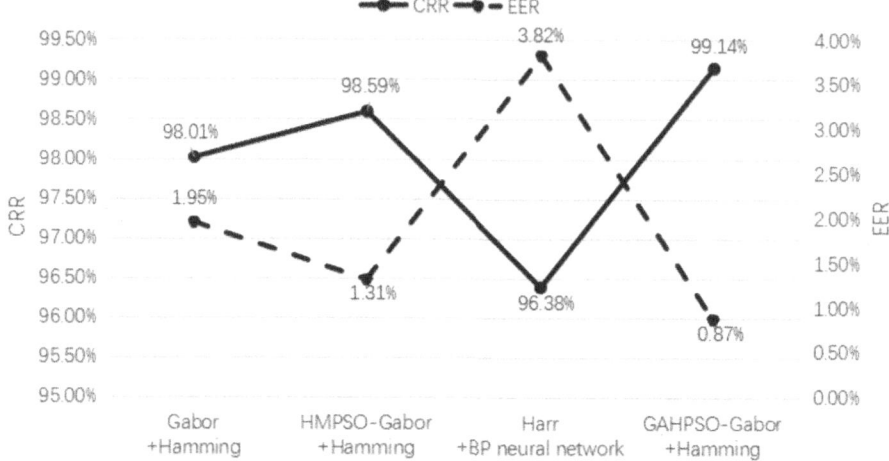

Fig. 6. CRR and EER values of different methods in JIU experiment.

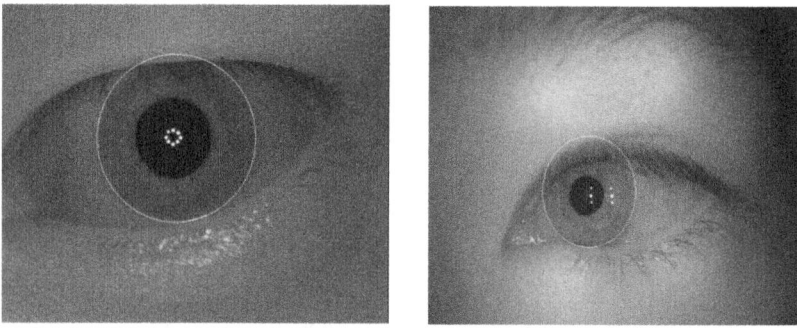

Fig. 7. Examples of correctly identified image (left) and incorrectly identified image (right).

It can be seen from the experimental data that the proposed method is more accurate than the filter method, because the asynchronous genetic particle swarm optimization in this paper can regulate the population diversity and search ability well. In the crossover operation, the particle speed is affected by the optimal position of the other parent, and in the mutation operation, the particle is affected by the optimal position of its own and the population, which can make full use of the cooperation between the groups and find the possible better solution with the help of memory. Moreover, the dynamic time-varying inertia weight and asynchronous learning factors reduce the possibility of the algorithm falling into the local optimal value, and improve its search accuracy and robustness. In addition, optimizing filter parameters can greatly improve its feature extraction capability without changing the number of filters. The proposed method is more accurate than the PSO optimization filter method, because the combination of genetic algorithm can simulate biological genetic evolution to produce new particles, has the ability to deviate from local values, and expands the population diversity without adding the number of particles. Compared with the method of optimizing filters with SAPSO and HMPSO,

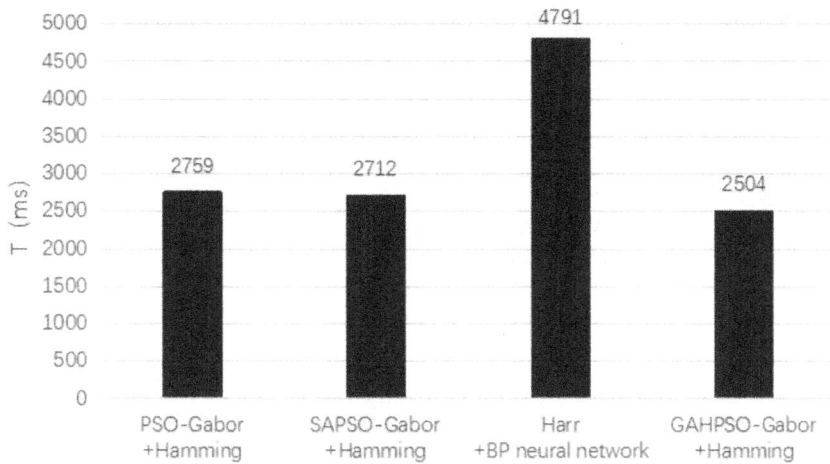

Fig. 8. Identification time of different methods in CASIA V4.0 experiment.

the accuracy of this method is higher, and the optimization ability of this algorithm has also been proved. GAHPSO algorithm is used to adapt filter parameters, which greatly improves its ability to extract iris features, and the proposed method is effective in distance class methods.

The accuracy of the proposed method is higher than that of the neural network method, because the neural network method has only a few training samples in small-scale iris recognition scenarios, which is difficult to give full play to the maximum performance of the neural network, and the training results are not optimal. Besides, the neural network method takes a long time, and small-scale recognition is generally used for routine operations such as attendance checking and enterprise access control, so the accuracy requirements are not so rigorous. At the same time, people desire convenience and speed, so the method in this paper is more in line with the needs.

Experiments have been carried out on CASIA V4.0 iris library and JIU series iris library respectively, and satisfactory results have been obtained, which proves the universality of this method. Different numbers of iris images have been used in experiments, which can prove that this method is suitable for ultra-small-scale and small-scale iris recognition. In summary, the method proposed in this paper can perform iris recognition quickly and accurately, which is better than the existing methods.

References

1. Malgheet, J.R., Manshor, N.B., Affendey, L.S., et al.: Iris recognition development techniques: a comprehensive review. Complexity **2021**, 1–32.9 (2021)
2. Zadnik, D., Žemva, A.: Image acquisition device for smart-city access control applications based on iris recognition. Sensors **21**(18), 6185 (2021)
3. Zhongliang, L., Zhuohua, D., Jingguo, D.: Research on Iris recognition in meat food traceability based on internet of things. Math. Pract. Cogn. **43**(05), 102–108 (2013)
4. Feng, H., Yupeng, D.: Design of mobile phone iris recognition lens. J. Appl. Opt. **41**(01), 37–42 (2020). (in Chinese)

5. Yu, X., Guo, B., Li, X., Shen, T.: Design of small distortion mobile phone lens based on iris recognition. Adv. Laser Optoelectron. **57**(01), 228–233 (2020). (in Chinese)
6. Ammour, B., Boubchir, L., Bouden, T., et al.: Face–iris multimodal biometric identification system. Electronics **9**(1), 85 (2020)
7. Xiong, Q., Zhang, X., Xu, X., et al.: A modified chaotic binary particle swarm optimization scheme and its application in face-iris multimodal biometric identification. Electronics **10**(2), 217 (2021)
8. Chen Ying, W., Liang, W.X., Shubin, G.: Non-cooperative identity authentication with iris-periocular features. J. Image Graph. **28**(05), 1462–1476 (2023). (in Chinese)
9. Luo, Z., Li, J., Zhu, Y.: A deep feature fusion network based on multiple attention mechanisms for joint iris-periocular biometric recognition. IEEE Signal Process. Lett. **28**, 1060–1064 (2021)
10. Zhao, T., Liu, Y., Huo, G., et al.: A deep learning iris recognition method based on capsule network architecture. IEEE Access **7**, 49691–49701 (2019)
11. Benalcazar, D.P., Bastias, D., Perez, C.A., et al.: A 3D iris scanner from multiple 2D visible light images. IEEE Access **7**, 61461–61472 (2019)
12. Dronky, M.R., Khalifa, W., Roushdy, M.: Using residual images with BSIF for iris liveness detection. Expert Syst. Appl. **182**, 115266 (2021)
13. Huijie, G.: An enhanced iris image quality assessment algorithm. J. Optoelectron. Technol. **39**(01), 26–29 (2019). https://doi.org/10.19453/j.cnki.1005-488x.2019.01.006
14. Gu, Z., Wang, C., Tian, Q., Zhang, Q.: Noise iris image segmentation method based on transformer and symmetric codec. J. Comput. Aided Des. Graph. **34**(12), 1887–1898 (2022)
15. Sun, Z., Heran, W.L., et al.: Development report of biometrics. J. Image Graph. **26**(06), 1254–1329 (2021). (in Chinese)
16. Rehman, A.: Light microscopic iris classification using ensemble multi-class support vector machine. Microscopy Res. Tech. **84**(5), 982–991 (2021)
17. Li, J., Sun, X., Li, S., Li, R.: Improved particle swarm optimization Algorithm based on genetic cross factor. Comput. Eng. (02), 181–183 (2008). (in Chinese)
18. Huang Yang, L., Kaibo, H.X., Shijuan, H.: Adaptive particle swarm optimization algorithm based on S-type function. Comput. Sci. **46**(01), 245–250 (2019). (in Chinese)
19. He, L., Yao, Y., Li, S., Feng, J.: Multi-parameter coincidence calculation of artillery interior trajectory based on improved particle swarm optimization. Fire Control Command Control **46**(11), 165–169 (2021). (in Chinese)
20. He, F.: Research on multi-feature extraction and fusion recognition of iris based on Gabor filter. J. Jilin Univ. (2015)
21. Thakkar, S., Patel, C.: Iris recognition supported best gabor filters and deep learning cnn options. In: 2020 International Conference on Industry 4.0 Technology (I4Tech). IEEE, pp. 167–170, 2020
22. Liu, S.: Research on improved Iris feature extraction and recognition algorithm based on particle swarm optimization. Jilin University, 2019

Joint Torque Sensor-Based Distributed Coordinated Control of Dual-Arm Reconfigurable Manipulators

Bing Ma[1], Hangwei Zhang[1], Ximing Yao[2], Yuan Yao[1], Tianjiao An[1], and Qiang Pan[1(✉)]

[1] Department of Control Science and Engineering, Changchun University of Technology, Changchun, China
{mabing,antianjiao}@ccut.edu.cn, qqp5482314@163.com
[2] Jilin Province Key Laboratory of Measuring Instrument and Technology, Jilin Institute of Metrology, Changchun, China

Abstract. This paper proposes a distributed coordinated control strategy for a dual-arm reconfigurable manipulator (DRM) based on joint torque sensors to complete grasping and transportation tasks. Through kinematic and force analysis of the manipulators, the dynamic models of DRM and object are established under the joint torque feedback (JTF) technique and Newton-Euler algorithm, respectively. A fusion error function is designed to reflect the tracking performance of object and the influence of internal forces. The adaptive observer is established to estimate the uncertain dynamic terms of DRM. Consequently, a novel coordinated controller is proposed for DRM. The asymptotic stability of DRM is demonstrated utilizing Lyapunov stability theory. Finally, the effectiveness of the proposed method is verified by experiments.

Keywords: Reconfigurable manipulators · Dual-arm coordinated control · Load distribution · Sliding mode control

1 Intruduction

Dual-arm manipulator mimics human hands to adapt to the requirements of different working environments through coordinated control. The dual-arm reconfigurable manipulator (DRM) are composed of several standardized interface modules, allowing for different tasks by adding or removing modules [1]. An et al. proposed an optimal control method for reconfigurable robotic manipulator based on cooperative game [2] and dynamic event-triggered strategies [3]. Compared with traditional manipulators, DRM significantly enhance manipulation performance through flexible modular design and adaptive capabilities, with wide applicability in manufacturing, medical, and aerospace fields [4]. Therefore, it is essential to develop a suitable controller for DRM to effectively accomplish coordinated handling tasks within complex environments.

According to the current state of research, coordinated control is generally divided into master-slave control [5], impedance control [6] and force/position control [7,8]. Research on coordinated manipulator control has existed since the 1990s, For example, T. Sugar et al. discussed the coordinated control of two or three manipulator platforms [9]. Erhart et al. introduced an impedance-based coordinated control scheme for multi-manipulator operations to limit unnecessary internal forces when there is kinematic uncertainty [10]. The aforementioned methods typically depend on the manipulator's configuration, where the structure of the manipulator dictates the controller architecture. However, for reconfigurable manipulators, the configuration needs to be adjusted flexibly according to the adapting tasks. To utilize this characteristic, some researchers have developed dynamic models for reconfigurable manipulators based on joint torque sensors to accommodate changing task requirements. Liu et al. proposed a distributed control method for reconfigurable manipulators based on joint torque feedback (JTF) technique for solving the decomposition and compensation problem of model uncertainty [11]. They also studied the design and control method of spring-assisted modular robot manipulators [12]. However, the majority of existing research on JTF technique focuses solely on single manipulator, with little attention given to studying coordinated control for dual-arm manipulator. Furthermore, due to the variability of grasped objects leading to uncertain dynamic models, and the fact that the impact of an object's dynamic model on manipulator performance has not been fully considered. Therefore, this paper proposes a novel dual-arm distributed coordinated control method based on JTF technique to complete the transportation tasks.

2 System Description and Some Preliminaries

2.1 Kinematics

Denoting $x_e = [x_{e1}^T, x_{e2}^T]^T \in \mathbb{R}^{2m}$ as the pose vector of the two end-effectors, the relationship between joint velocity and end-effector velocity is:

$$\dot{x}_e = J_D \dot{q}_D \tag{1}$$

where $q_D = [q_1^T, q_2^T]^T \in \mathbb{R}^{(n_1+n_2) \times 1}$ and $q_i \in \mathbb{R}^{n_i \times 1}$ for $i = 1, 2$ is the joint angle vector of the ith manipulator; $J_D = blockdiag\,[J_1, J_2] \in \mathbb{R}^{2m \times (n_1+n_2)}$, where J_i denotes the Jacobian matrix. $x_o \in \mathbb{R}^p$ is defined as the coordinate vector of the object's center of mass (COM) and is related to by the grasp matrix. Hence, we have

$$\dot{x}_e = J_o \dot{x}_o \tag{2}$$

where $J_o^T = [J_{o1}^T, J_{o2}^T] \in \mathbb{R}^{p \times 2m}$ denotes the grasp matrix and it can be calculated by $J_{oi}^T = [I_3, 0; P_i, I_3] \in \mathbb{R}^{p \times m}$ for $i = 1, 2$. Also, the matrix P_i is defined as: $P_i = S(r_{ori}) = [0, -p_{iz}, p_{iy}; p_{iz}, 0, -p_{ix}; -p_{iy}, p_{ix}, 0]$, $r_{ori} = [p_{ix}, p_{iy}, p_{iz}]^T$ denotes the description of the vector from the COM of being grasped to the end of the manipulators in the world frame.

Considering two arms acting on the object at the same time, we have

$$\dot{x}_o = J_o^+ J_D \dot{q}_D = J_\phi \dot{q}_D \tag{3}$$

where $J_\phi = J_o^+ J_D$ and J_o^+ is the pseudoinverse of J_o. The time derivative of Eq. (3) leads to

$$\ddot{x}_o = \frac{d(J_\phi)}{dt} \dot{q}_D + J_\phi \ddot{q}_D \tag{4}$$

2.2 Force Analysis of DRM System

Based on the principle of virtual work, the connection between the net force F_o acting on the end-effector and the contact force F_e applied to it can be represented as:

$$F_o = J_o^T F_e \tag{5}$$

where $F_o = [F_{o1}, F_{o2}]^T \in \mathbb{R}^p$, $F_e = [F_{e1}, F_{e2}]^T \in \mathbb{R}^{2m}$ and the force F_e consists of two parts, one is the external force used to drive the motion of the object, and the other is the internal force used to hold the object stable while grasping it.

$$F_e = F_M + F_I \tag{6}$$

where $F_M = \left(J_o^T(x_o)\right)^+ F_o$, $F_I = \mathbb{N} F_e$ and $\mathbb{N} = I - \left(J_o^T(x_o)\right)^+ J_o^T(x_o)$ is the null-space matrix of $J_o^T(x_o)$. Inspired by Walker et al. [13], the relationship between the resultant force F_o of the object and the component F_{oi} can be written as:

$$F_{oi} = \frac{1}{2} F_o \tag{7}$$

2.3 Dynamics

The dynamic models of the dual-arm and object can be expressed separately as follows:

$$I_m \gamma \ddot{q}_D + f(q_D, \dot{q}_D) + \frac{\tau_s}{\gamma} = \tau - \frac{J_D^T F_e}{\gamma} \tag{8}$$

$$M_o(x_o)\ddot{x}_o + C_o(x_o, \dot{x}_o)\dot{x}_o + g_o(x_o) = F_o \tag{9}$$

where $I_m \in \mathbb{R}^{(n_1+n_2)\times(n_1+n_2)}$ is the moment of inertia of actuator rotor; $\gamma \in \mathbb{R}^{(n_1+n_2)\times(n_1+n_2)}$ denotes the reduction ratio; $f(q_D, \dot{q}_D) \in \mathbb{R}^{(n_1+n_2)}$ represents the joint lumped frictional torque; $\tau_s = [\tau_{s1}, \tau_{s2}]^T \in \mathbb{R}^{(n_1+n_2)}$ represents the coupling torque at the joint torque sensor; $\tau = [\tau_1, \tau_2]^T \in \mathbb{R}^{(n_1+n_2)}$ indicates the control torque at the motor end. $M_o(x_o) \in \mathbb{R}^{p\times p}$ is object's symmetric positive definite inertial matrix; $C_o(x_o, \dot{x}_o) \in \mathbb{R}^{p\times p}$ is object's Coriolis and centrifugal matrix; $g_o(x_o) \in \mathbb{R}^p$ is the gravitational force vector.

Dividing both sides of Eq. (8) by γ, we can obtain

$$I_m \gamma^2 \ddot{q}_D + \gamma f(q_D, \dot{q}_D) + \tau_s = \gamma \tau - J_D^T F_e \tag{10}$$

The control torque at the motor end τ multiplied by γ can be approximated as the torque at the joint end, i.e. $u = \gamma\tau$. Let $A = I_m\gamma^2$, $F = \gamma f(q_D, \dot{q}_D)$. Moreover, we have

$$A\ddot{q}_D + F + \tau_s = u - J_D^T F_e \tag{11}$$

Considering Eq. (6) and Eq. (7), we can obtain the dynamic model of a single manipulator through load distribution.

$$A_i \ddot{q}_i + F_i + \tau_{si} = u_i - J_i^T \left[\frac{1}{2}(J_{oi}^T)^+ F_o + F_{Ii} \right] \tag{12}$$

Considering Eq. (9), Eq. (12) and kinematics Eq. (3), Eq. (4), the dynamic of the single manipulator combined with object is rewritten as:

$$[A_i + \Gamma M_o(x_o) J_{\phi i}] \ddot{q}_i + \left[\Gamma M_o(x_o) \frac{dJ_{\phi i}}{dt} + \Gamma C_o(x_o, \dot{x}_o) J_{\phi i} \right] \dot{q}_i \\ + F_i + \tau_{si} + \Gamma g_o(x_o) + \tau_{fi} = u_i \tag{13}$$

where $\Gamma = \frac{1}{2} J_i^T (J_{oi}^T)^+$, $\tau_{fi} = J_i^T F_{Ii}$. According to Eq. (13), the state space equation of the dual-arm system can be described as:

$$\dot{x}_i : \begin{cases} \dot{x}_1 = x_2 \\ \dot{x}_2 = \Delta + \Phi(u_i - \tau_{si} - \tau_{fi}) \end{cases} \tag{14}$$

where $x_i = [x_1, x_2]^T = [q_i, \dot{q}_i]^T$.

$$\Delta = -\Phi \left[\left(\Gamma M_o(x_o) \frac{dJ_{\phi i}}{dt} + \Gamma C_o(x_o, \dot{x}_o) J_{\phi i} \right) \dot{q}_i + F_i + \Gamma g_o(x_o) \right]$$

$$\Phi = (A_i + \Gamma M_o(x_o) J_{\phi i})^{-1}$$

3 Design of Dual-Arms Coordinated Control Scheme

3.1 Sliding Mode Controller Design

Define F_{Ii} as:
$$F_{Ii} = \xi^T \lambda_{Ii} \tag{15}$$

where ξ^T is to describe the direction of force vector, λ_{Ii} is value of F_{Ii}.

Define internal force tracking error:

$$e_{\lambda_{Ii}} = \lambda_{Ii} - \lambda_{Iid} \tag{16}$$

where λ_{Iid} is the magnitude of the expected internal force. Define internal force error signal and reference desired velocity $\dot{q}_{ird}$ as:

$$\Lambda e_{\lambda_{Ii}} = \int_0^t e_{\lambda_{Ii}} d\lambda \tag{17}$$

$$\dot{q}_{ird} = J_{\phi i}^{+}\left[\dot{x}_{od} - \delta\left(x_{o} - x_{od}\right)\right] \tag{18}$$

where δ is a positive constant.

Define joint angle tracking error e_{qi} as:

$$e_{qi} = J_{\phi i}^{+}\left(x_{o} - x_{od}\right) = q_{i} - q_{ird} \tag{19}$$

The reference joint angle velocity $\dot{q}_{ir}$ as:

$$\dot{q}_{ir} = \dot{q}_{ird} - K_{q}e_{qi} - K_{I}J_{i}^{+}\xi^{T}\Lambda e_{\lambda_{li}} \tag{20}$$

where K_q and K_I are positive constants.

Define a joint sliding variable s_i as:

$$\begin{aligned} s_i &= \dot{q}_i - \dot{q}_{ir} \\ &= \dot{q}_i - \dot{q}_{ird} + K_q e_{qi} + K_I J_i^{+}\xi^{T}\Lambda e_{\lambda_{li}} \end{aligned} \tag{21}$$

Substituting Eq. (14) into Eq. (21), we get

$$\begin{aligned} \dot{s}_i &= \ddot{q}_i - \ddot{q}_{ird} + K_q \dot{e}_{qi} + K_I \frac{d\left(J_i^{+}\xi^{T}\right)}{dt}\Lambda e_{\lambda_{li}} + K_I J_i^{+}\xi^{T} e_{\lambda_{li}} \\ &= \Delta + \Phi\left(u_i - \tau_{si} - \tau_{fi}\right) + \nu_i \end{aligned} \tag{22}$$

where $\nu_i = -\ddot{q}_{ird} + K_q \dot{e}_{qi} + K_I \frac{d\left(J_i^{+}\xi^{T}\right)}{dt}\Lambda e_{\lambda_{li}} + K_I J_i^{+}\xi^{T} e_{\lambda_{li}}$.

Then, the coordinated control strategy is designed as:

$$u_i = -\Phi^{-1}\left(d_i s_i + \Delta + \nu_i\right) + \tau_{si} + \tau_{fi} \tag{23}$$

where d_i is a positive constant.

3.2 Adaptive Observer Design

For the model uncertainty in Eq. (14), we design an adaptive observer as:

$$\dot{\hat{x}}_i : \begin{cases} \dot{\hat{x}}_1 = \hat{x}_2 + k_1\left(x_1 - \hat{x}_1\right) \\ \dot{\hat{x}}_2 = \hat{\Delta} + \hat{\Phi}\left(u_i - \tau_{si} - \tau_{fi}\right) + k_2\left(x_2 - \hat{x}_2\right) \end{cases} \tag{24}$$

where $\dot{\hat{x}}_i = \left[\dot{\hat{x}}_1, \dot{\hat{x}}_2\right]^T$ is the observation of $\dot{x}_i$, k_1, k_2 are positive definite observation gain values. $\hat{\Delta}$ and $\hat{\Phi}$ are the estimation of model uncertainty which can be updated by:

$$\begin{aligned} \dot{\hat{\Delta}} &= \alpha_1 E_2 \\ \dot{\hat{\Phi}} &= \alpha_2 E_2\left(u_i - \tau_{si} - \tau_{fi}\right) \end{aligned} \tag{25}$$

where α_1, α_2 are positive definite constants, and $e_F = [E_1, E_2]^T = [x_1 - \hat{x}_1, x_2 - \hat{x}_2]^T$ is the model observation error.

Combining Eq. (14) with Eq. (24), we have

$$\dot{e}_F = \begin{cases} \dot{E}_1 = \dot{x}_1 - \dot{\hat{x}}_1 = E_2 - k_1 E_1 \\ \dot{E}_2 = \dot{x}_2 - \dot{\hat{x}}_2 = e_\Delta + e_\Phi (u_i - \tau_{si} - \tau_{fi}) - k_2 E_2 \end{cases} \quad (26)$$

where $e_\Delta = \Delta - \hat{\Delta}$ and $e_\Phi = \Phi - \hat{\Phi}$ are the observation error of nonlinear terms Δ and Φ, respectively.

Theorem 1. *For Eq. (14) with model uncertainty, the model observation error is guaranteed to be uniformly ultimately bounded (UUB) under the developed adaptive observer with the adaptive updated law Eq. (25).*

Proof. Define the Lyapunov function candidate as follows:

$$V_1(t) = \frac{1}{2} e_F^T e_F + \frac{1}{2} e_\Delta^T \alpha_1^{-1} e_\Delta + \frac{1}{2} e_\Phi^T \alpha_2^{-1} e_\Phi \quad (27)$$

The derivative of Eq. (27) with respect to time t is given by

$$\begin{aligned}
\dot{V}_1(t) &= e_F^T \dot{e}_F + e_\Delta^T \alpha_1^{-1} \dot{e}_\Delta + e_\Phi^T \alpha_2^{-1} \dot{e}_\Phi \\
&= E_1 \dot{E}_1 + E_2 \dot{E}_2 - \dot{\hat{\Delta}}^T \alpha_1^{-1} e_\Delta - \dot{\hat{\Phi}}^T \alpha_2^{-1} e_\Phi \\
&= E_1 (E_2 - k_1 E_1) + E_2 (e_\Delta + e_\Phi (u_i - \tau_{si} - \tau_{fi}) - k_2 E_2) \\
&\quad - \dot{\hat{\Delta}}^T \alpha_1^{-1} e_\Delta - \dot{\hat{\Phi}}^T \alpha_2^{-1} e_\Phi \\
&= E_1 E_2 - k_1 E_1^2 - k_2 E_2^2 + \left(E_2 - \dot{\hat{\Delta}}^T \alpha_1^{-1} \right) e_\Delta + \\
&\quad \left(E_2 (u_i - \tau_{si} - \tau_{fi}) - \dot{\hat{\Phi}}^T \alpha_2^{-1} \right) e_\Phi \\
&\leq - \left(k_1 - \frac{1}{2} \right) E_1^2 - \left(k_2 - \frac{1}{2} \right) E_2^2 - \left(\dot{\hat{\Delta}}^T \alpha_1^{-1} - E_2 \right) e_\Delta \\
&\quad - \left(\dot{\hat{\Phi}}^T \alpha_2^{-1} - E_2 (u_i - \tau_{si} - \tau_{fi}) \right) e_\Phi
\end{aligned} \quad (28)$$

Substituting Eq. (25) into Eq. (28), we can observe that $\dot{V}_1(t) \leq 0$ holding $k_1 \geq 1/2$, $k_2 \geq 1/2$. Therefore, the model observation error is UUB. This completes the proof.

Define approximation as: $e_i = e_\Delta + e_\Phi(u_i - \tau_{si} - \tau_{fi})$, estimation error as: $\tilde{e}_i = e_i - \hat{e}_i$.

The control strategy can be rewritten as:

$$u_i = -\hat{\Phi}^{-1} \left(d_i s_i + \hat{\Delta} + v_i + \hat{e}_i \right) + \tau_{si} + \tau_{fi} \quad (29)$$

The adaptive law for error estimation is designed as:

$$\dot{\hat{e}}_i = \alpha_3 s_i \quad (30)$$

where α_3 is a positive constant.

4 Proof of Stability

In this section, the Lyapunov function is employed to analyze the stability of DRM.

Theorem 2. *Given DRM operating in free space, the dynamic model of the ith manipulator subsystem is established in Eq. (14). Considering the model uncertainties presented in Eq. (13), under the combined control laws of Eq. (25) and Eq. (30), the angle tracking error of each manipulator joint are UUB.*

Proof. Choosing the Lyapunov candidate as:

$$V(t) = \frac{1}{2}s_i^T s_i + \frac{1}{2}\alpha_3 \tilde{e}_i^2 \tag{31}$$

The time derivative of $V(t)$ is taken as:

$$\begin{aligned}
\dot{V}(t) &= s_i \dot{s}_i - \alpha_3^{-1}\tilde{e}_i\dot{\hat{e}}_i \\
&= s_i\left(\Delta + \Phi(u_i - \tau_{si} - \tau_{fi}) + \nu_i\right) - \alpha_3^{-1}\tilde{e}_i\dot{\hat{e}}_i \\
&= s_i\left(\Delta + \hat{\Phi}(u_i - \tau_{si} - \tau_{fi})\right) + s_i\left(\left(\Phi - \hat{\Phi}\right)(u_i - \tau_{si} - \tau_{fi}) + \nu_i\right) - \alpha_3^{-1}\tilde{e}_i\dot{\hat{e}}_i \\
&= s_i\left(\Delta + \hat{\Phi}\left(\hat{\Phi}^{-1}\left(-d_i s_i - \hat{\Delta} - \nu_i - \hat{e}_i\right)\right)\right) - \alpha_3^{-1}\tilde{e}_i\dot{\hat{e}}_i \\
&\quad + s_i\left(e_\Phi(u_i - \tau_{si} - \tau_{fi}) + \nu_i\right) \\
&= s_i\left(\Delta - \hat{\Delta} - d_i s_i - \nu_i - \hat{e}_i\right) + s_i\left(e_\Phi(u_i - \tau_{si} - \tau_{fi}) + \nu_i\right) - \alpha_3^{-1}\tilde{e}_i\dot{\hat{e}}_i \\
&= s_i\left(e_\Delta + e_\Phi(u_i - \tau_{si} - \tau_{fi}) - \hat{e}_i - d_i s_i\right) - \alpha_3^{-1}\tilde{e}_i\dot{\hat{e}}_i \\
&= -d_i s_i^2 + \tilde{e}_i\left(s_i - \alpha_3^{-1}\dot{\hat{e}}_i\right)
\end{aligned} \tag{32}$$

Substituting Eq. (30) into Eq. (32) and yields the expression:

$$\dot{V}(t) = -d_i \|s_i\|^2 \leq 0 \tag{33}$$

Aiming at DRM with task constraints, it can be concluded that $\dot{V}(t) \leq 0$ through Eq. (33). According to the Lyapunov stability theorem, the angle and internal force tracking errors are asymptotically stable.

5 Experiment Study

To demonstrate the effectiveness of the proposed controller, we experimented with DRM rigidly transport object (see Fig. 1). Due to space limitation, this paper only provides experimental data of a reconfigurable manipulator to verify the proposed control algorithm.

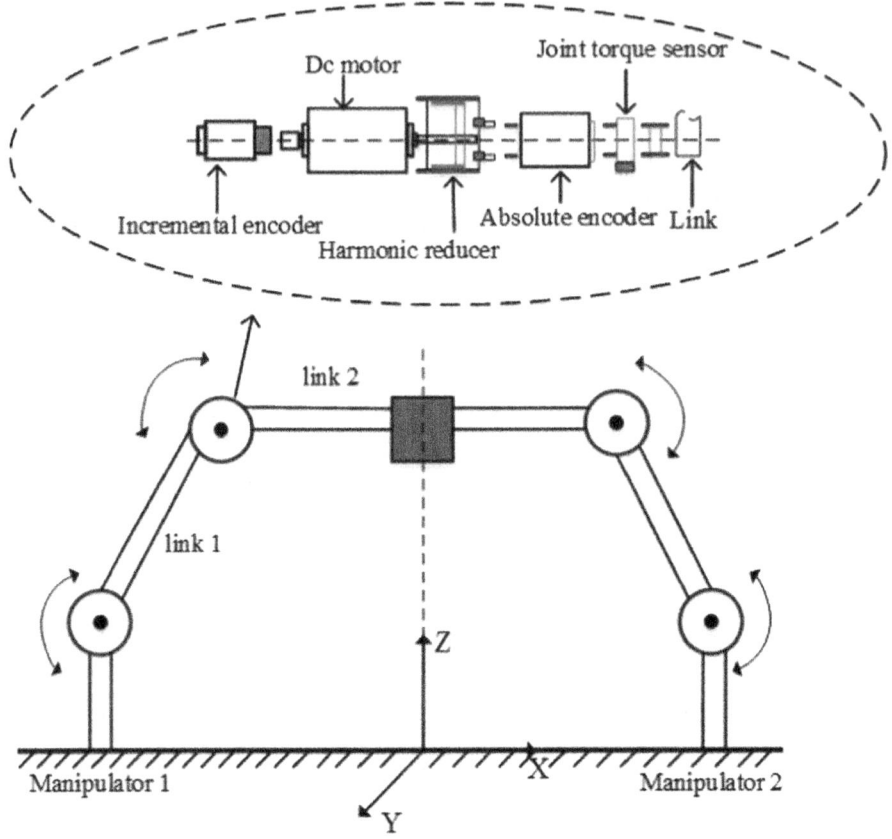

Fig. 1. Structure diagram of DRM.

5.1 Experiment Settings

The physical parameters of DRM and object are enumerated in Table 1. The parameters of the reconfigurable manipulator and object are shown in Table 1. Choose a suitable position between the two manipulators to establish the world coordinate system, the object coordinate system is located at COM of object. In this article, we assume that the object is constrained with only one degree of freedom. The ideal value of the internal force acting on both hands is $F_{Id1} = F_{Id2} = 20N$.

5.2 Experiment Analysis

Experiment curves are showed in Fig. 2–4. These curves show the trajectory tracking effect, error size and internal force tracking effect of the robot joint under the proposed controller. The detailed analysis and description of the curves are as follows.

Table 1. Parameters of reconfigurable manipulator system.

Parameters	Designation	Value
Length of link	$l1,l2$	0.265 m, 0.315 m
Mass of link	$m1,m2$	3.1 kg, 3.7 kg
Mass of object	M	0.5 kg
Gravitational acceleration	g	9.8 m/s^2

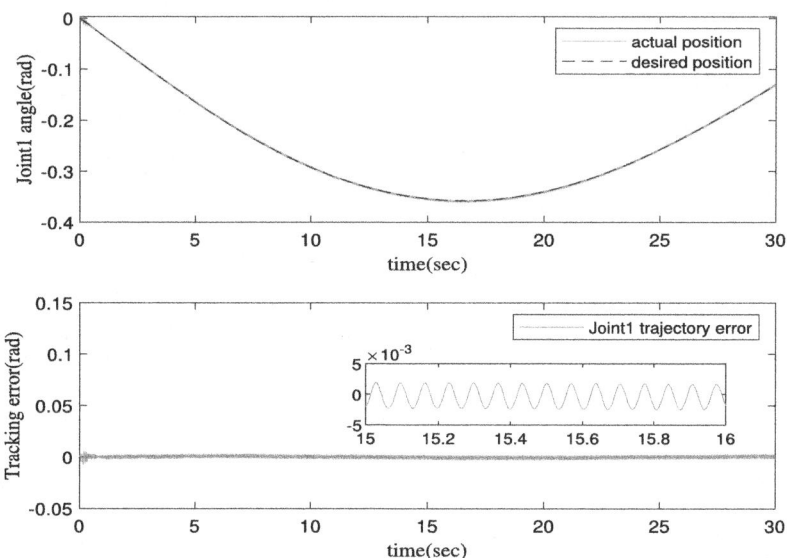

Fig. 2. Trajectory tracking and tracking errors curves of joint 1.

Figure 2 shows the joint position tracking and tracking error of the first joint of DRM under the proposed controller, where the black dashed and red lines represent the actual and desired positions, respectively. And the controller can quickly realize the asymptotic tracking between the actual trajectory and the desired trajectory. The trajectory tracking error of the joint fluctuated slightly in the initial stage, yet remained within an acceptable range. Subsequently, it stabilized and stayed within a range of ±0.005 rad.

Figure 3 also shows the joint position tracking and tracking error of the second joint of DRM under the proposed controller, and the representation rules are the same as above. The controller can quickly realize the asymptotic tracking between the actual trajectory and desired trajectory. However, the trajectory tracking performance is slightly inferior to that of the first joint, with a slightly larger tracking error. This is due to the physical properties of the reconfigurable robot, where the motion of the first joint affects the second joint, but the motion of the second joint does not affect the first joint. The error quickly stabilized and remained within the range of ±0.01 rad.

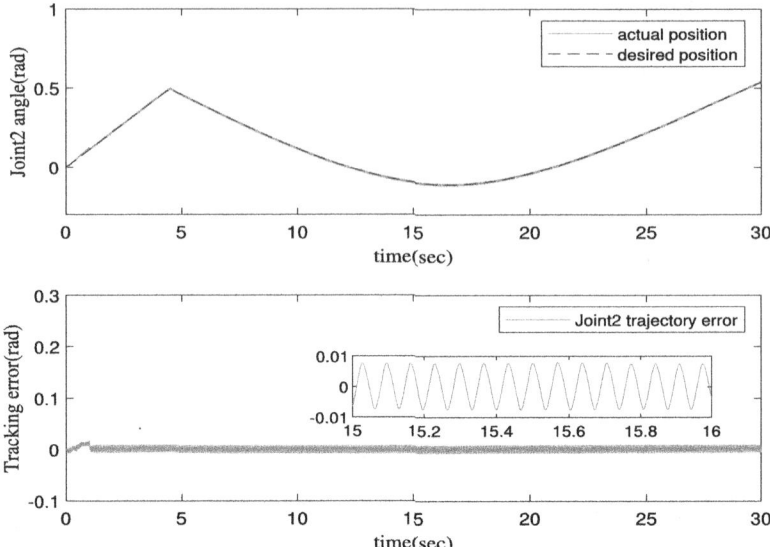

Fig. 3. Trajectory tracking and tracking errors curves of joint 2.

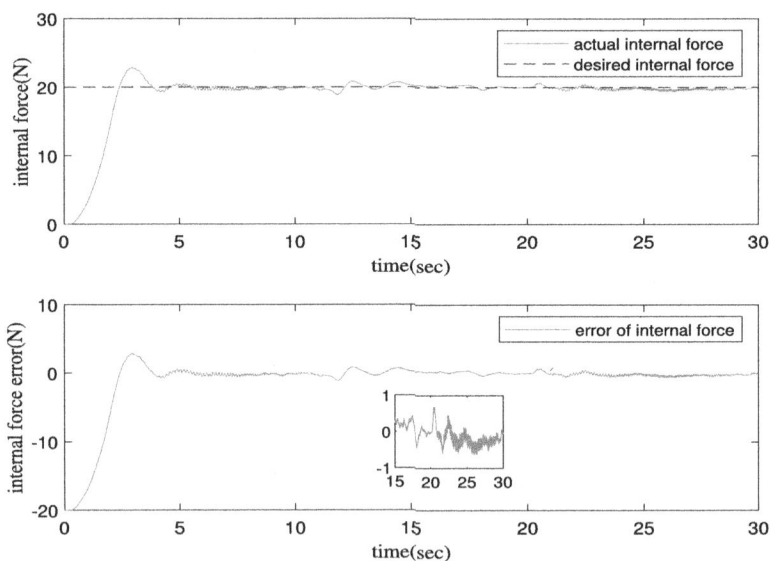

Fig. 4. Internal force tracking and tracking error curve of DRM.

Figure 4 shows the internal force tracking curve and error curve of DRM. The black dashed lines indicate the expected internal forces, and the red indicates the actual internal forces. It can be seen from the figure that when the DRM end effector contacts the grasped object, the internal force can quickly reach the expected value and maintain dynamic stability above and below the expected value. After the internal force tends to be stable, the internal force error remains within ±1 N.

Results of the experiments show that the distributed coordinated control method based on the joint torque sensor is effective for the dual-arm reconfigurable manipulator system.

6 Conclusion

A distributed coordinated strategy for DRM based on the JTF technique is proposed in this paper to facilitate handling tasks. The dynamic models of the reconfigurable manipulator and object are constructed using JTF technique and load distribution. An novel fusion error function, incorporating joint tracking error and internal force error, is designed. With the assistance of an adaptive observer, a coordinated controller for DRM is derived. The system's asymptotic stability is verified in accordance with Lyapunov stability theory. Finally, experiments demonstrate the effectiveness of the proposed control method.

Acknowledgments. This work was supported by the National Natural Science Foundation of China (Grant no. 62173047, 62473063), the Scientific Technological Development Plan Project in Jilin Province of China (Grant no. 20230508159RC).

References

1. Paredis, C.J., Brown, H.B., Khosla, P.K.: A rapidly deployable manipulator system. In: IEEE International Conference on Robotics and Automation, vol. 21, pp. 289–304 (1997)
2. An, T., Dong, B., Yan, H., Liu, L., Ma, B.: Dynamic event-triggered strategy-based optimal control of modular robot manipulator: a multiplayer nonzero-sum game perspective. IEEE Trans. Cybern. (2024). https://doi.org/10.1109/TCYB.2024.3468875
3. An, T., Wang, Y., Liu, G., Li, Y., Dong, B.: Cooperative game-based approximate optimal control of modular robot manipulators for human-Robot collaboration. IEEE Transactions on Cybernetics **53**(7), 4691–4703 (2023)
4. Kalaycioglu, S., de Ruiter, A.: Dual arm coordination of redundant space manipulators mounted on a spacecraft. Robotica **41**(8), 2489–2518 (2023)
5. Chen, C., Pan, J., Liu, S., Wan, L.: Task space bilateral teleoperation of co-manipulators using power-based tdpc and leader-follower admittance control. In: IECON 2021-47th Annual Conference of the IEEE Industrial Electronics Society, pp. 1–6 (2021)
6. Ye, D., Yang, C., Jiang, Y., Zhang, H.: Hybrid impedance and admittance control for optimal robot-environment interaction. Robotica **42**(2), 510–535 (2024)
7. Mohajerpoor, R., Rezaei, S.M., Talebi, H.A., Monfaredi, R.: A robust adaptive hybrid force/position control scheme of two planar manipulators handling an unknown object interacting with an environment. In: IEEE International Conference on Robotics and Biomimetics, vol. 226, pp. 509–522 (2012)

8. Ma, B., Dong, B., Zhou, F., Li, Y.: Adaptive dynamic programming-based fault-tolerant position-force control of constrained reconfigurable Manipulators. IEEE Access **8**, 183286–183299 (2020)
9. Sugar, T., Kumar, V.: Multiple cooperating mobile manipulators. In: IEEE International Conference on Robotics and Automation, vol. 2, pp. 1538–1543 (1999)
10. Erhart, S., Sieber, D., Hirche, S.: An impedance-based control architecture for multi-robot cooperative dual-arm mobile manipulation. In: IEEE/RSJ International Conference on Intelligent Robots and Systems, pp. 315–322 (2013)
11. Liu, G., Abdul, S., Goldenberg, A.A.: Distributed control of modular and reconfigurable robot with torque sensing. Robotica **26**(1), 75–84 (2008)
12. Liu, G., Liu, Y., Goldenberg, A.A.: Design, analysis, and control of a spring-assisted modular and reconfigurable robot. IEEE/ASME Trans. Mechatron. **16**(4), 695–706 (2011)
13. Walker, I.D., Freeman, R.A., Marcus, S.I.: Analysis of motion and internal loading of objects grasped by multiple cooperating manipulators. Int. J. Robot. Res. **10**(4), 396–409 (1991)

MAMF-ResNet: Multi-Attention Mechanisms Fusion Network for Brain Tumor Classification

Qingyun Huo[1], Meng Yang[2], Yuxuan Li[1], Yunyu Wang[1], Xin Zhang[1], and Mingtao Liu[1](✉)

[1] School of Information Science and Engineering, Linyi University, Linyi 276000, Shandong, China
liumingtao@lyu.edu.cn
[2] College Library, Jinan University, Guangzhou 250022, Shandong, China

Abstract. Early diagnosis of brain tumors enables earlier, more targeted treatment, significantly improving the patient's chances of recovery. As deep learning rapidly advances in medical image processing, brain tumor classification algorithms based on it have gained significant attention. Convolutional neural network (CNN), known for their powerful feature learning abilities, can automatically extract tumor-related features from brain images for classification by training on large-scale medical image datasets. While VGG's successful application in the ImageNet competition is well-known, simply deepening the network structure does not always result in significant performance improvements for medical image tasks. As the network depth increases, the risk of overfitting and extracting irrelevant features may grow, resulting in a lack of robustness in the model's learned features. Networks that are excessively deep may experience gradients that become either too small or too large during backpropagation, leading to instability in weight updates. This paper addresses the above issues by using the ResNet network as the backbone for brain tumor image classification and proposes a way based on the dual attention mechanism: MAMF-ResNet (Multi-Attention Mechanisms Fusion ResNet). The structure of the ResNet network is optimized and adapted as a way to improve the classification accuracy. This chapter uses the publicly available BraTS 2018 and BraTS 2019 datasets for training and testing. Experimental results show that the improved model proposed in this chapter outperforms the base model in all evaluation metrics.

Keywords: Brain Tumor Classification · ResNet · Non-Local · Se-Block

1 Introduction

There are three types of brain tumors: benign, malignant and pituitary tumors. For example, according to the World Health Organization Guidelines for the Classification of Tumors of the Central Nervous System, gliomas are classified as grades I to IV, with the severity of the tumor increasing with the grade [1]. These four tumor grades can be grouped into two types: (1) Low-grade glioma (LGG), comprising grade I and II gliomas, is less malignant, slow-growing, and less aggressive, with mild tumor cell atypia. Due to their more defined margins, low-grade gliomas are relatively easier to remove. (2) High-grade glioma (HGG), comprising grade III and IV gliomas, is typically fast-growing, highly atypical, and aggressive, often spreading to surrounding normal brain tissue [2]. Treatment options for brain tumors vary widely depending on the grade, including surgery, radiation therapy, chemotherapy, targeted therapy, and oncolytic virus therapy [3]. Early diagnosis can significantly improve treatment outcomes, and the use of medical imaging technology in brain imaging analysis, diagnosis, and adjuvant therapy is a current research focus.

The goal of brain tumor classification is to automatically identify the region of the medical image containing the tumor and determine its type, thereby guiding subsequent diagnosis and treatment. Popular brain tumor classification techniques include support vector machines, random forests, K-means clustering, etc. Chen et al. [4] A method of EKF-SVM combining EKF and SVM is proposed to improve the accuracy of brain tumor image classification. It performs better for positive brain tumor images, but has limitations in classifying negative cases. Aggarwal et al. [5] used the grayscale co-occurrence matrix of brain tumor MR images to extract statistical texture features and combined them with a random forest classifier, achieving an accuracy of 83.3%. Khan et al. [6] used K-means clustering to preprocess brain tumor images and a fine-tuned VGG19 model to classify them as benign or malignant, improving classification accuracy through data augmentation. The machine learning-based classification method is highly flexible and customizable, with parameters that can be adjusted to suit datasets with similar characteristics. However, this method has poor generalization and is highly sensitive to dataset quality. Thus, researchers have begun to use deep learning-based methods for brain tumor classification.

So far, many deep learning models have been applied to brain tumor classification. Mzoughi et al. A Multiscale 3D convolutional neural network for glioma classification is proposed, using MR images covering the entire volume of T1 modalities. Multi-scale 3D CNNs allow the model to capture features at various scales and depths. Deepak et al. [6] used Siamese networks to compare similarities and differences between brain images, while proximity analysis methods offer insights into the relationships between data points and their contexts, helping to better understand and interpret the data characteristics. Rasool et al. [7] proposed a novel brain tumor classification method based on a CNN-SVM hybrid approach, combining GoogleNet and SVM for improved classification performance. Nayak et al. [8] proposed a dense EfficientNet network for classifying three types of brain tumors. This method combines minimal normalization with a dense, efficient network to enhance data and improve training accuracy with deeper network layers. Ali et al. [9] A proposed novel network architecture for tumor detection and related category classification. The method uses a serial approach to extract features from two pre-trained AlexNet and Inception-V3 models and fuses them for final classification. Kang et al. [10] A new approach was proposed to classify brain tumors by extracting deep features using a pre-trained convolutional neural network and combining it with a machine learning classifier. This approach overcomes the limitations of a single CNN model and achieves excellent performance on multiple datasets, providing an effective tool for early brain tumor diagnosis. Masood et al. [11] proposed a custom Mask Re-gion-based Convolutional Neural Network (Mask RCNN), which achieved better classification results and more accurate delineation of tumor regions, making it a potential new automated diagnostic tool.

Currently, existing deep learning-based brain tumor classification methods can automatically extract complex tumor features through training, enabling tumor classification. However, as network depth increases, overfitting and the extraction of unnecessary features may occur, Reducing models robustness of learned features.

In this paper, we improve the ResNet network structure and propose a deep learning-based brain tumor classification method, MAMF-ResNet, with the following key contributions:

(1) Integrating the channel attention mechanism module (Se-Block) after the residual structure, by learning the weights of each channel, the network focuses more on the channel information highlighted by the feature excitation part while preserving the original information. This helps optimize feature representation and enhances the model's ability to extract tumor features.
(2) The position attention mechanism module (Non-Local) was embedded between the residual modules to address the spatial information gaps that the channel attention mechanism could not fully utilize in the input brain tumor image feature map. By calculating the autocorrelation of features, the channel attention mechanism multiplies the convolution results of the input feature map and applies linear activation to obtain weight values between 0 and 1, which are used as self-attention weights.

2 Method

This section begins with a brief description of the data preprocessing used in this work, followed by a detailed explanation of the ResNet network, including the spatial and location attention mechanism modules.

2.1 Data Preprocessing

Due to the significant variation in the intensity range of MR images of brain tumors across different cases within the same modality, normalizing the maximum and minimum values [12] is an effective image preprocessing method. This operation maps the pixel values of the image to a uniform range. Minus the minimum value and divide by pixel intensity range, each image is adjusted to a normalized 0 to 1 intensity range. Depending on the application, maximum and minimum normalization is typically performed using the following formula:

$$X_{normalized} = \frac{X - \min(X)}{\max(X) - \min(X)} \quad (1)$$

where X stands for pixel value of the original input brain tumor MR image, and $X_{normalized}$ represents the normalized pixel value of the tumor image. This ensures that the input data is within a relatively small range, which helps accelerate convergence and reduce training instability.

2.2 NetWork

During this experiment, we propose a brain tumor classification method called MAMF-ResNet, using the ResNet network as the base structure, specifically de-signed for processing 3D brain tumor MR images to perform automatic classification. Figure 1 illustrates the detailed network structure of MAMF-ResNet. First, feature extraction is performed using a 3D convolutional layer with a size of $7 \times 7 \times 7$ and a stride of 2, with zero padding applied to maintain consistent feature map size. Next, batch normalization and ReLU activation functions were applied to prevent gradient vanishing, maximum pooling operations are then performed to shrink the input brain tumor feature map and improve computational efficiency. Next, the processed brain tumor feature map passed through a series of residual modules. This includes the channel focus mechanism, which makes the network pay more attention to important channel information and enhances the model's ability to recognize key features. After the second residual module, both the channel attention mechanism and the position attention mechanism are introduced to fully utilize the global context information and enhance the model's understanding of the global structure of the input brain tumor images. Finally, the spatial dimension of the processed feature map was reduced to a scalar value using global average pooling [13], followed by fully connected layers to output the final brain tumor classification results.

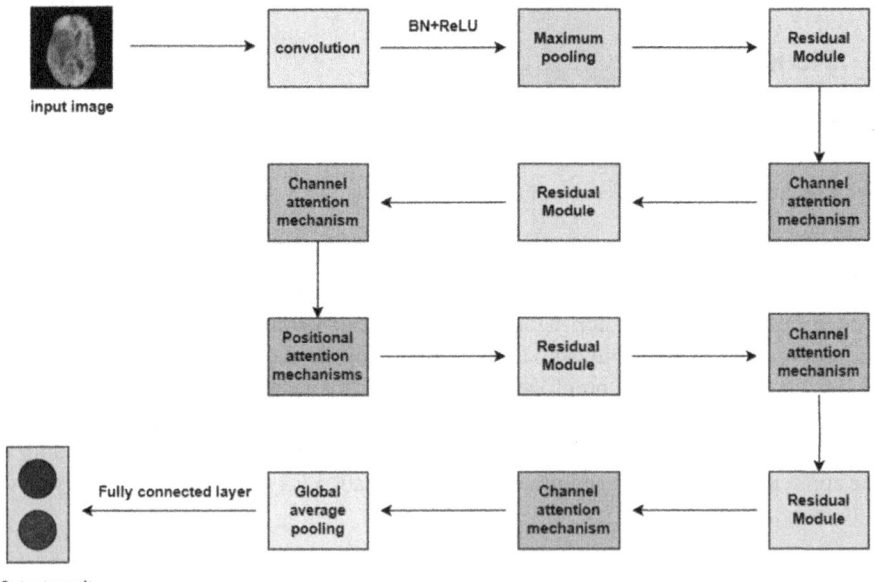

Fig. 1. MAMF-ResNet network structure

(1) Channel Attention Mechanisms.

By learning the weights of each channel, the mechanism of channel focus al-lows the network to pay more attention to channel information highlighted by the feature excitation part while retaining the original data, which helps optimize feature representation, improve model performance, and enhance generalization ability. The module consists of three main steps: first, the incoming feature map is compressed. second, the features are stimulated. and third, the features are fused.

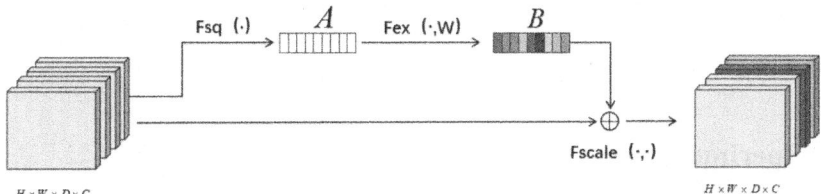

Fig. 2. Channel-wise Attention Mechanism

The specific structure is shown in Fig. 2, where the first feature map of size H × W × D × C is input, with H for height, W for width, D for depth, and C for number of channels. First, the input feature map is processed by global average pooling, where each feature channel is squeezed along the spatial dimension and compressed into a channel feature map A with global context. The weight matrix W of learning parameters is then introduced to weight the compressed feature map A, resulting in the excitation feature map B. This process enables the net-work to adaptively emphasize or suppress the feature responses of different channels. The final output feature map is obtained by weighting and summing the input feature map with the feature map B from the previous process.

(2) Positional Attention Mechanisms.

Channel attention mechanisms do not fully utilize the spatial information of feature maps, so the position attention mechanism is embedded between the residual modules of the ResNet network. Figure 3 shows the structure of the positional attention mechanism. First, a 1 × 1 × 1 convolutional linear mapping was applied to the input feature map A, compressing it to obtain three new feature maps: B, C, and D. These feature maps were then merged along all dimensions except the channel dimension through a remodeling operation. Next, matrix dot multiplication is performed on feature maps B and C to calculate the autocorrelation of the features. The resulting values are then linearly activated to obtain weight values between 0 and 1, which serve as the self-attention weights. Finally, the self-attention weight is multiplied by feature map D, and the result undergoes a 1 × 1 × 1 convolution to produce feature map E. Feature map E is then added to the original input feature map A via residual linking to generate the final output feature map F. The feature map E is then added to the original input feature map A via residual linking.

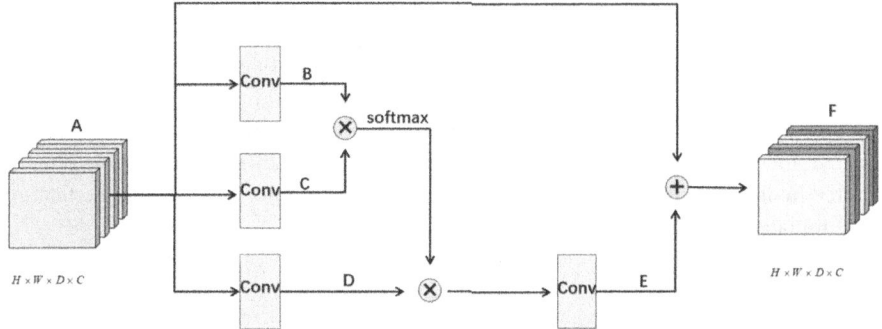

Fig. 3. Spatial attention mechanism

3 Experiment

3.1 Dataset

This chapter uses two datasets, the same as those in the BraTS 2018 and BraTS 2019 challenges [14], to distinguish between the grades of HGG and LGG. The classification of HGG and LGG is crucial for patients, as it informs both treatment decisions and post-recovery evaluations. HGG is typically linked to worse outcomes and is harder to treat, while LGG is relatively easier to manage. Accurate classification enables physicians to plan treatment more effectively, improve outcomes, and predict survival and recurrence risks with greater accuracy.

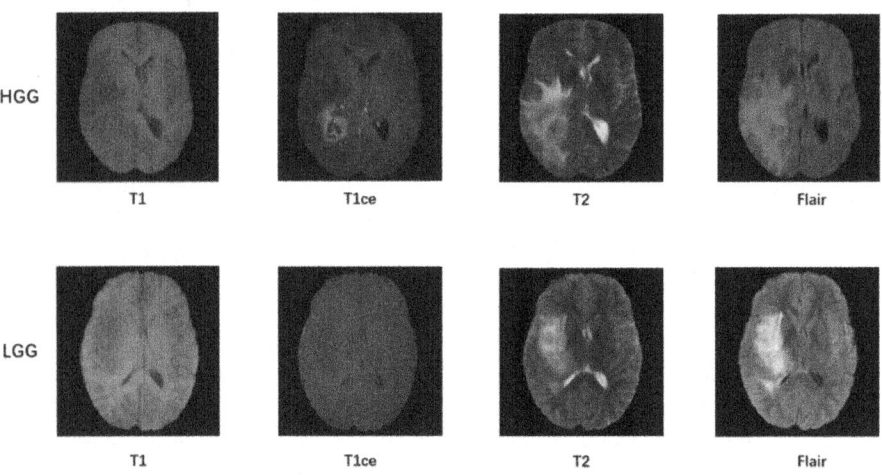

Fig. 4. In the dataset example, visualizations of various modalities under HGG/LGG states

As shown in Fig. 4, the dataset includes 469 HGG images across four modalities and 151 LGG images, also covering the same modalities. In constructing the training set, 543 samples were randomly selected from the overall dataset, including 421 HGG) and

122 LGG. This ensures that the training set is sufficiently diverse and representative, improving the training of the deep learning model. To evaluate the model's performance, 48 HGG and 29 LGG were selected, resulting in 77 samples for testing. Table 1 presents the sources and numbers of images used in these experiments.

Table 1. Dataset information

lable	source	quantities
HGG	BraTS 2018	210
LGG	BraTS 2018	75
HGG	BraTS 2019	259
LGG	BraTS 2019	76

3.2 Experimental Settings

In this experiment, high-performance computing hardware supported the in-depth study and accurate evaluation of brain tumor classification algorithms on medical image datasets. Table 2 provides the hardware and version details used in this experiment.

Table 2. The hardware facilities and corresponding parameters involved in the experimental use

Hardware configuration and version of the platform	
GPU	A5000
random access memory (RAM)	24G
operating system	Ubuntu
development environment	Tensorflow-gpu = 2.5
Development IDE	VSCode
programming language	Python
CUDA version	11.2

3.3 Evaluation Indicators

To fully evaluate the performance of ResNet neural network in the brain tumor classification task, the confusion matrix is used as the key evaluation metric in this chapter.

Consider the classification task of HGG and LGG, where the positive class is high-grade glioma and the negative class is low-grade glioma. In the experiments of this chapter, TP represents the correct classification of HGG samples as high-grade, FP refers

Table 3. Example of confusion matrix

confusion matrix		Real Labeling	
		positive	negative
Predictive Labeling	positive	True positive (TP)	False positive (FP)
	negative	False negative (FN)	True negative (TN)

to the incorrect classification of LGG samples as high-grade, TN denotes the correct classification of LGG samples as low-grade, and FN refers to the incorrect classification of HGG samples as low-grade. Analyzing the confusion matrix helps gain a deeper understanding of the ResNet model's performance across different brain tumor classification scenarios and offers valuable insights for further model optimization and enhancement. The confusion matrix examples in Table 3 provide clear, detailed information to interpret the network's classification performance.

(1) Accuracy is an index calculated by determining the proportion of brain tumors correctly categorized cases requiring classification. It is the most intuitive and easy-to-understand evaluation metric. Accuracy is presented as a percentage and provides a clear view of the model's overall performance across the entire dataset. The formula is presented in Eq. 2.

$$ACC = \frac{TP + TN}{TP + TN + FP + FN} \quad (2)$$

(2) In the brain tumor classification task, sensitivity measures the model's ability to correctly identify high-grade gliomas. High sensitivity indicates that the model can better capture the presence of high-grade gliomas, reducing the risk of missed diagnoses and suggesting that the model has learned more characteristic features of high-grade gliomas. Sensitivity is also referred to as recall. The formula is presented in Eq. 3.

$$SEN = \frac{TP}{TP + FN} \quad (3)$$

(3) Specifications measure the ability of the model to correctly identify low-grade gliomas in the dataset. Improved specificity helps reduce misclassification, ensuring more accurate results and better treatment options for patients. The formula is presented in Eq. 4.

$$SPE = \frac{TN}{TN + FP} \quad (4)$$

(4) In the brain tumor classification task, PPV refers to the proportion of actual high-grade gliomas among all samples predicted as high-grade gliomas, also known as precision. The formula is presented in Eq. 5.

$$PPV = \frac{TP}{TP + FP} \quad (5)$$

(5) In the brain tumor classification task, negative predictive value (NPV) refers to the proportion of actual low-grade gliomas among all samples predicted as low-grade gliomas, as presented in Eq. 6.

$$NPV = \frac{TN}{TN + FN} \tag{6}$$

(6) The F1 score is a composite indicator that includes evaluates a classification model's performance by balancing positive predictive value and sensitivity. This ensures optimal performance in minimizing misdiagnoses while correctly identifying all true positive instances in brain tumor classification. The formula is presented in Eq. 7.

$$F1 = \frac{2 \times PPV \times SEN}{PPV + SEN} \tag{7}$$

3.4 Experimental Results

The T1ce modality is an MRI of a brain tumor obtained using contrast enhancement, which enhances tissue visibility and allows for more accurate differentiation between normal tissue and abnormal lesions compared to other modalities. Therefore, the T1ce modality from the MRI dataset is chosen as the focus of this experiment.

(1) Hyperparameter experiment

To ensure the effectiveness of model training, this chapter's experiments focus on setting the initial learning rate to avoid both overly rapid convergence due to a large learning rate and excessively slow training due to a small learning rate. To address these issues, this experiment uses a learning rate decay strategy to balance the model's speed and stability during training. In the early stages, a large learning rate was chosen to rapidly approach the global optimal solution. As training progresses, the learning rate is gradually reduced to help stabilize the model weights, thereby accelerating convergence. The experiment consists of 50 iterations, divided into 4 stages to reduce the learning rate: the first stage is 1–5 rounds, the second stage is 5–25 rounds, the third stage is 25–35 rounds, and the fourth stage is 35–50 rounds. To verify the optimality of the initial learning rate, a ResNet-based network structure is used in this chapter, and hyperparameter experiments are conducted. The performance differences between various initial learning rates are compared by evaluating the Acc and F1 scores under different learning rate combinations. The optimal learning rate combination, [0.01, 0.0005, 0.0001, 0.00005], is selected. The specific experimental data are shown in Tables 4 and 5.

To accelerate model convergence, this chapter adopts a batch processing method. The data is divided into several subsets, and gradients are calculated for each subset to update the parameters, allowing the model to find a local optimal solution more quickly and improve convergence speed. However, several factors must be considered when setting the batch size. An inappropriate batch size may lead to insufficient memory or cause the model to overfit, negatively affecting its generalization ability on the brain tumor test dataset. On the other hand, setting the batch size too small may underutilize hardware resources and result in unstable gradient estimation. To address the above

Table 4. Confusion matrix under different learning rates in T1ce mode

Initial setting of the learning rate	TP	TN	FN	FP
0.01,0.005,0.001,0.0005	43	19	5	10
0.001,0.0005,0.0001,0.00005	47	18	1	11
0.0001,0.00005,0.00001,0.000005	47	13	1	16

Table 5. Comparison of experimental results of T1ce mode under different initial learning rates

Initial setting of the learning rate	Acc	F1 score
0.01,0.005,0.001,0.0005	80.52%	85.15%
0.001,0.0005,0.0001,0.00005	**84.42%**	**88.68%**
0.0001,0.00005,0.00001,0.000005	77.92%	84.68%

issues, the classification accuracy and F1 score were compared under different batch sizes to evaluate their effect. The initial batch size was then selected as 4. The specific experimental data are shown in Tables 6 and 7.

Table 6. Confusion matrix under different learning rates in T1ce mode

Initial setting of batch size	TP	TN	FN	FP
2	47	13	1	16
3	45	18	3	11
4	47	18	1	11
5	42	20	6	9

Table 7. Comparison of experimental results of T1ce mode under different initial learning rates

Initial setting of batch size	Acc	F1 score
2	77.92%	84.68%
3	81.82%	86.54%
4	**84.42%**	**88.68%**
5	80.51%	84.85%

(2) Ablation Study

To visually assess the impact of multiple improvement strategies on the performance of the algorithm, this chapter demonstrates through ablation experiments that the channel-focused and location-focused mechanisms improve the classification performance of the underlying network to varying degrees. Different network models were trained independently using the same experimental parameters to ensure the reliability of the comparison results. Our experimental results show that combining the channel attention mechanism and the location attention mechanism can improve the brain tumor classification performance of the network to different degrees. The specific results are shown in Table 8, where "Se Block" refers to the channel attention mechanism module, and "MAMF-ResNet" refers to the network structure that includes both the channel attention and position attention modules.

Table 8. Confusion Matrix for T1ce Modality

Method	TP	TN	FN	FP
ResNet	47	18	1	11
Resnet + Se Block	46	22	2	7
MAMF-ResNet	46	23	2	6

Additionally, this chapter provides a detailed record of the performance evaluation indicators for the T1ce modality. For the F1 evaluation metric, the channel attention mechanism module increased from 88.68% to 91.09%, while the position attention module further increased to 92%. Similarly, for the accuracy (ACC) metric, the channel attention module improved from 84.42% to 88.31%, and the position attention module increased to 89.61%. This clearly demonstrates the positive impact of introducing the channel attention and position attention modules on improving the model's overall accuracy. The results are summarized in Table 9, where "Se Block" represents the channel attention mechanism module, and "MAMF-ResNet" refers to the network structure that includes both the channel attention mechanism and the position attention module.

Table 9. Model Performance Metrics for T1ce Modality

Method	F1	ACC	SEN	SPE	PPV	NPV
ResNet	88.68%	84.42%	**97.92%**	62.07%	81.03%	**94.74%**
ResNet + Se Block	91.09%	88.31%	95.83%	75.86%	86.79%	91.67%
MAMF-ResNet	**92%**	**89.61%**	95.83%	**79.31%**	**88.46%**	92%

4 Conclusion

This chapter presents a deep learning-based way for classifying brain tumor images. First, the basic process of the method is described, and the selected ResNet-based underlying network architecture is explained in detail. To enable the net-work to establish finer correlations between features at different levels and dynamically adjust the feature weights of each residual block, this chapter improves the network and proposes an enhanced method for brain tumor image classification based on the ResNet model, MAMF-ResNet. In the improved method, a channel attention module is introduced between each residual module to enable multi-scale context information fusion, thereby improving the model's perceptual ability. Additionally, a spatial attention mechanism is added to the decoder module to assign different weights to features, enabling accurate selection of the most important features and improving the segmentation model's precision and accuracy. The performance of the improved method was then thoroughly evaluated through ablation experiments, which confirmed its significant impact on enhancing the accuracy of brain tumor image segmentation. In the training process, only the T1ce modality from the four available brain tumor MR image modalities is input into the network, without leveraging the complementary information be-tween modalities. In future work, we could explore feeding all four modalities into the network simultaneously during training to extract their corresponding features.

References

1. Komori, T.: Grading of adult diffuse gliomas according to the 2021 WHO classification of tumors of the central nervous system. Lab. Invest. **102**(2), 126–133 (2022)
2. Nadeem, M.W., Ghamdi, M.A.A., Hussain, M., et al.: Brain tumor analysis empowered with deep learning: a review, taxonomy, and future challenges. Brain Sci. **10**(2), 118 (2020)
3. Zaccagna, F., Grist, J.T., Quartuccio, N., et al.: Imaging and treatment of brain tumors through molecular targeting: recent clinical advances. Eur. J. Radiol. **142**, 109842 (2021)
4. Chen, B., Zhang, L., Chen, H., et al.: A novel extended Kalman filter with support vector machine based method for the automatic diagnosis and segmentation of brain tumors. Comput. Methods Programs Biomed. **200**, 105797 (2021)
5. Aggarwal, A.K.: Learning texture features from GLCM for classification of brain tumor mri images using random forest classifier. Trans. Signal Process. **18**, 60–63 (2022)
6. Khan, A.R., Khan, S., Harouni, M., et al.: Brain tumor segmentation using K-means clustering and deep learning with synthetic data augmentation for classification. Microscopy Res. Tech. **84**(7), 1389–1399 (2021)
7. Rasool, M., Ismail, N.A., Boulila, W., et al.: A hybrid deep learning model for brain tumour classification. Entropy **24**(6), 799 (2022)
8. Nayak, D.R., Padhy, N., Mallick, P.K., et al.: Brain Tumor Classification Using Dense Efficient-Net. Axioms 2022, 11, 34. Mathematical Fuzzy Logic in the Emerging Fields of Engineering, Finance, and Computer Sciences, 2022: 151
9. Ali, M., Shah, J.H., Khan, M.A., et al.: Brain tumor detection and classification using pso and convolutional neural network. Comput. Mater. Contin. **73**(3), 4501–4518 (2022)
10. Kang, J., Ullah, Z., Gwak, J.: Mri-based brain tumor classification using ensemble of deep features and machine learning classifiers. Sensors **21**(6), 2222 (2021)
11. Masood, M., Nazir, T., Nawaz, M., et al.: A novel deep learning method for recognition and classification of brain tumors from MRI images. Diagnostics **11**(5), 744 (2021)
12. Dubey, A.K., Jain, V.: Comparative study of convolution neural network's relu and leaky-relu activation functions. In: Applications of Computing, Automation and Wire-less Systems in Electrical Engineering: Proceedings of MARC 2018, pp. 873–880. Springer, Singapore, 2019
13. Kumar, R.L., Kakarla, J., Isunuri, B.V., et al.: Multi-class brain tumor classification using residual network and global average pooling. Multimed. Tools Appl. **80**, 13429–13438 (2021)
14. Caelen, O.: A Bayesian interpretation of the confusion matrix. Ann. Math. Artif. Intell. **81**(3–4), 429–450 (2017)

Noise Sensitive Relation Aware Cross-Lingual Entity Alignment

Yunlong Tian[1], Shuanglong Yao[2], Liqiu Shan[3], and Xing Wang[2](✉)

[1] Qingdao Haier Technology Co., Ltd., QingDao, China
[2] School of Information Science and Engineering, Linyi University, Linyi, China
wangxing@lyu.edu.cn
[3] School of Electronic and Information Engineering, Liaoning Technical University, Huludao, China

Abstract. The goal of cross-language entity alignment is to connect the same entities in the cross-language knowledge graph to help downstream applications that need cross-language knowledge. At present, the main methods to realize cross-language entity alignment are knowledge graph representation learning method and graph convolution network method. The former uses the knowledge graph representation learning method to learn the embedded vectors of knowledge in different knowledge graphs and aligns entities by calculating the similarity between vectors, but this method cannot well capture the complex relationships commonly existing in multi-relational knowledge graphs. In contrast, the latter shows obvious advantages. However, pre-aligned entity pairs from manual annotation usually contain errors, which will affect the final alignment results. Therefore, this paper proposes a noise-sensitive relationship-aware dual graph convolution network (NSRDGCN), which is composed of noise detection and relationship-aware entity alignment. The noise detection part uses GAN to detect the noise in the training set, and the relationship-aware entity alignment part learns a better entity representation through the influence between the triples of (entity, relationship, entity) and its corresponding new triples (relationship, entity, relationship). Experiments on three cross language entity data sets show that our method is significantly better than the existing entity alignment methods. In particular, based on the benchmark data DBP15K$_{JA-EN}$, the entity alignment model NSRDGCN achieves a relative improvement of about 1.6% for Hits@10 compared to the RDGCN model.

Keywords: Knowledge Graph · Dual Relation Graph · Noise Detection · Entity Alignment

1 Introduction

The progress of science and technology has led to the increasingly prominent role of knowledge graph in different fields. Recently, many knowledge graphs

[1,2] have been constructed to provide structural knowledge for different applications. The languages used to construct knowledge graphs are likely to be different. These separately constructed knowledge graphs contain heterogeneous but complementary contents. Therefore, it is meaningful to integrate different knowledge graphs across languages into a unified knowledge graph, which is very helpful for downstream applications that need cross language knowledge, such as recommendation system [3], information extraction [4,5] and question and answer [6].

In order to integrate the knowledge graph, an effective method is to align two or more different information sources but the same entity, that is, entity alignment. Early entity alignment methods mainly relied on manpower [7] or additional resources [7,8] to find new aligned entity pairs. Since 2013, the rise of knowledge graph representation learning [9–11] has provided a new idea for entity alignment. These methods [12–17] use the existing knowledge graph representation learning methods to learn the embedded vectors of knowledge in different knowledge graphs, and align entities by calculating the similarity between vectors in the same vector space. Compared with the early methods, this method only requires less people to participate in the feature construction, which can not only retain the structure of the knowledge graph, but also implicitly complete the missing links in the existing knowledge of the knowledge graph [18], but this method can not well capture the common complex relationship information in the multi-relationship knowledge graph. It will be guided by the error of the hypothesis of "head entity + relationship $\approx$ tail entity" of the knowledge graph representation learning method, so as to reduce the efficiency of the model in acquiring relationships in multiple relationship graphs [19].

Recent work [20] adopts a different method. They use the emerging graph convolution network model (GCN) [21] to form a new entity vector by adding the vector information of its neighbors to the entity vector in the knowledge graph, which opens a new idea of entity alignment. However, this method cannot correctly model the relationship information. Because the general graph convolution network operates on undirected graph and unmarked graph, the model based on graph convolution network will ignore the edge relationship information of knowledge graph, and the edge relationship information is likely to be useful. Relational graph convolution network (R-GCN) [22] can be used to model multiple relational graphs, but the number of parameters required by relational graph convolution network for knowledge graph containing many relationships is very large, which will undoubtedly slow down the learning speed of a relational graph convolution network model. The dual primitive graph convolution network (DPGCNN) [23] provides a new solution to this problem. The dual primitive graph convolution network alternately convolutes the primitive graph and the dual graph (whose vertices correspond to the edges of the primitive graph), and applies the graph attention mechanism to learn the primitive edge representation. Therefore, the dual primitive graph convolution network is better for the learning of complex multilateral structure knowledge graph, and can also get a better representation of knowledge graph [19]. However, the training of the above models is inseparable from the clean training set, which makes the models more vulnerable to the influence of noise [24]. In particular, the influence is more obvious because the relationship aware dual graph convolution entity alignment model [19] has more in-depth mining of entity information and relationship information.

In order to overcome the limitation of relational aware dual graph convolution entity alignment model in dealing with noisy entity pairs, this paper proposes a noise sensitive relational aware cross language entity alignment model (NSRDGCN). The basic idea of NSRDGCN is to combine the noise detection module with the relational aware entity alignment module, and use the relational aware entity alignment module to model the cross-language knowledge graph and learn the representation of nodes. Generative Adversarial Network (GAN) [25], an adversarial training based method, is used for noise detection in the noise detection module. Since the training set uses only clean entities during training, the noise generator receives inputs of clean entity pairs, samples them and generates noisy entity pairs. The noise discriminator takes the clean entity pairs and the noisy entity pairs as inputs and trains a model capable of recognizing the noise. The noise generator and the noise discriminator conduct confrontation training to improve the generation and recognition ability of noise data. At the same time, the noise discriminator generates a trust score for the input aligned entity pair and supplements the aligned entity pair with higher score to the clean aligned entity pair training set. It is used to represent and learn the nodes of the relationship aware dual graph convolution entity alignment module. The two modules are trained iteratively to solve the problem of correct acquisition and integration of relational information and the problem of noise in the training set at the same time.

The contribution of this paper is mainly reflected in the following three points:

1. By combining the noise detection module with the relation aware dual graph convolutional entity alignment model, we propose a noise sensitive relation aware cross-lingual entity alignment model.
2. We solve the problem of higher requirements for data sets in relation aware cross-lingual entity alignment model, and improve the robustness of relation aware cross-lingual entity alignment.

2 Method

2.1 Problem Formulation

We use $G_1 = (E_1, R_1, T_1)$ and $G_2 = (E_2, R_2, T_2)$ to represent two aligned knowledge graphs of different languages, where E, R, and T represent G entities, relations, and triples, respectively. The entities in G_1 may have corresponding entities of different languages in G_2. $AS = \{(e_1, e_2) | e_1 \in E_1, e_2 \in E_2\}$, where e_1 and e_2 are aligned entity pairs; AS is the seed set. The entity alignment task is to use the alignment seed to automatically find more equivalent entities. When noise is involved, the AS seed set is not completely correct because there may be errors in the process of labeling alignment entity pairs. Thus, we use AS^T as a reliable and noise-free alignment entity pair to train entity alignment, and $AS^U = AS \setminus AS^T$ as an unreliable tag entity pair. Given AS^U with noisy entity pairs and AS^T with only trusted entity pairs, our model finds and aligns entities without noise in AS^U.

2.2 Approach: NSRDGCN

The framework can be divided into two parts: relationship aware entity alignment part and noise detection part, as shown in Fig. 1. Firstly, the relationship aware entity alignment part is used for training to minimize the distance between entities in the entity pair in the training set; Secondly, we use the noise detection part to distinguish between real entity pairs and noise entity pairs. At the same time, we need the interaction between generator and discriminator to guide our own optimization.

2.3 Relation Aware Dual Graph Convolutional Entity Alignment

Inspired by the higher utilization of knowledge in the knowledge graph by the relationship aware dual graph convolution entity alignment model [19], our model draws on the idea of the relationship aware entity alignment model, but the result is better, because our model can make the entity information and relationship information in the knowledge graph cleaner, so it improves the final entity alignment effect. In order to integrate the relationship into the entity representation more fully, in the case of a given input knowledge graph, the dual relationship graph is constructed, and the vertices of the dual relationship graph represent the edge relationships in the original knowledge graph. Then, we use the attention mechanism of the graph to generate weights to stimulate the interaction between the original knowledge graph and the dual graph, The vertex vector in the original knowledge graph is then sent to the GCN [21] layer to increase the information of vertex neighbors, and the final entity information will be provided to the noise detection part.

Dual Graph. G_1 and G_2 together constitute the original knowledge graph $G_e = (V_e, E_e)$, $V_e = E_1 \cup E_2$ includes all entity vertices in the original knowledge graph, $E_e = T_1 \cup T_2$ includes all relationship edges in the original knowledge graph. Given a primal graph G_e, its dual graph $G_r = (V_r, E_r)$ can be constructed according to the following method:

1. For relation r in G_e, there are vertices v^r, $v^r \in V_r$, so $V_r = R_1 \cup R_2$.
2. Create the edge u_{ij}^r connecting v_i^r and v_j^r in G_r if two relations r_i and r_j have the same entity in G_e.

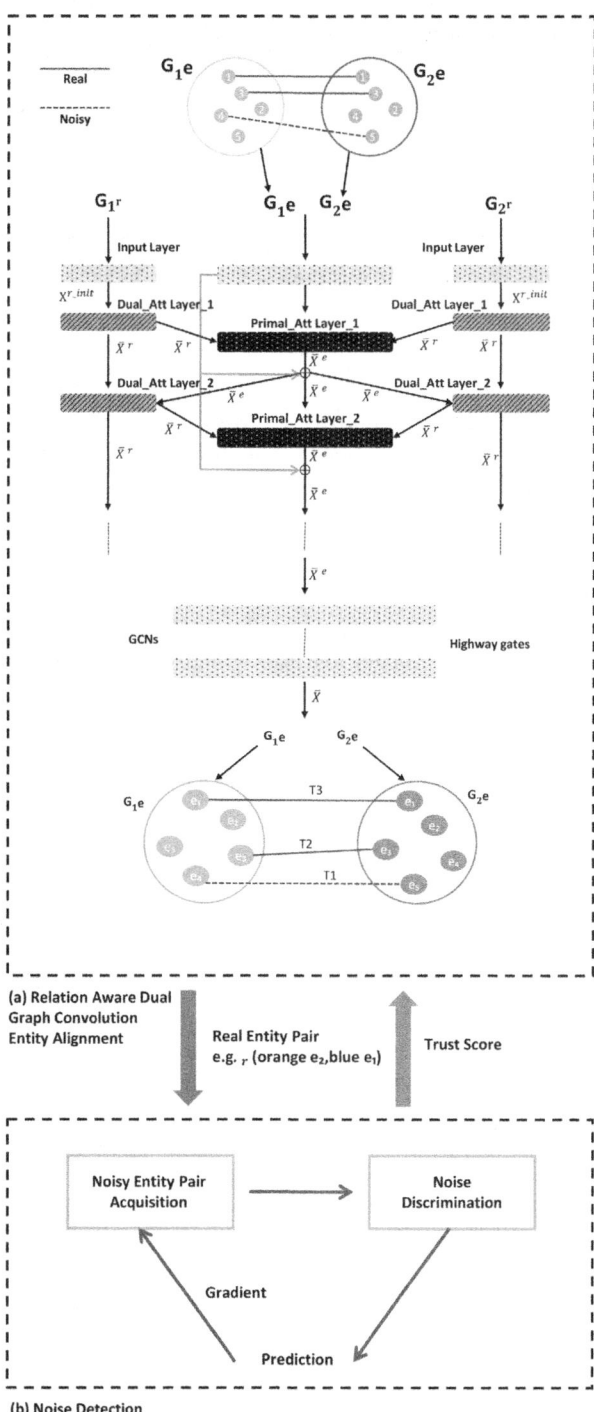

Fig. 1. The framework of noise sensitive relationship aware cross language entity alignment model.

In order to make the nodes in the dual graph more expressive, we define the weight w_{ij}^r for the edge u_{ij}^r in the dual graph according to the situation of sharing head and tail entities,

$$w_{ij}^r = H(r_i, r_j) + T(r_i, r_j), \tag{1}$$

$$H(r_i, r_j) = \frac{H_i \cap H_j}{H_i \cup H_j}, \quad T(r_i, r_j) = \frac{T_i \cap T_j}{T_i \cup T_j}. \tag{2}$$

In the formula H_i and T_i are the head entity set and tail entity set of relation r_i in G_e.

The Function of Dual Graph and Primitive Knowledge Graph. The attention [26] of the graph to generate weights is used to obtain the vertex vectors of the original knowledge graph and the dual graph in this paper. The interaction between the original knowledge graph and the dual graph includes two parts, namely, the dual attention part and the primal attention part.

Dual Attention Part
$X^r \in \mathbb{R}^{m \times 2d}$ is the vertex vector matrix of the dual graph, represents vertices in a dual graph. The primitive vertex feature, which is generated by the primitive attention part (Eq. 8), is used to calculate the dual attention score:

$$\widetilde{x}_i^r = \sigma^r \left(\sum_{j \in N_i^r} a_{ij}^r x_j^r \right), \tag{3}$$

$$\alpha_{ij}^r = \frac{exp(\eta(w_{ij}^r \alpha^r [c_i \parallel c_j]))}{\sum_{k \in N_i^r} exp(\eta(w_{ij}^r \alpha^r [c_i \parallel c_k]))}, \tag{4}$$

where $\widetilde{x}_i^r$ is the output representation of d'-dimensional of dual vertex v_i^r, x_j^r is the dual representation of vertex v_j^r, N_i^r is the set of neigh bors of v_i^r, α_{ij}^r is the attention score, α^r is a fully connected layer mapping 2d'-dimensional input to scalar, σ^r is ReLU, and η is Leaky ReLU, $\parallel$ is the join operation, and c_i is the relationship vector representation obtained from the original knowledge graph.

In the framework based on graph embedding, we cannot directly provide relational representation, because there is too little training data. Therefore, in order to approximate the relation vector of r_i, we connect the average vector representation of the head entity and the tail entity of r_i in G_e.

$$c_i = \left[\frac{\sum_{k \in H_i} \hat{X}_k^e}{|H_i|} \parallel \frac{\sum_{l \in T_i} \hat{X}_l^e}{|T_i|} \right], \tag{5}$$

where $\hat{X}_k^e$ and $\hat{X}_l^e$ are the vector of the k-th layer head entity and the L-th layer tail entity of the relationship in the attention part of the original knowledge graph.

Primal Attention Part

When applying GAT [27] to the primitive knowledge graph, we can use the dual vertex representation in to calculate the primal attention score. So we can use the relation representation generated by double attention layer to influence the primal vertex embedding.

Specifically, we use $X^r \in \mathbb{R}^{n \times d}$ to represent the vertex vector matrix of original knowledge graph. For the entity e_q in the primal G_e, its expression $\widetilde{x}_q^e$ can be calculated by the following formula:

$$\widetilde{x}_q^e = \sigma^e \left(\sum_{t \in N_q^e} \alpha_{qt}^e x_t^e \right), \quad (6)$$

$$\alpha_{qt}^e = \frac{exp\left(\eta\left(a^e\left(\widetilde{x}_{qt}^r\right)\right)\right)}{\sum_{k \in N_q^e} exp\left(\eta\left(a^e\left(\widetilde{x}_{qk}^r\right)\right)\right)}, \quad (7)$$

where $\widetilde{x}_{qt}^r$ is the dual representation of r_{qt} in G_r, N_q^e is a set of neighbor indexes of e_q in G, α_{qt}^e is the primal attention score, a^e is a fully connected layer mapping, and σ^e is the primal attention part activation function.

We preserve information by combining the initial vector representation with the output of the original attention layer:

$$\hat{X}_q^e = \beta_s * \widetilde{x}_q^e + x_q^{e_{init}}. \quad (8)$$

where $\widetilde{x}_q^e$ represents the final output of the interaction module of the entity e_q in G_e, and β_s is the weighting parameter of the s-th layer of the primal attention part.

Incorporating Information. After multiple rounds of interaction between dual graph and original knowledge graph, we can collect relevant perceptual entity representations. Next, we apply two high-speed gate controlled GCN [25] to the final primal graph to further collect the information of its adjacent structure.

In each GCN layer l, the entity representation $X^{(l)}$ is used as the input, and the output representation $X^{(l+1)}$ can be calculated as:

$$X^{(l+1)} = \xi \left(\widetilde{D}^{-\frac{1}{2}} \widetilde{A} \widetilde{D}^{-\frac{1}{2}} X^{(l)} W^{(l)} \right). \quad (9)$$

where I is an identity matrix, $\widetilde{A} = A + I$ is the adjacency matrix, $\widetilde{D}_{jj} = \sum_k \widetilde{A}_{jk}$ and $W^{(l)} \in \mathbb{R}^{d^{(l)} \times d^{(l+1)}}$ are layer specific training weight matrices; ξ is the ReLU activation function.

The Object Of Relation Aware Dual Graph Convolutional Entity Alignment. After the entities of knowledge graph G_1 and G_2 are embedded from the relation aware dual graph convolutional entity alignment module, the entities in the marked entity pair in AS should be closer in the vector space because of the same semantics. For example, when (e_x, e_y) established, that is $e_x \approx e_y$, the marked pair of entities can be used as training sets to train the relation aware dual graph convolutional entity alignment module. But most of the existing works assume that the data in all training sets are correct, there is inevitably noise in the training set, inspired by REA model [24], we introduce a trust score to describe the possibility of the marked entity pair (e_x, e_y) as true. Therefore, the objective function based on marginal ranking for the alignment of noise sensing entities can be defined as follows by trust score:

$$\tau_{EA} = \sum_{(e_x,e_y)\in AS} \sum_{(e'_x,e'_y)\in AS'_{(e_x,e_y)}} TS(e_x,e_y)\left[\gamma + f_d(e_x,e_y) - f_d(e'_x,e'_y)\right]_+, \tag{10}$$

where $TS(e_x, e_y)$ is the trust score of entity pair (e_x, e_y) and $[x]_+ = max\{0, x\}$ is the part where x is greater than 0, γ is margin hyper-parameter greater than 0, $f_d(\cdot)$ is a distance function

$$f_d(e_x, e_y) = \|e_x - e_y\|_1, \tag{11}$$

AS' refers to the negative sample set of entity pair (e_x, e_y):

$$AS'_{(e_x,e_y)} = \{(e'_x, e_y) \mid e'_x \in \varepsilon_i\} \cup \{(e_x, e'_y) \mid e'_y \in \varepsilon_j\}. \tag{12}$$

where (e'_x, e_y) and (e_x, e'_y) are the negative sampling entity pairs obtained by randomly replacing e_x or e_y with (e_x, e_y).

2.4 Noise Detection

Firstly, we need to design a strategy to get the pair of noise entities to overcome the shortcomings of the pair of noise marking entities. Then we need to design a discriminant model to distinguish between noisy data and labeled real data. After convergence, the discriminant model can be used to classify the entities in AS. Inspired by GAN [9], the noise generator and noise discriminator are unified by the minimax game, and a noise detection model is formed. The formal model formula is as follows:

$$L_{ND} = \max_\varphi \min_\theta \sum_{(e_x,e_y)\in AS^T} E_{(e_x,e_y)} \sim AS^T \left[\log D\left((e_x,e_y);\varphi\right)\right]$$
$$+ E_{(e'_x,e'_y)\sim G(\cdot|(e'_x,e'_y);\theta)}\left[\log(1 - D((e'_x,e'_y);\varphi))\right]. \tag{13}$$

where $D(;\varphi)$ is noise discriminator, which takes real entity pair (e_x, e_y) or generated noise entity pair (e'_x, e'_y) as input, $G(\cdot \mid (e'_x, e'_y);\theta)$ represents noise entity pair generator, which generates noise entity pair (e'_x, e'_y) by using given real entity pair (e_x, e_y), and in addition, real entity pair samples AS^T from a set of trusted entities.

Noise Generator. The noise generator should be able to fit the potential correlation distribution in the real data, and extract the most "real" entity pair from the fitted set of negative sample entity pairs for each entity pair in the AS^T as the noise sample to deceive the noise discriminator. More specifically, the noise generator uses the (e_x, e_y) entity embedding learned from the RDGCN entity alignment module to generate noise pairs (e'_x, e'_y) as close to the real distribution as possible. The formula of noise generator is as follows:

$$\theta^* = arg \min_{\theta} \sum_{(e_x,e_y) \in AS^T} \frac{E_{(e'_x,e'_y) \sim G(\cdot \mid (e'_x,e'_y);\theta)} \left[log(1 - D((e'_x,e'_y);\varphi)) \right]}{L_G(e_x, e_y)}, \tag{14}$$

where $L_G(e_x, e_y)$ is the loss function of entity to (e_x, e_y).

The probability formula of generating noise (e'_x, e'_y) as follows:

$$G((e'_x, e'_y) \mid (e_x, e_y); \theta) = \frac{exp(f_\theta(e'_x, e'_y))}{\sum exp(f_\theta(e^*_x, e^*_y))}, \tag{15}$$

$$(e^*_x, e^*_y) \in N(e_x, e_y) \subset AS'_{(e_X, e_y)}, \tag{16}$$

where $f_\theta(x, y)$ is a two-layer neural network. The probability of generating all (e^*_x, e^*_y) in $N(e_x, e_y)$ determines the probability of generating (e'_x, e'_y). we use a subset of $AS'_{(e_X, e_y)}$ as a set of negative samples. Polynomial sampling is used to collect sample (e'_x, e'_y) from $G(\cdot \mid (e'_x, e'_y); \theta)$.

The entity pair (e'_x, e'_y) generated by the generator is discrete, and the gradient descent based optimizer cannot optimize it. At present, the commonly used solution is to use reinforcement learning algorithm based on strategy gradient [28,29]. The gradient of the solid pair (e_x, e_y) can be derived as follows:

$$\nabla_\theta L_G(e_X, e_y)$$
$$= \nabla_\theta E_{(e'_x,e'_y) \sim G(\cdot \mid (e'_x,e'_y);\theta)} \left[log(1 - D((e'_x, e'_y); \varphi)) \right]$$
$$= E_{(e'_x,e'_y) \sim G(\cdot \mid (e'_x,e'_y);\theta)} \left[\nabla_\theta log G(\cdot \mid (e'_x, e'_y); \theta) log(1 - D((e'_x, e'_y); \varphi)) \right] \tag{17}$$
$$\simeq \frac{1}{k} \sum^1_k \nabla_\theta log G(\cdot \mid (e'_x, e'_y); \theta) log(1 - D((e'_x, e'_y); \varphi)),$$

(e_x, e_y) can be regarded as a stage, $G(\cdot \mid (e_x, e_y))$ as a strategy, (e'_x, e'_y) as an action, $log(1 - D((e'_x, e'_y); \varphi)$ and as a reward. The generator is an agent in reinforcement learning, which can perform interaction with the discriminator and environment according to the above state and strategy, and then update it by maximizing the reward. In addition, in order to reduce variance, we introduced baseline [33] into the reward function, and the formula is as follows:

$$R_G = \left[log(1 - D((e'_x, e'_y); \varphi))\right] - E_{(e'_x, e'_y) \sim G(\cdot|(e_x, e_y); \theta)} \left[log(1 - D((e'_x, e'_y); \varphi))\right]. \tag{18}$$

where $E_{(e'_x, e'_y) \sim G(\cdot|(e_x, e_y); \theta)} \left[log(1 - D((e'_x, e'_y); \varphi))\right]$ is the baseline function in the policy gradient. In addition, the noise entity pair generated from the generator can be used as negative samples to improve the performance of the noise aware entity alignment part.

Noise Discrimination. The noise discriminator is used to separate the real data and noise in AS^U. Firstly, the samples generated by the noise generator are taken as negative samples, and then the entity pairs in AS^T are taken as positive samples. The objective function of the discriminator is to maximize the log likelihood of correctly distinguishing positive and negative pairs when $G(\cdot \mid (e_x, e_y); \theta)$ is fixed. The formula of noise discriminator is as follows:

$$\varphi^* = argmax_\varphi \sum_{(e_x, e_y) \in AS^T} E_{(e_x, e_y)} \in AS^T \left[logD((e_x, e_y); \varphi)\right] \tag{19}$$
$$+ E_{(e'_x, e'_y) \sim G(\cdot|(e'_x, e'_y); \theta)} \left[log(1 - D((e'_x, e'_y); \varphi))\right],$$

$$D((e'_x, e'_y); \varphi) = \sigma(f_\varphi(e_x, e_y)) = \frac{exp(f_\varphi(e_r, e_y))}{exp(f_\varphi(e_x, e_y)) + 1}, \tag{20}$$

where $f_\varphi(x, y)$ is a two-layer neural network with input $\|x - y\|_1$, which uses ReLU as the activation function, and $\sigma(x)$ is sigmoid function. And use random gradient descent to update the objective function:

$$TS_{(e_x, e_y)} = \begin{cases} 1 & \sigma(f_\varphi(e_x, e_y)) \geq \delta \\ 0 & \sigma(f_\varphi(e_x, e_y)) < \delta \end{cases}. \tag{21}$$

where $\sigma(f_\varphi(e_x, e_y))$ is the output of the discriminator, σ is th e threshold that divides AS^U into two groups, the one is real entity pair, and the other is noise entity pair. Therefore, we set $TS_{(e_x, e_y)}$ as a binary score to indicate whether the entity pair is real or not. AS^T will expand according to the change of trust score of the entity pair in AS^U.

3 Experiments

3.1 Experimental Setup

Datasets. To validate the model, we evaluated our method in three large cross-lingual datasets of DBP15K [14]. These data sets are based on Chinese, English, Japanese and French versions of DBpedia. Each data set contains data from two knowledge graphs of different languages, and provides 15K pre aligned entity

Table 1. Summary of the DBP15K datasets.

Datasets		Entities	Relations	triples
$DBP15K_{ZH-EN}$	Chinese	66469	2830	153929
	English	98125	2317	237674
$DBP15K_{JA-EN}$	Japanese	65744	2043	164373
	English	95680	2096	233319
$DBP15K_{FR-EN}$	French	66858	1379	192191
	English	105889	2209	278590

pairs. Table 1 shows the statistical information of the data set. We use 30% for training and 70% for testing.

Experimental Setup. $\beta_1 = 0.1$, $\beta_2 = 0.3$, $\gamma_1 = 1.0$. The size of hidden layer in dual primal attention layer is $d = 300$, $d' = 600$ and $\sim d = 300$. The size of all hidden layers in GCN layer is 300, the learning rate is set to 0.001, and the error entity pair $k = 125$. The vector initialization of entity name in knowledge graph is the same as RDGCN [19].

Evaluation Metrics. we use the classic evaluation index Hits@N By measuring the proportion of correctly aligned entities ranked in the top N list Hits@N The higher the score, the better the performance of the model.

3.2 Contrast Experiment

In order to prove the effectiveness of NSRDGCN model, we compare it with several better entity alignment models, they are JE [12], MTransE [13], JAPE [14], IPTransE [15], BootEA [16], GCN-Align [20], KECG [18], SEA [30], REA [24], RDGCN [19]. The results of comparative experiments are shown in Table 2, and the bold font is the best data result.

As can be seen from the data in Table 2, the NSRDGCN model achieved significant improvements in all datasets and evaluation scores. Compared with other models, RDGCN shows better performance. Only in the $DBP15K_{ZH-EN}$ dataset, Hits@10 score of RDGCN is slightly lower than BootEA. However, all scores of the NSRDGCN model not only outperform RDGCN but also outperform BootEA. Especially in the baseline data $DBP15K_{JA-EN}$, the entity alignment model NSRDGCN achieves a relative improvement of about 1.6% in Hits@10 compared to the RDGCN model. Additionally, NSRDGCN requires less training data to learn better entity representation.

3.3 Ablation Experiment

In order to verify the influence of noise detection module and GCN layer on the performance of NSRDGCN model, ablation experiments are carried out on three data sets. The comparison results are shown in Table 3, and the bold font is the

Table 2. The overall alignment performance of all models on $DBP15K_{ZH-EN}$, $DBP15K_{JA-EN}$ and $DBP15K_{FR-EN}$. Among them, best published scores in underlined and best scores in **bold**. The results [⋆] are taken from [19], the results [†] are taken from [18], and others are taken from [24].

Model	$DBP15K_{ZH-EN}$		$DBP15K_{JA-EN}$		$DBP15K_{FR-EN}$	
	Hits@1	Hits@10	Hits@1	Hits@10	Hits@1	Hits@10
JE [⋆]	21.27	42.77	18.92	39.97	15.38	38.84
MtransE [⋆]	30.83	61.41	27.86	57.45	24.41	55.55
IPTransE [⋆]	40.59	73.47	36.69	69.26	33.30	68.54
JAPE [⋆]	41.18	74.46	36.25	68.50	32.39	66.86
BootEA [⋆]	62.94	84.75	62.23	85.39	65.30	87.44
GCN-Align [†]	41.25	74.38	39.91	74.46	37.29	74.49
KECG [†]	47.9	83.6	49.0	84.4	48.6	85.1
SEA	42.4	79.6	38.5	78.3	40.0	79.7
REA	34.80	61.71	27.33	57.87	28.73	68.04
RDGCN [⋆]	70.81	84.61	76.74	89.50	88.62	95.71
NSRDGCN	**71.20**	**85.56**	**77.84**	**90.89**	**88.88**	**96.06**

Table 3. Alignment performance of ablation models on DBP15k datasets.

Model	$DBP15K_{ZH-EN}$		$DBP15K_{JA-EN}$		$DBP15K_{FR-EN}$	
	Hits@1	Hits@10	Hits@1	Hits@10	Hits@1	Hits@10
RD	61.81	73.83	68.54	80.22	84.64	91.98
NSRD	61.89	73.85	68.59	80.23	84.62	91.99
RDGCN	70.81	84.61	76.74	89.50	88.62	95.71
NSRDGCN	**71.20**	**85.56**	**77.84**	**90.89**	**88.88**	**96.06**

best data result. Among them, RD is the primal RDGCN model to remove the GCN layer, RDGCN is to add a noise detection module on the basis of RD, and RDGCN is an extension of the RD model that incorporates a GCN layer.

As shown in Table 3, when comparing RD and NSRD, it can be observed that incorporating the noise detection module leads to small improvements in all metrics. This indicates that adding the noise detection module can enhance the effectiveness of cross-lingual entity alignment to some extent. Furthermore, comparing RD and DRGCN, significant improvements in all metrics are observed after adding the GCN layer. Particularly in the $DBP15K_{ZH-EN}$ dataset, Hits@10 shows a notable increase of 11.73% (approximately 15.9% relative improvement). Comparing RDGCN and NSRDGCN, it can be seen that incorporating the noise detection module on top of RDGCN further improves all metrics. Especially in the $DBP15K_{JA-EN}$ dataset, Hits@10 increases by 13.3% relative to RD, 13.2% relative to NSRD, and 1.5% relative to RDGCN. This indicates that the noise detection module is also a significant factor affecting the performance of cross-lingual entity alignment. Comparing RD and NSRD, it can be observed that without adding the GCN layer, the impact of the noise detection module on cross-lingual entity alignment models is minimal. This is because the GCN layer effectively captures the neighboring structural information of the knowledge graph, and the interaction between the GCN layer and the dual original graph is complementary. By combining them, better relation-aware representations can be learned, providing more assistance to the noise detection module.

4 Conclusion

In this paper, a noise-sensitive relation-aware cross-lingual entity alignment model is proposed. The method overcomes the limitations of the relation-aware cross-lingual entity alignment method in dealing with noisy entity pairs and improves the robustness of the method. The NSRDGCN model learns better entity representations for cross-lingual entity alignment by connecting the relational information to the neighboring structural information through the interaction of the original knowledge graph and the pairwise relational graph. The noise detection part detects possible noisy data in the training data, thus making the training data clean, which improves the entity alignment effect and the robustness of the model. Compared with existing methods, our model achieves the best cross-lingual entity alignment performance on three real datasets.

However, the modeling effect of the model mainly depends on the effect of relationship-aware modeling. In the follow-up work, we will improve the model by increasing the effect of the relationship-aware model, which will improve the overall alignment effect of the new cross-lingual entity alignment model.

Declarations

Availability of data and materials. The data that support the findings of this study are openly available.

Funding. This work was supported by the Natural Science Foundation of China (Grant NO. 62341603) and the Introduction and Cultivation Program for Young Innovative Talents of Universities in Shandong Province under grant No. 2021QC-YY003 and National Natural Science Foundation Special Project, 62341603.

Conflicts of interest. The authors have no competing interests to declare that are relevant to the content of this article.

References

1. Lehmann, J., et al.: Dbpedia–a large-scale, multilingual knowledge base extracted from wikipedia. Semantic Web **6**(2), 167–195 (2015)

2. Rebele, T., Suchanek, F., Hoffart, J., Biega, J., Kuzey, E., Weikum, G.: YAGO: a multilingual knowledge base from wikipedia, wordnet, and geonames. In: Groth, P., et al. (eds.) ISWC 2016. LNCS, vol. 9982, pp. 177–185. Springer, Cham (2016). https://doi.org/10.1007/978-3-319-46547-0_19
3. Zhang, F., Yuan, N.J., Lian, D., Xie, X., Ma, W.Y.: Collaborative knowledge base embedding for recommender systems. In: Proceedings of the 22nd ACM SIGKDD International Conference on Knowledge Discovery and Data Mining, pp. 353–362, 2016
4. Han, X., Liu, Z., Sun, M.: Neural knowledge acquisition via mutual attention between knowledge graph and text. In: Proceedings of the AAAI Conference on Artificial Intelligence, 2018
5. Cao, Y., Hou, L., Li, J., Liu, Z.: Neural collective entity linking. In: Proceedings of the 27th International Conference on Computational Linguistics, pp. 675–686, 2018
6. Cui, W., Xiao, Y., Wang, H., Song, Y., Hwang, S., Wang, W.: Kbqa: learning question answering over qa corpora and knowledge bases. Proc. VLDB Endow. **10**(5), 565–576 (2017)
7. Mahdisoltani, F., Biega, J., Suchanek, F.M.: Yago3: a knowledge base from multilingual wikipedias. In: Conference on Innovative Data Systems Research, 2015
8. Wang, Z., Li, J., Tang, J.: Boosting cross-lingual knowledge linking via concept annotation. In: Twenty-Third International Joint Conference on Artificial Intelligence. Citeseer, 2013
9. Bordes, A., Usunier, N., Garcia-Duran, A., Weston, J., Yakhnenko, O.: Translating embeddings for modeling multi-relational data. Adv. Neural Inf. Process. Syst. **26**, 2787–2795 (2013)
10. Lin, Y., Liu, Z., Sun, M., Liu, Y., Zhu, X.: Learning entity and relation embeddings for knowledge graph completion. In: Proceedings of the AAAI Conference on Artificial Intelligence, vol. 29, 2015
11. Ji, G., He, S., Xu, L., Liu, K., Zhao, J.: Knowledge graph embedding via dynamic mapping matrix. In: Proceedings of the 53rd Annual Meeting of the Association for Computational Linguistics and the 7th International Joint Conference on Natural Language Processing, pp. 687–696, 2015
12. Hao, Y., Zhang, Y., He, S., Liu, K., Zhao, J.: A joint embedding method for entity alignment of knowledge bases. In: Chen, H., Ji, H., Sun, L., Wang, H., Qian, T., Ruan, T. (eds.) CCKS 2016. CCIS, vol. 650, pp. 3–14. Springer, Singapore (2016). https://doi.org/10.1007/978-981-10-3168-7_1
13. Chen, M., Tian, Y., Yang, M., Zaniolo, C.: Multilingual knowledge graph embeddings for cross-lingual knowledge alignment, pp. 1511–1517, 2017
14. Sun, Z., Hu, W., Li, C.: Cross-Lingual entity alignment via joint attribute-preserving embedding. In: d'Amato, C., et al. (eds.) ISWC 2017. LNCS, vol. 10587, pp. 628–644. Springer, Cham (2017). https://doi.org/10.1007/978-3-319-68288-4_37
15. Zhu, H., Xie, R., Liu, Z., Sun, M.: Iterative entity alignment via joint knowledge embeddings. In: IJCAI, vol. 17, pp. 4258–4264 (2017)
16. Sun, Z., Hu, W., Zhang, Q., Qu, Y.: Bootstrapping entity alignment with knowledge graph embedding. In: IJCAI, vol. 18, 2018
17. Chen, M., Tian, Y., Chang, K.W., Skiena, S., Zaniolo, C.: Co-training embeddings of knowledge graphs and entity descriptions for cross-lingual entity alignment, 2018
18. Li, C., Cao, Y., Hou, L., Shi, J., Li, J., Chua, T.S.: Semi-supervised entity alignment via joint knowledge embedding model and cross-graph model. In: Proceedings of

the 2019 Conference on Empirical Methods in Natural Language Processing and the 9th International Joint Conference on Natural Language Processing (EMNLP-IJCNLP), pp. 2723–2732. Association for Computational Linguistics, 2019
19. Wu, Y., Liu, X., Feng, Y., Wang, Z., Yan, R., Zhao, D.: Relation-aware entity alignment for heterogeneous knowledge graphs, pp. 5278–5284, 2019
20. Wang, Z., Lv, Q., Lan, X., Zhang, Y.: Cross-lingual knowledge graph alignment via graph convolutional networks. In: Proceedings of the 2018 Conference on Empirical Methods in Natural Language Processing, pp. 349–357, 2018
21. Kipf, T.N., Welling, M.: Semi-supervised classification with graph convolutional networks, 2017
22. Schlichtkrull, M., Kipf, T.N., Bloem, P., van den Berg, R., Titov, I., Welling, M.: Modeling relational data with graph convolutional networks. In: Gangemi, A., Gangemi, A., et al. (eds.) ESWC 2018. LNCS, vol. 10843, pp. 593–607. Springer, Cham (2018). https://doi.org/10.1007/978-3-319-93417-4_38
23. Monti, F., Shchur, O.: Aleksandar Bojchevski. Stephan Günnemann, and Michael M Bronstein. Primal-dual mesh convolutional neural networks, Or Litany (2020)
24. Pei, S., Yu, L., Yu, G., Zhang, X.: Rea: robust cross-lingual entity alignment between knowledge graphs. In: Proceedings of the 26th ACM SIGKDD International Conference on Knowledge Discovery & Data Mining, pp. 2175–2184, 2020
25. Goodfellow, I., et al.: Generative adversarial nets, vol. 27, 2014
26. Veličković, P., Cucurull, G., Casanova, A., Romero, A., Liò, P., Bengio, Y.: Graph attention networks. stat **1050**(20), 10–48550 (2017)
27. Qu, M., Bengio, Y., Tang, J.: Weakly supervised entity alignment with positional inspiration, pp. 814–822, 2023
28. Radford, A., Metz, L., Chintala, S.: Unsupervised representation learning with deep convolutional neural network for remote sensing images, pp. 97–108, 2015
29. Karras, T., Laine, S., Aila, T.: A style-based generator architecture for generative adversarial networks. In: 2019 IEEE/CVF Conference on Computer Vision and Pattern Recognition (CVPR), pp. 4396–4405, 2019
30. Lin, X., Yang, H., Wu, J., Zhou, C., Wang, B.: Guiding cross-lingual entity alignment via adversarial knowledge embedding. In: 2019 IEEE International Conference on Data Mining (ICDM), pp. 429–438. IEEE (2019)

Origin Classification of Rice Wine Based on an Electronic Nose and Convolutional Neural Network

Wenqi Sun, Ancai Zhang, Guangyuan Pan, and Wenbo Zheng(✉)

Linyi University, Linyi 276000, China
zhengwenbo@lyu.edu.cn

Abstract. Research on the origin of rice wine has significant commercial value for agricultural markets. Inspired by the advantages of fast, non-destructive and high sensitivity of electronic nose (e-nose) in rice wine analysis, this paper applies an e-nose system based on headspace sampling and combines with a convolutional neural network (CNN) to achieve effective identification for rice wine origins. The results show that, first, the odor information of rice wine from 10 origins is obtained by the e-nose system. Second, odor features of the samples from different origins are obtained by the convolutional and pooling layers. Finally, the combination of the e-nose and CNN achieved the best performance, including the accuracy of 98.00%, the kappa coefficient of 97.51%, and the F_1-score of 98.00%, in compared with multiple classification models. Effective identification of rice wine origin results is acquired by the e-nose and CNN, providing a tool for quality assessment of food products.

Keywords: Rice wine · Origin identification · E-nose · CNN

1 Introduction

Rice wine, a popular traditional fermented beverage, not only has a long history and cultural significance in Asia [1], but it is also gaining traction in the international market. The quality and flavor of rice wine are significantly influenced by the raw materials, brewing process, and the environment of its origin. Therefore, studying the origins of rice wine holds great commercial value for preserving this traditional product, enhancing the consumer experience, and increasing market competitiveness. Zheng *et al.* revealed the impact of origin on the flavor of rice wine by analyzing the chemical composition of rice wine from various regions [2]. Shen *et al.* explored how geographical indications affect consumers' purchase intentions [3]. Yu *et al.* investigated the differences in the nutritional and functional components of yellow wine products produced in different regions of China, revealing that rice wine from diverse origins significantly influences its nutritional components [4].

Currently, the primary methods for studying the origin of rice wine include sensory evaluation and physicochemical analysis. Sensory assessment relies on trained tasters to evaluate the flavor of rice wine; however, this approach has several disadvantages, including high subjectivity, lack of repeatability, and significant costs [5]. Physicochemical analysis methods, such as gas chromatography-mass spectrometry (GC-MS), can provide detailed information on chemical composition, but they typically require complex sample pre-treatment, expensive equipment, and specialized operating skills [6]. Additionally, these traditional methods are inefficient for processing large numbers of samples and struggle to monitor flavor changes in rice wine in real time [7]. The following is a summary of the results of the study.

The electronic nose (e-nose) technology offers significant advantages in the analysis of rice wine compared to traditional sensory evaluation and physicochemical analysis. E-nose can quickly and non-destructively detect the volatile components of rice wine, requiring minimal operator skill [8]. The principle of the e-nose is based on a gas sensor array that recognizes odor patterns. The data collected are analyzed using pattern recognition techniques to identify the origin of the rice wine [9]. Typically, the structure of an e-nose includes a sampling system, a sensor array, a signal processing unit, and data processing software [10]. In the field of rice wine analysis, Zheng *et al.* used an e-nose to classify rice wines from different origins, achieving promising results [11]. Gliszczyńska-Świgło *et al.* investigated the aging process of rice wine using e-nose technology [12], while Zhang *et al.* employed the e-nose to verify the authenticity of rice wine [13].

Therefore, this paper uses an e-nose system based on headspace sampling to effectively identify the origin of rice wine. The system enhances the detection sensitivity and selectivity of the volatile components in rice wine by optimizing the sampling conditions and selecting appropriate sensor arrays. The e-nose system offers advantages such as ease of operation, rapid response, and low cost. Building on this, a convolutional neural network (CNN) is designed in this paper to process the output data from the e-nose system, achieving excellent discrimination performance.

The contribution of this paper is the development of a novel detection system that combines an e-nose with a CNN. This system not only enhances the accuracy and efficiency of rice wine origin recognition but also serves as an effective tool for food quality assessment. Through extensive experimental validation, the system achieves an accuracy of 98.00%, a kappa coefficient of 97.51%, and an F_1-score of 98.00% in the rice wine origin recognition task, demonstrating strong potential for practical application. These research findings not only provide technical support for the advancement of the rice wine industry but also offer new insights for origin analysis and quality control of other food products.

2 Material and Methods

2.1 Sample

Samples of rice wine from ten provinces in China were selected to validate the proposed origin identification method. Detailed information about the samples is presented in Table 1. As shown in Table 1 and Fig. 1, different geographical locations can result in variations in the alcohol content of rice wines with identical compositions. All samples were measured directly using an e-nose system without any pre-processing. Additionally, to slow down the aging of the rice wine and preserve its quality, all samples were refrigerated prior to testing.

Table 1. The specific information on the samples.

No	Brand	Origin	Alcohol content (% vol)
1	Tingyuan	Fujian, China	5.0
2	Qiansanbai	Guizhou, China	18.0
3	Shenglong	Hubei, China	13.0
4	Qinjiu	Jilin, China	1.0
5	Suzhouqiao	Jiangsu, China	5.0
6	Nongxiangwang	Jiangxi, China	20.0
7	Huatianxiangzi	Shanxi, China	11.0
8	Qingcaosha	Shanghai, China	12.0
9	Mise	Chongqing, China	6.0
10	Aixiaoxi	Sichuan, China	4.0

Fig. 1. The distribution of origin of rice wine samples.

2.2 Instrument

The e-nose system comprises the following components: sampling tube (ST), active carbon (AC), thermostatic controller (TC), flow rate valve (FRV), air pump (AP), sensor chamber (SC), signal amplifier (SA), digital to analog converter (DAC), and data acquisition system (DAS).

ST is not only capable of functioning under various temperature and pressure conditions, but it also effectively prevents sample contamination and deterioration, maintaining the stability of the sample to ensure the accuracy of subsequent analysis results. The ST controls the temperature of the sample to produce volatile gases, which facilitates sensor detection and further eliminates impurities in the gas through the AC. The FRV can be utilized to adjust the flow characteristics of the fluid as needed, precisely regulating the flow rate to prevent overflow and reduce the risk of failure. The SC is a sensor array composed of metal oxide sensors (MOS). The gas sample data is obtained through the conductivity generated between the sensor and the gas, establishing a relationship between gas concentration and conductivity.

There is a relationship between gas concentration and conductivity: the higher the gas concentration, the greater the conductivity. In this study, the sensor output is expressed as S1/S2, where S1 represents the conductivity of the sensor in contact with the sample gas and S2 represents the conductivity of the sensor in contact with clean air. The characteristics of the sample are analyzed based on the difference in conductivity. The data collected by the sensor is further processed and optimized by the SA to enhance the signal for improved digital conversion by the ADC.

2.3 Experiment

The methodology for testing gas samples is outlined as follows: Select a well-ventilated, undisturbed laboratory where the temperature is maintained between 20 °C and 25 °C, and the humidity is kept between 40% and 60%. The testing procedure is as follows:

First, the e-nose was warmed up for one hour before testing to ensure the stability of the gas sensor output.

Next, the air chamber is cleaned and treated by opening AP1 and closing AP2, as shown in Fig. 2. The sensor chamber is treated with AC-treated clean air at a flow rate of 450 ml/min for 200 s.

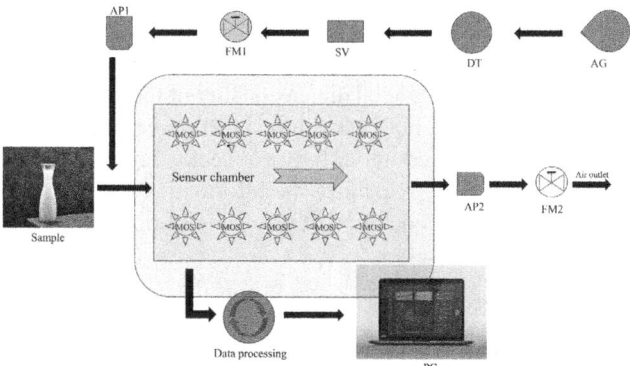

Fig. 2. The detection processes of the e-nose system.

To initiate gas absorption, 30 ml of the sample was placed in a 100 ml sampling bottle, which was then heated to approximately 30 °C using the thermostat controller

(TC) to ensure complete volatilization of the sample. This process increased the gas concentration. Subsequently, AP2 was activated while AP1 was deactivated to commence detection. The sampling frequency was set to 1 Hz, the airflow rate to 450 ml/min, and the sensor's sampling duration was established at 90 s. The gas absorption and desorption times were set at 60 s and 30 s, respectively.

At last, the above procedure was repeated for each sample. A total of 30 samples were prepared for each brand, and each sample was tested 10 times, resulting in a total of 300 samples for the 10 rice wines.

2.4 CNN

CNN is a deep learning model widely utilized in various fields, including image recognition, natural language processing, and bioinformatics. The core component of a CNN is the convolutional layer, which extracts features from the input data through convolutional operations. The mathematical expression for the convolution operation is as follows:

$$(f * g)(x) = \int_{-\infty}^{\infty} f(t)g(x - t)dt \tag{1}$$

where f is the input signal, g is the convolutional kernel, and * denotes the convolutional operation. The advantage of CNN lies in their ability to automatically learn a hierarchical feature representation of the data, ranging from low-level edges and textures to high-level semantic information. Additionally, CNNs reduce model parameters through weight sharing and local connectivity, which improves computational efficiency and enhances the model's generalization ability [14–16]. CNNs have achieved remarkable results in several fields, including image classification, target detection, and speech recognition.

In this thesis, we design a CNN architecture specifically for processing rice wine origin data generated by an e-nose system. The CNN model comprises multiple convolutional and pooling layers, followed by a fully connected layer. Each convolutional layer contains several convolutional kernels of size (3, 3), which are utilized to extract features at various scales. After each convolutional layer, we apply a maximum pooling layer with a size of (2, 2) to reduce the spatial dimensionality of the features while preserving the most significant information. Following these layers, we obtain a set of features that are subsequently flattened and fed into a fully connected layer for classification. This CNN architecture is designed to maximize the information extracted from the e-nose data, enabling highly accurate recognition of the origin of rice wine.

3 Results and Discussion

3.1 CNN Processing Results

In this study, the original dataset consisted of dimensions (1, 900), representing the volatile component signals of each rice wine sample. Before being input into the CNN model, the data was reshaped into a 15 × 60 matrix format to meet the model's input requirements. Through the multi-layer convolution and pooling processes of the CNN model, we successfully extracted the key features of the rice wine samples. Specifically, the model comprises two convolutional layers, each followed by a max pooling layer. The first convolutional layer employs 32 convolutional kernels of size 3 × 3, while the second layer utilizes 64 convolutional kernels of the same size. After two pooling operations, the size of the feature map is effectively reduced, resulting in a final extraction of 64 and 128 features from each convolutional layer, with feature sizes of 7 × 7 and 5 × 5, respectively. These features are then flattened and fed into the fully connected layer for classification.

For the parameter settings of the CNN model, we selected an initial learning rate of 0.001 and conducted training over 600 iterations. The optimizer employed was stochastic gradient descent (SGD), which facilitated the model's convergence. The training set, including the validation set, comprised 200 samples, while the test set contained 100 samples. This dataset provided the model with sufficient information to learn how to differentiate between rice wines from various origins.

3.2 Performance Evaluation of e-nose and CNN

Table 2 presents the results of rice wine origin recognition and emphasizes the superiority of the e-nose when combined with the CNN model, as compared to the performance of six models (including the fully connected neural network (FCNN), recurrent neural network (RNN), extreme learning machine (ELM), radial basis function neural network (RBFNN), probabilistic neural network (PNN), and graph convolutional network (GCN)). Our comparison revealed that the FCNN exhibited the poorest performance, likely due to its excessive number of parameters, which can lead to overfitting and a loss of information regarding spatial structure. In contrast, while the RNN is adept at processing sequential data, it struggles with image data that possesses spatial structure [17]. This limitation may stem from its susceptibility to the vanishing or exploding gradient problem when handling long sequences. Conversely, the CNN model effectively reduces both the number of parameters and computational load, enhancing the model's generalization ability through its local connectivity and weight-sharing properties, while simultaneously capturing local features and preserving spatial structure information. Similar findings were reported in the study by Guo *et al.*, which demonstrated that CNN outperforms RNN [18].

With the e-nose and CNN, we achieved optimal performance, including an accuracy of 98.00%, a kappa coefficient of 97.51%, and an F_1-score of 98.00%. These results highlight the advantages of CNN models in performing such tasks, as well as the efficiency and accuracy of e-nose in the analysis of rice wine.

In our error analysis, we conducted a comprehensive examination of the models' prediction results and created error bar charts. These charts illustrate the distribution of prediction accuracy, kappa coefficient, and F_1-score for various models on the test set, along with their variances. The comparative analyses (shown in Fig. 3) provide a clearer understanding of the advantages of the CNN model across each metric. The accuracy of the CNN model is significantly higher than that of the other models, and its variance is smaller, indicating high stability and reliability. These analyses further validate the effectiveness of the CNN model in the task of rice wine origin recognition.

Table 2. The identification results of the sample origins.

No	Model	Accuracy	F_1-score	Kappa
1	FCNN	89.46%	89.88%	88.42%
2	RNN	92.86%	93.03%	92.13%
3	ELM	92.12%	91.06%	91.32%
4	RBFNN	93.66%	92.25%	93.00%
5	PNN	93.16%	93.32%	92.46%
6	GCN	96.58%	96.57%	95.73%
7	**CNN**	**98.00%**	**98.00%**	**97.51%**

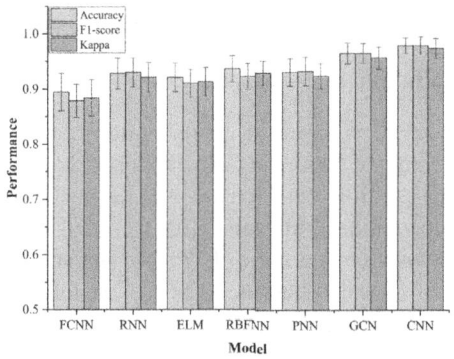

Fig. 3. The analytical results of the model stability.

In summary, the e-nose system and CNN presented in this paper achieved impressive results in identifying the origin of rice wine, with an accuracy of 98.00%, a kappa coefficient of 97.51%, and an F_1-score of 98.00%.

4 Conclusion

In this paper, we propose a method that utilizes an e-nose system in conjunction with a CNN model to effectively identify the origin of rice wine. The main conclusions of this paper are as follows:

(1) Compared to the results of various recognition models, the e-nose and CNN models demonstrated the highest recognition performance, achieving an accuracy of 98.00%, an F_1-score of 97.51%, and a kappa coefficient of 98%.
(2) The optimal classification stability of the CNN model for rice wine samples was validated using error histograms, further demonstrating the effectiveness of the proposed method for classifying the origin of rice wine.

In conclusion, the e-nose and CNN model successfully identified the origin of rice wine. In the next phase, the e-nose and deep learning model will be employed to assess the quality of agricultural products, including rice wine.

Acknowledgments. This work was supported by the Taishan Scholar Program of Shandong Province under Grant No. Tsqn202211240, the National Natural Science Foundation of Shandong Province under Grant No. ZR2024QC153 and ZR2024MF041, the Development Plan of Youth Innovation Team of University in Shandong Province under Grant No. 2023KJN049, and the Yunnan Provincial Science and Technology Plan Project under Grant No. 202302AD080006.

Disclosure of Interests. The authors declare that they have no known competing financial interests or personal relationships that could have appeared to influence the work reported in this paper.

References

1. Zhang, J.X., Li, T., Zou, G., Wei, Y.J., Qu, L.B.: Advancements and future directions in yellow rice wine production research. Fermentation **10**(1), 40–55 (2024)
2. Zheng, W.B., Sun, W.Q., Liang, X., Yuan, Q., Zhang, A.C.: Identification of the quality for rice with different storage humidity: an electronic nose combined with multiblock feature integration method. Measurement **237**, 115236 (2024)
3. Shen, X.C., Sun, Q.H., Guo, C.: Research on the influence of brand image on consumers' purchase intention of hometown geographical indication agricultural products. J. Xi'an Coll. Arts Sci. (Soc. Sci. Edit.) **27**(4), 93–99 (2024)
4. Yu, J.Z.: Comparative analysis of nutritional and functional components in yellow wine from different origins. J. Anhui Agric. Sci. **37**(42), 15989–15991 (2009)
5. Balhorn, L.S., Weber, J.M., Buijsman, S., Hildebrandt, J.R., Ziefle, M., Schweidtmann, A.M.: Empirical assessment of ChatGPT's answering capabilities in natural science and engineering. Sci. Rep. **14**, 4998–5008 (2024)
6. Hao, J.H., Xu, F.F., Yang, D., Wang, B., Qiao, Y.Y., Tian, Y.Y.: Analytical pyrolysis of biomass using pyrolysis-gas chromatography/mass spectrometry. Renew. Sustain. Energy Rev. **208**, 115090 (2025)
7. Sánchez-Quezada, V., Luzardo-Ocampo, I., Gaytán-Martínez, M., Loarca-Piña, G.: Physicochemical, nutraceutical, and sensory evaluation of a milk-type plant-based beverage of extruded common bean (*Phaseolus vulgaris* L.) added with iron. Food Chemistry **453**, 139602 (2024)
8. Dang, Y., Reddy, Y.V.M., Cheffena, M.: Facile e-nose based on single antenna and graphene oxide for sensing volatile organic compound gases with ultrahigh selectivity and accuracy. Sens. Actuators B Chem. **419**, 136409 (2024)
9. Yurdakos, O.B., Cihanbegendi, O.: System design based on biological olfaction for meat analysis using e-nose sensors. ACS Omega **9**(30), 33183–33192 (2024)
10. Ma, X., Wu, F., Yan, J., Duan, S.K., Peng, X.Y.: TF-TCN: a time-frequency combined gas concentration prediction model for e-nose data. Sens. Actuators A **376**, 115654 (2024)
11. Zheng, W.B., Wang, Y.W., Liang, X., Zhang, A.C.: Origin identification for rice wines based on an electronic nose and convolution dot-product attention mechanism. Sens. Actuators A **375**, 115521 (2024)
12. Gliszczyńska-Świgło, A., Chmielewski, J.: Electronic nose as a tool for monitoring the authenticity of food. a review. Food Anal. Methods **10**, 1800–1816 (2017)

13. Zhang, H.H., Shao, W.Q., Qiu, S.S., Wang, J., Wei, Z.B.: Collaborative analysis on the marked ages of rice wines by electronic tongue and nose based on different feature data sets. Sensors **20**(4), 1065–1082 (2020)
14. Bird, J.J., Lotfi, A.: CIFAKE: image classification and explainable identification of AI-generated synthetic images. IEEE Access **12**, 15642–15650 (2024)
15. Xu, S.F., Geng, S.J., Xu, P.F., Chen, Z.H., Gao, H.M.: Cognitive fusion of graph neural network and convolutional neural network for enhanced hyperspectral target detection. IEEE Trans. Geosci. Remote Sens. **62**, 1–15 (2024)
16. Diep, Q.B., Phan, H.Y., Truong, T.C.: Crossmixed convolutional neural network for digital speech recognition. PLoS ONE **19**(4), e0302394 (2024)
17. Khatti, J., Grover, K.S.: CBR prediction of pavement materials in unsoaked condition using LSSVM, LSTM-RNN, and ANN approaches. Int. J. Pavement Res. Technol. **17**, 750–786 (2024)
18. Guo, Y.M., Liu, Y., Bakker, E.M., Guo, Y.H., Lew, M.S.: CNN-RNN: a large-scale hierarchical image classification framework. Multimed. Tools Appl. **77**, 10251–10271 (2018)

Parkinglot Obstacle Detection System Using Infastrue-Based Cameras

Yuesheng He[1]( ), Fei Wang[1], Zihan Zong[2], and Ming Yang[1]

[1] Department of Automation, Shanghai Jiao Tong University, Shanghai, China
heyuesh@sjtu.edu.cn
[2] University of Michigan - Shanghai Jiao Tong University Joint Institute, Shanghai, China

Abstract. This paper presents a parking lot obstacle detection system that utilizes infrastructure-based cameras to detect and track obstacles within a parking lot. The system begins by capturing images from ceiling-mounted cameras and employs object detection algorithms to identify 2D detection boxes and estimate 3D bounding boxes' ground points. By leveraging camera intrinsics and extrinsics, along with the assumption that all objects are on a plane, we transform these points into the world coordinate system. The system then fuses detection results from multiple cameras using spatial and temporal fusion techniques, including the mean shift algorithm and the Expanded Kalman Filter, to optimize and refine the detection results. Our approach effectively addresses the challenge of depth estimation in monocular camera setups and provides a robust solution for parking lot obstacle detection.

Keywords: Obstacle Detection · Monocular Camera · 3D Bounding Box · Spatial Fusion · Temporal Fusion

1 Introduction

Parking lots are intricate environments where vehicles and pedestrians frequently navigate in close proximity, necessitating reliable obstacle detection systems to ensure safety and efficiency. Traditional object detection methods, such as YOLO and DETR, have been widely employed for various applications but encounter challenges when applied to parking lot scenarios due to the lack of depth information in monocular camera images. This paper introduces a novel system that addresses these challenges by reformulating the problem into keypoint estimation and employing both spatial and temporal fusion techniques to enhance detection accuracy. Our system not only detects obstacles but also estimates their 3D bounding boxes, providing a comprehensive solution for parking lot monitoring.

The primary contributions of this paper are twofold:

1. Efficient and Accurate Detection of Ground Bounding Boxes: We reformulate the problem of detecting object bounding boxes in the parking lot by estimating the distribution of key points of the objects. This innovative approach

allows us to efficiently and precisely detect the ground-bound boxes of targets within the parking lot. By assuming all targets are on the ground, we focus on detecting the four points of the 3D bounding box on the ground and estimate their distribution through a network. This method effectively addresses the challenge of depth estimation in monocular camera setups, which is a significant advancement over traditional methods that struggle with the lack of depth information.
2. Real-time Obstacle Detection System Based on Edge-side Monocular Cameras: We have developed a real-time obstacle detection system based on edge-side monocular cameras for underground parking lots. This system captures images from ceiling-mounted cameras and employs object detection algorithms to identify 2D detection boxes and estimate 3D bounding boxes' ground points. By leveraging camera intrinsics and extrinsics, along with the assumption that all objects are on a plane, we transform these points into the world coordinate system. The system then fuses detection results from multiple cameras using spatial and temporal fusion techniques, including the mean shift algorithm and the Expanded Kalman Filter, to optimize and refine the detection results. This system is designed to operate in real time, providing a robust solution for parking lot obstacle detection.

In summary, this paper presents a cutting-edge approach to parking lot obstacle detection that leverages keypoint estimation and fusion techniques to achieve high accuracy and real-time performance. Our system's ability to estimate 3D bounding boxes and its real-time capabilities make it a valuable tool for enhancing safety and efficiency in parking lot environments. The following sections will detail our methodology, experimental setup, and results, demonstrating the effectiveness of our approach in various scenarios.

2 Related Works

The field of parking lot obstacle detection has seen significant advancements with the integration of intelligent vision systems and advanced algorithms. This section reviews the latest research pertinent to our parking lot obstacle detection system.

Deep Learning in Object Detection – The application of deep learning, particularly convolutional neural networks (CNNs), has revolutionized object detection. He et al. provide a comprehensive survey on the use of transformers in vision tasks, highlighting their potential in object detection [14]. Dosovitskiy et al. demonstrate the effectiveness of transformers for image recognition at scale, offering a new paradigm for detecting objects in images [10].

3D Object Detection Techniques – The transition from 2D to 3D object detection is crucial for accurate obstacle detection in parking lots. Qi et al. discuss deep learning methodologies for 3D object detection and segmentation, which are essential for understanding the spatial context of objects [27]. Chen et al. present a multi-view 3D object detection network tailored for autonomous driving, emphasizing the importance of multiple perspectives in detecting objects [7].

Monocular Camera Depth Estimation – One of the challenges in monocular camera setups is the lack of depth information. Zhou et al. survey the field of monocular 3D object detection for autonomous driving, addressing the depth estimation problem [8]. Wang et al. (2022) provide an overview of depth estimation techniques for monocular object detection, which is vital for our system's accuracy [13].

Fusion Techniques in Obstacle Detection – The integration of data from multiple sensors is key to improving detection reliability. Geiger et al. introduce the KITTI vision benchmark suite, which includes multi-sensor fusion for 3D object detection [12]. Li et al. review multi-sensor fusion techniques for 3D object detection, underlining the significance of combining data for enhanced accuracy [24].

Extended Kalman Filter in Object Tracking – The Extended Kalman Filter (EKF) is a popular method for fusing detection results over time. Arulampalam et al. offer a tutorial on particle filters for online nonlinear/nongaussian Bayesian tracking, which includes the application of EKF [4]. Simon discusses optimal state estimation techniques, including EKF, which is integral to our system's object tracking capabilities [18].

Parking Lot Monitoring Systems – Specific to parking lots, Zhang et al. explore the detection and occupancy estimation of parking spaces using aerial images, providing insights into monitoring parking lots [2]. Li et al. survey deep learning applications for parking space detection and status estimation, which is directly relevant to our system [20].

Camera-Based Obstacle Detection Systems – The effectiveness of camera-based systems in obstacle detection is demonstrated by Bochkovskiy et al. with YOLOv4, an optimal speed and accuracy object detection framework [5]. Redmon et al. further improve upon this with YOLOv5, an incremental improvement over previous versions [3].

Multi-View Object Detection and Fusion – Wang et al. survey multi-view fusion for 3D object detection, emphasizing the importance of integrating information from different viewpoints [1]. Li et al. introduce a graph-based message passing and attention mechanism for multi-view 3D object detection, which could apply to our system [25].

Mean Shift Algorithm in Clustering – For spatial fusion, the mean shift algorithm is a robust clustering method. Comaniciu et al. discuss the mean shift approach to feature space analysis, which applies to our system's spatial fusion of key points [9]. Chen et al. review mean shift clustering techniques, providing further insights into its application [26].

Real-Time Obstacle Detection Systems – Real-time detection is crucial for practical applications. Wang et al. present YOLOv7, which sets a new state-of-the-art for real-time object detectors [22]. Bochkovskiy et al. introduce YOLOv8, focusing on scalability and performance for real-time detection [19].

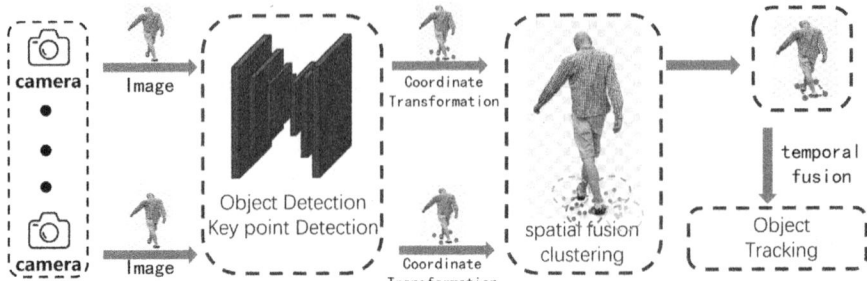

Fig. 1. System Overview.

These works provide a solid foundation for the design and implementation of our parking lot obstacle detection system. By building upon these studies, we aim to enhance the safety and efficiency of parking lot monitoring through the integration of advanced vision systems and algorithms.

3 Method

3.1 System Overview

An overview of the parking lot obstacle detection system is shown in Fig. 1. The system starts by receiving images from the ceiling-mounted cameras in the parking lot and then performs object detection to obtain the 2D detection box of the object and its four ground points of the 3d bounding box, i.e., front-left, front-right, back-left, and back-right points. After obtaining the detection results, first, since these four points are on the ground, we project them from the camera coordinate system to the world coordinate system using the camera's intrinsic and extrinsic, obtaining a coarse bounding box of the target in the world coordinates. Then, the spatial fusion of the detection results is required as multiple cameras may detect the same object. After obtaining the spatial fusion results, we perform a temporal fusion of the targets using Expanded Kalman filtering to optimize the target detection results and obtain the final parking lot obstacle detection results.

3.2 Obstacle Keypoint Detection

There are already many mature methods for object detection in images, such as YOLO [5,11,15–17,21–23] and DETR [6], which have been applied in numerous real-world tasks. However, for cameras mounted on parking garage ceilings, due to the lack of depth information in the images, the 2D bounding box of an object detected directly in the image cannot be readily converted into a 3D bounding box of the object in the world coordinate system.

Many works are also dedicated to 3D object detection with monocular cameras. However, due to the lack of depth information, this problem often reduces to monocular depth estimation of the object.

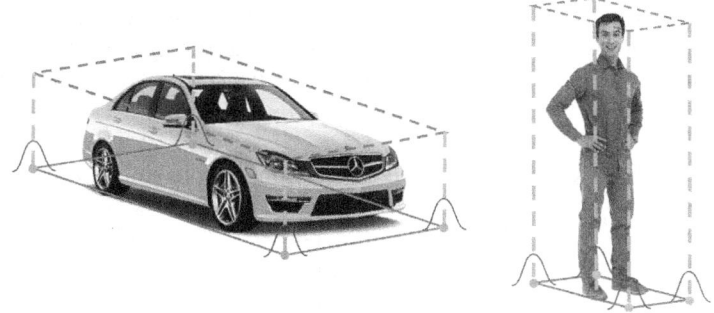

Fig. 2. The illustration of keypoint detection result

We note that in the parking lot, the objects are all on a plane, which is a useful constraint. Therefore, we reformulate the problem of detecting object bounding boxes in the parking lot as estimating the distribution of keypoints of the objects. As shown in Fig. 2, each object has a 3D bounding box that defines its boundaries. However, we assume that all targets in the parking lot are on the ground, so we only detect the four points of the 3D bounding box on the ground(marked as orange points in Fig. 2) and estimate the distribution of these four points through the network.

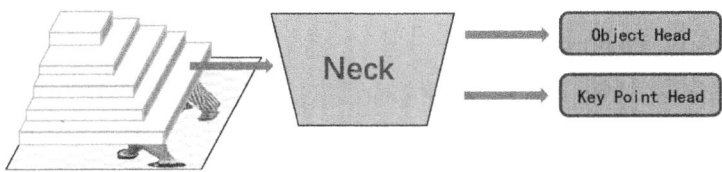

Fig. 3. The network structure

Network. For object and projection point detection, we propose a key point detection method. The network architecture we designed is shown in Fig. 3. The network uses a feature extraction network to extract multi-level features and then fuses these features using a feature pyramid. Finally, the network has two heads to obtain the obstacle detection results and ground key point estimation results separately.

Through the network, we can obtain the 2D bounding box of the object in the image, its category, and the four key points on the ground. And these four points also have corresponding classes, i.e., front-left, front-right, back-left, and back-right points.

Transformation. After detecting the object and its four points of the object's 3D bounding box on the ground, and then using the camera's intrinsic and extrinsic along with the ground plane constraint, we obtain the distribution of these four points in the world coordinate system. The projection formula is as follows:

First, transform the ground plane to the camera coordinate system.

$$\begin{bmatrix} \hat{a} \\ \hat{b} \\ \hat{c} \\ \hat{d} \end{bmatrix} = T \begin{bmatrix} a \\ b \\ c \\ d \end{bmatrix} \quad (1)$$

where a, b, c, d are the plane parameters($ax + by + cz = d$).

Then, for each point

$$depth = \frac{-\hat{d}}{\hat{a}(x - c_x)/f_x + \hat{b}(y - c_y/f_y) + \hat{c}} \quad (2)$$

where c_x, c_y, f_x, f_y is camera's intrinsic.

Thus, the point is

$$\begin{bmatrix} x \\ y \\ z \end{bmatrix} = T^{-1} \begin{bmatrix} \frac{depth(x-c_x)}{f_x} \\ \frac{depth(y-c_y)}{f_y} \\ depth \end{bmatrix} \quad (3)$$

3.3 Result Fusion

Because multiple cameras from different perspectives may simultaneously detect the same object, and because the estimated ground key points are assumed to follow a normal distribution, fusing results from multiple views can provide a better estimation of the object's ground bounding box.

Since the estimated ground key points have class information, we only need to cluster the key points of the same class for all detection results. Considering that the detection results will have outliers, the mean shift algorithm is a suitable algorithm to compute the clustering algorithm that removes outliers.

The mean shift algorithm finds stable outlier-free keypoint means by repeatedly calculating the direction of shift of the mean of points within the Euclidean distance threshold to achieve spatial fusion of the same keypoint class.

3.4 Object Tracking

An Extended Kalman Filter (EKF) is adopted for tracking and fusing the obstacle detection results of object 3D bounding box in coordinate system of the world. The EKF is an ideal filter for removing noise between the detection noise using temporal fusion.

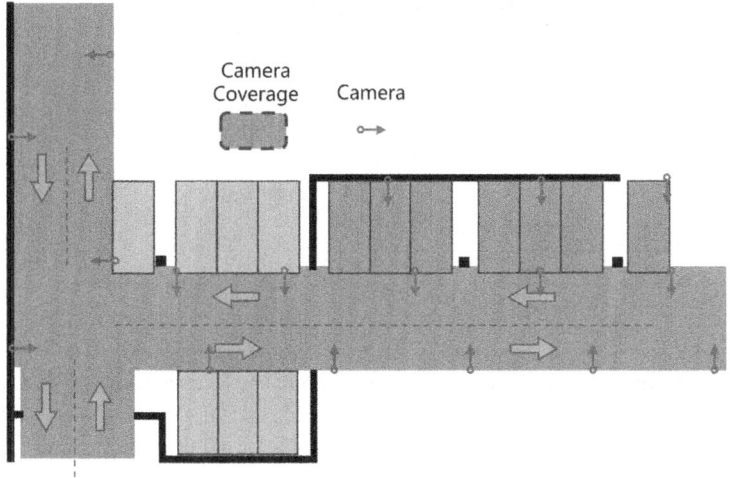

Fig. 4. The sensor network.

The obstacle state aims to estimate is

$$x = \begin{bmatrix} x \\ y \\ \theta \\ w \\ h \end{bmatrix}, \quad (4)$$

representing obstacle's position and its rotation angle, the bounding box's width and height. Then, the standard procedure of EKF propagation procedure is applied to update the obstacle's state.

First, use the prediction equation to predict the state of the obstacle,

$$\hat{x}_{\bar{k}} = A\hat{x}_{k-1} + Bu_{k-1} \quad (5)$$

$$P_{\bar{k}} = AP_{k-1}A^T + Q \quad (6)$$

Then after obtaining the observations z_k, the state of the Kalman filter is updated to obtain the spatial fusion results.

$$K_k = \frac{P_{\bar{k}} H^T}{H P_{\bar{k}} H^T + R} \quad (7)$$

$$\hat{x}_k = \hat{x}_{\bar{k}} + K_k \left(z_k - H\hat{x}_{\bar{k}} \right) \quad (8)$$

$$P_k = (I - K_k H) P_{\bar{k}} \quad (9)$$

Table 1. Model Detection Result

Method	Bounding Box		Key Point	
	AP_{50}	AP_{50-95}	AP_{50}	AP_{50-95}
Ours	0.98	0.938	0.97	0.955

4 Experiments

4.1 Sensor Network and Data Prepare

We deployed the system proposed in Sect. 3 in an indoor parking lot. We installed a total of 17 cameras on the roadside and connected these cameras with network cables to form a sensor network. As shown in Fig. 4, the blue area represents the coverage of the camera, containing roughly 50 m of the driveway and 7 parking spaces, the red arrows represent the camera's mounting position and orientation, with a field of view of 120° the camera was selected to be mounted at about 45° downwards.

To prepare data for training the detection network, we also installed LiDARs in the parking lot to assist in accurately annotating 3D detection boxes for objects in the images. After calibrating the LiDAR and camera, we only need to annotate the 3D detection boxes for objects in the LiDAR point cloud. Then, we can convert these detection boxes into the camera's perspective to obtain the 3D detection boxes of objects in the images.

4.2 Detection Experiment

First, we labeled approximately 800 images for training the model and we trained and validated the proposed network using the self-collected dataset. After training, our model achieved the accuracy shown in Table 1. The evaluation metrics of the model primarily consist of two parts: object detection and detection of object's ground points, i.e. bounding box and key points. For evaluation metrics, Intersection over Union (IoU) is used for bounding boxes. For key points, we used the object keypoint similarity(OKS) from COCO. Therefore, the AP50 for key point represents the average precision when keypoint similarity is greater than 0.5, considering those as true positives.

Due to the small training set, we achieved high detection accuracy and keypoint estimation accuracy on this dataset. The detection results are shown in the Fig. 5. The results indicate that our proposed network can detect objects in the parking lot and estimates the distribution of key points for their ground bounding boxes.

4.3 System Experiment

In the previous section, we demonstrated that the proposed network can accurately detect obstacles in the parking lot. Our proposed obstacle detection system

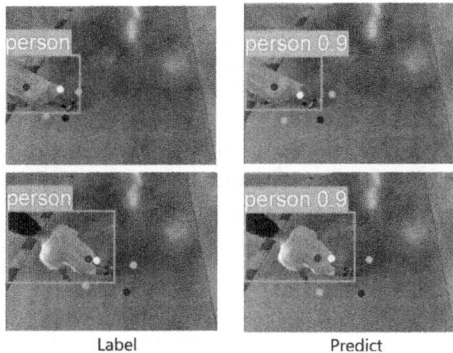

Fig. 5. Object Detection and Keypoint Results

also needs to evaluate the accuracy of detected obstacles in the world coordinate system.

Therefore, we deployed the trained model into the proposed system and used ground truth bounding boxes in the world coordinate system to evaluate the accuracy of our system. The experiment results are shown in the Table 2. We mainly used two metrics to evaluate our proposed system: the Root Mean Square Error (RMSE) of the object trajectory and the mIoU of the object ground bounding box. These metrics assess the accuracy of the detected objects positions and bounding boxes, respectively. Our method achieves a lower RMSE and a higher IoU for ground bounding boxes compared to the direct 2D projection detection method.

Table 2. Detection System Result

Method	Trajectory(m)	Ground mIoU
2D Box	1.908	0.046
Ours	**0.162**	**0.561**

At the same time, since the obstacle detection system is an real-time system, we also evaluated the inference time of our proposed system. Given that our system includes multiple cameras, we perform inference with a batch size of 8 on RTX3090. The inference time is shown in the Table 3.

Table 3. Inference Time

Method	Per Batch(ms)	Per Image(ms)
Ours	37.6	4.7

5 Conclusion

In this study, we have successfully developed and evaluated a parking lot obstacle detection system that utilizes infrastructure-based cameras, addressing the critical need for accurate and real-time monitoring in complex parking environments. Our system's innovative approach of reframing the ground bounding box detection problem as a keypoint distribution estimation issue has resulted in a significant advancement in the field of computer vision for parking lot applications.

The primary contributions of this research are the effective transformation of the detection problem into a keypoint estimation challenge and the implementation of a real-time, edge-based monocular camera system for underground parking lot obstacle detection. Our system demonstrates the ability to not only detect obstacles but also to estimate their 3D bounding boxes with high precision, which is a substantial improvement over traditional monocular camera setups that struggle with depth estimation.

The employment of spatial and temporal fusion techniques, including the mean shift algorithm and the Expanded Kalman Filter, has proven to be effective in refining detection results and enhancing the robustness of our system. The experiments conducted within an indoor parking lot, equipped with a network of 17 cameras, have yielded promising results. The system achieved high accuracy in detecting objects and estimating key points, with an AP50 score of 0.98 for bounding boxes and 0.97 for key points. Furthermore, the system's real-time performance, with an inference time of 4.7 ms per image, underscores its practical applicability.

The results of our system experiment, which used ground truth bounding boxes in the world coordinate system for evaluation, showed a significant reduction in the Root Mean Square Error (RMSE) of object trajectory and an increase in the mean Intersection over Union (mIoU) of object ground bounding boxes compared to traditional 2D box methods. These improvements highlight the effectiveness of our approach in providing accurate spatial of obstacles within the parking lot.

In conclusion, our parking lot obstacle detection system offers a comprehensive solution for monitoring and ensuring safety in parking lots. The system's ability to estimate 3D bounding boxes in real time, coupled with its high accuracy and robustness, makes it a valuable asset for intelligent transportation systems and autonomous vehicle technologies. Future work will focus on expanding the system's capabilities, incorporating additional sensors, and testing the system in diverse parking lot scenarios to further enhance its performance and reliability.

Disclosure of Interests. The authors have no competing interests to declare that are relevant to the content of this article.

References

1. Review of multi-view 3d object recognition methods based on deep learning. Displays **69**, 102053 (2021)
2. Acharya, D., Yan, W., Khoshelham, K.: Real-time image-based parking occupancy detection using deep learning. Research@ Locate **4**, 33–40 (2018)
3. et. al., G.J.: ultralytics/yolov5: v6.0 - YOLOv5n 'Nano' Models, Roboflow Integration, TensorFlow Export, OpenCV DNN support (October 2021)
4. Arulampalam, M.S., Maskell, S., Gordon, N., Clapp, T.: A tutorial on particle filters for online nonlinear/non-gaussian bayesian tracking. IEEE Trans. Signal Process. **50**(2), 174–188 (2002)
5. Bochkovskiy, A., Wang, C.Y., Liao, H.Y.M.: Yolov4: optimal speed and accuracy of object detection. arXiv preprint arXiv:2004.10934 (2020)
6. Carion, N., Massa, F., Synnaeve, G., Usunier, N., Kirillov, A., Zagoruyko, S.: End-to-end object detection with transformers. In: European Conference on Computer Vision, pp. 213–229. Springer (2020)
7. Chen, X., et al.: Multi-view 3D object detection network for autonomous driving. IEEE Trans. Pattern Anal. Mach. Intell. (2022)
8. Chen, X., Kundu, K., Zhang, Z., Ma, H., Fidler, S., Urtasun, R.: Monocular 3D object detection for autonomous driving. In: Proceedings of the IEEE Conference on Computer Vision and Pattern Recognition, pp. 2147–2156 (2016)
9. Comaniciu, D., Meer, P.: Mean shift: a robust approach toward feature space analysis. IEEE Trans. Pattern Anal. Mach. Intell. **24**(5), 603–619 (2002)
10. Dosovitskiy, A., et al.: An image is worth 16×16 words: transformers for image recognition at scale. arXiv preprint arXiv:2010.11929 (2021)
11. Farhadi, A., Redmon, J.: Yolov3: an incremental improvement. In: Computer Vision and Pattern Recognition, vol. 1804, pp. 1–6. Springer, Berlin/Heidelberg, Germany (2018)
12. Geiger, A., Lenz, P., Stiller, C.: Are we ready for autonomous driving? The kitti vision benchmark suite. In: 2012 IEEE Conference on Computer Vision and Pattern Recognition, pp. 3365–3372. IEEE (2012)
13. Huang, K.C., Wu, T.H., Su, H.T., Hsu, W.H.: Monodtr: monocular 3d object detection with depth-aware transformer. In: Proceedings of the IEEE/CVF Conference on Computer Vision and Pattern Recognition (CVPR), pp. 4012–4021 (June 2022)
14. Khan, S., Naseer, M., Hayat, M., Zamir, S.W., Khan, F.S., Shah, M.: Transformers in vision: a survey **54**(10s) (2022)
15. Li, C., et al.: Yolov6: a single-stage object detection framework for industrial applications. arXiv preprint arXiv:2209.02976 (2022)
16. Redmon, J.: You only look once: unified, real-time object detection. In: Proceedings of the IEEE Conference on Computer Vision and Pattern Recognition (2016)
17. Redmon, J., Farhadi, A.: Yolo9000: better, faster, stronger. In: Proceedings of the IEEE Conference on Computer Vision and Pattern Recognition, pp. 7263–7271 (2017)
18. Simon, D.: Optimal State Estimation: Kalman, H Infinity, and Nonlinear Approaches. John Wiley & Sons, Hoboken (2019)
19. Varghese, R., M., S.: Yolov8: a novel object detection algorithm with enhanced performance and robustness. In: 2024 International Conference on Advances in Data Engineering and Intelligent Computing Systems (ADICS), pp. 1–6 (2024)
20. Vu, H.T., Huang, C.C.: Parking space status inference upon a deep cnn and multi-task contrastive network with spatial transform. IEEE Trans. Circuits Syst. Video Technol. **29**(4), 1194–1208 (2019)

21. Wang, A., et al.: Yolov10: real-time end-to-end object detection. arXiv preprint arXiv:2405.14458 (2024)
22. Wang, C.Y., Bochkovskiy, A., Liao, H.Y.M.: Yolov7: trainable bag-of-freebies sets new state-of-the-art for real-time object detectors. In: Proceedings of the IEEE/CVF Conference on Computer Vision and Pattern Recognition, pp. 7464–7475 (2023)
23. Wang, C.Y., Yeh, I.H., Liao, H.Y.M.: Yolov9: learning what you want to learn using programmable gradient information. arXiv preprint arXiv:2402.13616 (2024)
24. Wang, X., Li, K., Chehri, A.: Multi-sensor fusion technology for 3D object detection in autonomous driving: a review. IEEE Trans. Intell. Transp. Syst. **25**(2), 1148–1165 (2024)
25. Wu, S., et al.: Graph-based 3D multi-person pose estimation using multi-view images, pp. 11128–11137 (2021)
26. Yang, J., Rahardja, S., Fränti, P.: Mean-shift outlier detection and filtering. Pattern Recogn. **115**, 107874 (2021)
27. Yi, F., Jeong, O., Moon, I., Javidi, B.: Deep learning integral imaging for three-dimensional visualization, object detection, and segmentation. Opt. Lasers Eng. **146**, 106695 (2021)

Self-tuning Control of Manipulator Based on Fuzzy Linear Active Disturbance Rejection Control and Particle Swarm Optimization Algorithm

Bo Tao, Zhuxiang Chen(✉), Du Jiang, and Juntong Yun

Key Laboratory of Metallurgical Equipment and Control Technology of Ministry of Education, Wuhan University of Science and Technology, Wuhan 430081, China
2219045902@qq.com

Abstract. The manipulator, due to the presence of unfavorable factors such as joint coupling, unknown disturbances, parameter ingestion and variable load tasks when performing complex tasks, greatly affects the control performance of the control system such as fast convergence, high precision tracking and robust robustness. Therefore, this paper proposes a self-tuning control method for manipulator based on fuzzy linear active disturbance rejection control (Fuzzy-LADRC) and particle swarm optimization (PSO) algorithm. Firstly, the PSO algorithm is the search algorithm for the optimal parameters of the controller, which ensures the reasonableness of the controller parameters. Secondly, the LADRC controller is utilized to assess and offset the unknown interference in the manipulator control system, which improves the manipulator control system's anti-jamming capability. Then, the parameters of the linear error feedback control rate (LSEF) are dynamically adjusted using fuzzy control to improve the control accuracy and robustness of the control system. Finally, the effectiveness of the proposed method is proved by simulation experiments to improve the control accuracy and robustness of the manipulator control system.

Keywords: Manipulator · Self-tuning control · Particle swarm optimization algorithm · Linear active disturbance rejection control · Fuzzy control

1 Introduction

With the wide application of manipulators in the fields of industrial, medical and service robots [1], the control accuracy, response speed, and stability of manipulators have received more and more attention. Since the manipulator is a non-linear, highly coupled, complex system that is highly susceptible to uncertainties such as friction, parameter pickup and other external disturbances when performing variable load tasks, it is difficult to achieve fast, accurate and stable angular control for it. Therefore, how to make the manipulator perform different complex tasks with more accurate control accuracy, faster response speed and more stable motion has become a key problem in the research of manipulator control.

In response to these problems mentioned above, traditional control methods have certain limitations due to their limitations for nonlinear and strongly coupled control systems [2]. Therefore, various new control methods have emerged. Long et al. [3] proposed a method of variable structure PID control by introducing variable structure dynamic control parameters, which improved the process control approach rate. Shang et al. [4] used a neuro net to identify the nonlinear terms and unknown disturbances of the manipulator, and presented a sliding mode control (SMC) approach that relies on neural network identification to boost the manipulator's control accuracy. Qiu et al. [5] combined fuzzy control and neuronal net for the multi-constraint nonlinear optimization problem of the manipulator to boost the manipulators' control system stability. Kang et al. [6] utilized the property of radial basis function neural network (RBFNN) that can assess the model online to enhance the control accuracy of the manipulator system. Mary et al. [7] included an additional robust term in the SMC legislation, which improved the manipulator control's robustness. Li et al. [8] used RBFNN to assess the unknown nonlinear friction online, and combined SMC to control the manipulator, which realized the steady control of the manipulator. Even though the studies above have shown satisfactory performance in manipulator control, certain problems remain, such as the need for an accurate system model, long computing time, easy to produce overshooting and difficult to eliminate the unknown interference and coupling effects. The LADRC controller, on the other hand, does not rely on an accurate system model and is able to observe and compensate for the uncertainties in the manipulator system through a linear expansion state observer (LESO) [9]. By designing fuzzy control rules, the control quantities of the linear self-resilient controller can be compensated dynamically and in real time, which further improves the accuracy, responsiveness and ruggedness.

Since the controller parameters often require a lot of manual time to repeatedly adjust the parameters [10], the adjustment of only three parameters takes a lot of time. Moreover, when the controller parameters are inaccurate, it will affect the control accuracy and robustness of the manipulator control system. Therefore, how to adjust the parameters quickly, efficiently and accurately has also become a difficult problem for many engineers. Along with the meteoric development of computer technology, intelligent algorithms have been extensively utilized in the adjustment of controller parameters[11], and their features such as high efficiency, global search capability, adaptivity, and reduction of manual intervention greatly improve the efficiency of parameter optimization [12]. To adjust Fuzzy-LADRC's three parameters, this paper introduces the PSO algorithm in intelligent algorithms for parameter search [13]. The main considerations for choosing PSO algorithm for the parameter adjustment in Fuzzy-LADRC controller in this paper include: a) global search ability: PSO algorithm has a powerful global search ability, and it can find a better solution in the complex search space; b) rapidity: PSO algorithm has a fast convergence speed, and it can find a better solution in a short time; c) parameter tuning simple: compared with other intelligent optimization algorithms, the PSO algorithm has fewer parameters and relatively simple adjustment, which reduces the complexity of parameter adjustment; d) strong ability to deal with nonlinear problems: the PSO algorithm is able to effectively solve nonlinear issues by means of iterative search.

Combining the above studies, this paper proposes a self-tuning control of the manipulator based on PSO-Fuzzy-LADRC algorithms. By using MATLAB/Simulink as the simulation platform, the manipulator is controlled by the Fuzzy-LADRC controller tuned by PSO algorithm, LADRC controller and PID controller, respectively. The simulated outputs indicate that the control effect of Fuzzy-LADRC controller optimized by PSO algorithm is better than that of LADRC and PID controllers optimized by PSO algorithm.

2 Kinetic Modeling of a Double-Jointed Manipulator

For an n-joint rigid manipulator system, a Lagrangian modeling approach is usually used to establish its kinetic equations, which are expressed as:

$$K(q)\ddot{q} + L(q,\dot{q})\dot{q} + B(q) = \tau \tag{1}$$

However, the manipulator will be affected by uncertainties such as friction, modeling error, parameter uptake and external environment disturbance in actual operation. The modeling error and friction are usually considered as the internal disturbances of the manipulator, and the external environmental disturbances are considered as the external disturbances of the manipulator. When considering the effects of these internal and external disturbances, the dynamics equation of the manipulator can also be expressed as:

$$\tau = K(q)\ddot{q} + L(q,\dot{q})\dot{q} + B(q) + \tau_h$$
$$\tau_h = \tau_f + \tau_d \tag{2}$$

where $\tau_f \in R^{n \times 1}$ is the internal disturbance vector of the manipulator; $\tau_d \in R^{n \times 1}$ is the external disturbance vector of the manipulator; $\tau_h \in R^{n \times 1}$ is the complex disturbance vector of the manipulator.

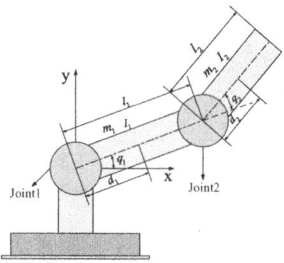

Fig. 1. Sketch of a double-joint rigid manipulator.

To validate the control efficacy of the controller presented in this research, a double-joint rigid manipulator was used as the study object, its sketch shown in Fig. 1. The kinetic equations of the double-jointed manipulator are represented by Eq. (2), and the expressions for the parameters are derived computationally:

$$K(q) = \begin{pmatrix} b_1 + b_2 + 2b_3 \cos q_2 & b_2 + b_3 \cos q_2 \\ b_2 + b_3 \cos q_2 & b_2 \end{pmatrix} \tag{3}$$

$$L(q, \dot{q}) = \begin{pmatrix} -b_3 \dot{q}_2 \sin q_2 & -b_3(\dot{q}_1 + \dot{q}_2) \sin q_2 \\ b_3 \dot{q}_1 \sin q_2 & 0 \end{pmatrix} \quad (4)$$

$$B(q) = \begin{pmatrix} b_4 g \cos q_1 + b_5 g \cos(q_1 + q_2) \\ b_5 g \cos(q_1 + q_2) \end{pmatrix} \quad (5)$$

$$\tau = \begin{bmatrix} \tau_1 & \tau_2 \end{bmatrix}^T \quad (6)$$

$$\tau_h = \begin{bmatrix} d_1 & d_2 \end{bmatrix}^T \quad (7)$$

where $b_1 = m_1 d_1^2 + m_2 l_1^2 + I_1$; $b_2 = m_2 d_2^2 + I_2$; $b_3 = m_2 l_1 d_2$; $b_4 = m_1 d_1 + m_2 l_1$; $b_5 = m_2 d_2$.

3 PSO-Fuzzy-LADRC Controller Design

3.1 LADRC Controller Design

The LADRC includes linear tracking differentiator (LTD), LESO and LSEF, whose core thinking is to consider the interior perturbation and the outside unknown perturbation as the total perturbation of the system. The transition process is arranged through LTD to solve the conflict between rapidity and overshooting amount. The total interference is evaluated and removed in real time through LESO, which in turn transforms the complex nonlinear controlled object into a simple integral series type and is not entirely dependent on the accurate theoretical modelling. Finally, the system is controlled and disturbance compensated by LSEF. Its basic structure is shown in Fig. 2.

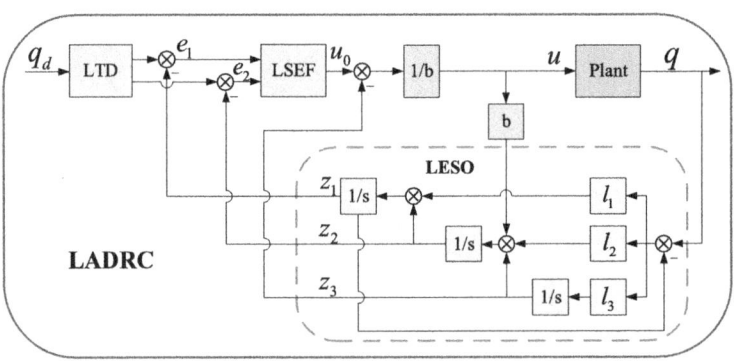

Fig. 2. Structure of LADRC controller.

The kinetic model of the double-jointed manipulator of Eq. (2) is derived as follows:

$$\ddot{q} = -K(q)^{-1} L(q, \dot{q}) \dot{q} - K(q)^{-1} B(q) - K(q)^{-1} \tau_h + K(q)^{-1} \tau \quad (8)$$

Due to its natural decoupling ability, LADRC can decouple the manipulator system into a single-input single-output system. Therefore, Eq. (8) can be simplified as:

$$\ddot{q} = f(q_1, q_2, \dot{q}_1, \dot{q}_2, d_1, d_2, t) + bu \tag{9}$$

where $f(q_1, q_2, \dot{q}_1, \dot{q}_2, d_1, d_2, t)$ is all internal and external system disturbances; b is the control parameter; u is the designed control rate.

Since the double-jointed manipulator system is a second-order system, the third-order LESO should be chosen for observing the joint angles, joint angular velocities, and total disturbances of the manipulator, respectively, then Eq. (9) can be written as:

$$\begin{cases} \dot{x}_1 = x_2 \\ \dot{x}_2 = x_3 + bu \\ \dot{x}_3 = \dot{f} \\ q = x_1 \end{cases} \tag{10}$$

The observer design of the LESO of the double-jointed manipulator is as follows:

$$\begin{cases} \dot{z}_1 = z_2 - \beta_1(z_1 - q) \\ \dot{z}_2 = z_3 - \beta_2(z_1 - q) + bu \\ \dot{z}_3 = -\beta_3(z_1 - q) \end{cases} \tag{11}$$

where z_1, z_2 and z_3 are estimates of the angle, angular velocity and total interference, respectively; β_1, β_2 and β_3 are the observer gains.

Based on the idea of bandwidth, let all the poles of the characteristic equation $\lambda(s)$ be $-w_o$, then the parameters of LESO take the value of:

$$\begin{cases} \lambda(s) = s^3 + \beta_1 s^2 + \beta_2 s + \beta_3 = (s + w_o)^3 \\ \begin{cases} \beta_1 = 3w_o \\ \beta_2 = 3w_o^2 \\ \beta_3 = w_o^3 \end{cases} \end{cases} \tag{12}$$

Since the study focuses on angle control, LSEF can be regarded as a PD controller. LSEF is designed to:

$$u_0 = k_p(r - z_1) - k_d z_2 \tag{13}$$

Similarly, the LSEF can be designed by the bandwidth method, which can be inferred from Eq. (13):

$$\begin{cases} s^2 + k_d s + k_p = (s + w_c)^2 \\ \begin{cases} k_p = w_c^2 \\ k_d = 2w_c \end{cases} \end{cases} \tag{14}$$

It can be noted that the observer and controller gains are only bandwidth dependent, which makes the design of the LESO and LSEF simplified and parameter tuning easier.

3.2 PSO Algorithm to Optimize LADRC Controller Parameters

The core idea of PSO algorithm is to determine the best way to solve the issue by simulating the foraging strategies of individuals in the flock [14]. The PSO algorithm's updated formulation is as follows [15]:

$$\begin{cases} v_{ij}^{t+1} = wv_{ij}^t + c_1 r_1 \left(pbest_{ij} - x_{ij}^t\right) + c_2 r_2 \left(gbest_j - x_{ij}^t\right) \\ x_{ij}^{t+1} = x_{ij}^t + v_{ij}^{t+1} \end{cases} \quad (15)$$

When using the PSO algorithm for LADRC controller parameter optimization, the three parameters (w_c, w_o, b) of the controller are taken as the target variables to be optimized. And it is transformed into a multidimensional search problem, and the best controller parameters are obtained by continuously iterating until the ideal solution is found. To enhance the optimization efficiency of the PSO algorithm, it is needed to limit the parameter range (w_c, w_o, b). The LADRC manipulator control system is constructed in the MATLAB/Simulink module, and the ranges of the w_c, w_o and b parameters are finally determined as [10,20], [800,1500], and [6, 12]. The LADRC controller parameter range is estimated by using debugging method:

(a) Reduce the search space, can avoid invalid search, reduce the complexity of finding the best parameters and the amount of calculation, enhance the optimization efficiency;
(b) Ensure the rationality of the controller parameters and ensure that the control system has good stability, fast response and accuracy.

The optimal solution problem of LADRC is to find a range of suitable parameters for the performance evaluation index optimal at the end of the algorithm operation. The commonly used error performance indexes are *ISE*, *IAE*, *ITAE* and *ISTE*, etc. The performance evaluation index used in this paper is ITAE, which is obtained by dividing the duration by the total value of the speed error and then integrating it, and this performance evaluation index reflects the accuracy and speed of the system control, and its expression is:

$$ITAE = \int_0^\infty t|e(t)|dt \quad (16)$$

After determining the *ITAE* performance indicators, the PSO algorithm is used to gradually assign values to the parameters to be optimized by LADRC, and the control system is run for parameter optimization. Under the premise of satisfying the constraints, the parameter with the smallest performance evaluation index function value is selected as the optimal solution of LADRC. The manipulator control system sketch of PSO optimization LADRC is illustrated in Fig. 3, and the process of PSO optimization LADRC is illustrated in Fig. 4.

3.3 Fuzzy Controller Design

When the manipulator is performing variable load tasks, dynamic adjustment of the LSEF parameters is realized by using the angular error e and the rate of change of the

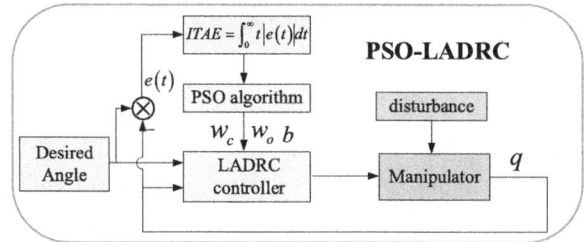

Fig. 3. PSO algorithm to optimize LADRC controller parameters.

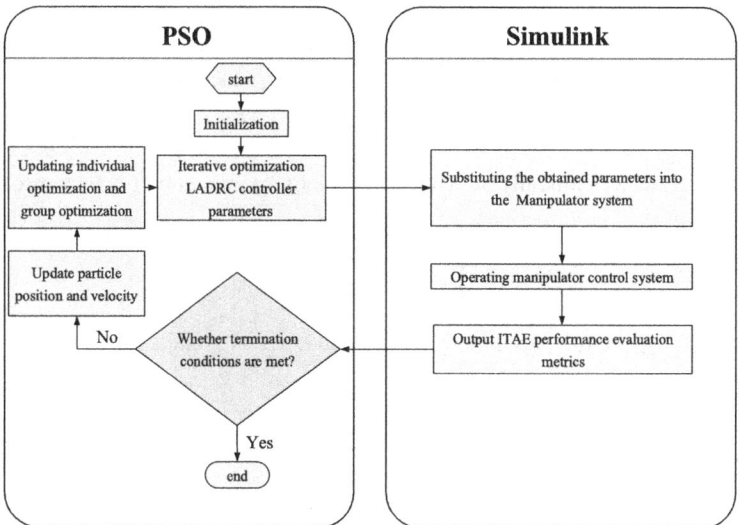

Fig. 4. Flowchart of PSO algorithm to optimize LADRC.

error ec as the inputs of the fuzzy controller. The fuzzy set rules are utilized to output the corrected values of the control parameters Δk_p and Δk_d, to enhance the control level and disturbance immunity. Based on the theoretical analysis, the fuzzy domains of the e and ec are designed as $[-3,3]$ and $[-3,3]$, respectively. The fuzzy areas of the correction values Δk_p and Δk_d are designed as $[-4,4]$ and $[-8,8]$, respectively. Seven levels of fuzzy subsets on the domain can be formed from the controller input and output variables: {NB, NM, NS, ZO, PS, PM, PB}.

The defuzzification method utilized in this article is using the central gravity method, which calculates all the input space combinations to derive the corresponding output control quantities to form the control table. The fuzzy control rules directly affect the control effect, while the fuzzy rules in Table 1 are formulated based on practical control experience and knowledge of LADRC theory.

Table 1. (a) Fuzzy control rules for Δk_p and Δk_d.

$\Delta k_p\ \Delta k_d$		ecec						
		NB	NM	NS	ZO	PS	PM	PB
e	NB	PB/PS	PB/PS	PM/NB	PM/NB	PS/NB	PS/NM	ZO/PS
	NM	PB/PS	PM/NS	PM/NB	PS/NM	PS/NM	ZO/NS	NS/ZO
	NS	PM/ZO	PM/NS	PS/NM	PS/NM	ZO/NS	NS/NS	NM/ZO
	ZO	PM/ZO	PS/ZO	PS/NS	ZO/ZO	NS/NS	NS/ZO	NM/ZO
	PS	PM/ZO	PS/ZO	ZO/ZO	NS/ZO	NS/ZO	NM/ZO	NM/ZO
	PM	PS/PB	ZO/NS	NS/PS	NS/PS	NM/PS	NM/PS	NB/PB
	PB	ZO/PB	NS/PM	NS/PM	NM/PM	NM/PS	NB/PS	NB/PB

Therefore, the dynamically adjusted controller parameters by fuzzy control are:

$$\begin{cases} k_{op} = k_{ip} + \Delta k_p \\ k_{od} = k_{id} + \Delta k_d \end{cases} \quad (17)$$

where k_{ip} and k_{id} are the parameters given in advance; k_{op} and k_{od} are the final output controller parameters.

4 Simulation Experiment

4.1 PSO-LADRC Controller Performance Test

To validate the control effect of PSO-LADRC controller on the manipulator, it is compared with PID and LADRC controllers. A simulation model of the manipulator control system is built on the MATLAB/Simulink simulation platform to confirm the controller's control effect. The PSO-LADRC controller parameters can be derived iteratively by the PSO algorithm, with the parameter ranges set to $w_c \in [10, 20]$, $w_o \in [800, 1500]$, and $b_0 \in [8, 12]$. The LADRC controller parameters are set to $w_c = 15$, $w_o = 1000$, and $b_o = 10$. The PID controller parameters are set to $k_p = 80$, $k_i = 5$ and $k_d = 10$. The parameters of the manipulator used in this paper are: $m_1 = 2\ kg$, $m_2 = 1\ kg$, $l_1 = 0.20\ m$, $l_2 = 0.15\ m$, $I_1 = 0.061\ kg \cdot m^2$, and $I_2 = 0.020\ kg \cdot m^2$.

The control performance of the PSO-LADRC controller is examined by applying a desired angle of 1 rad to both joints of the manipulator at a simulation time of 1 s. And it is compared with LADRC and PID controller under the same conditions. The simulation plot of the manipulator angle variation is shown in Fig. 5. Figure 5 (left) shows the simulation curves of joint 1. At the desired angle of 1 rad, PSO-LADRC reaches steady state without overshooting at 1.5 s. LADRC reaches steady state without overshooting at 1.7 s, and the PID reaches steady but not at the desired angle at 1.9 s. Figure 5 (right) shows the simulation curves of joint 2. At the desired angle of 1 rad, PSO-LADRC reaches steady state without overshooting at 1.5s, LADRC reaches steady state without overshooting at 1.7 s, and PID reaches steady state but beyond the desired angle at 1.9 s.

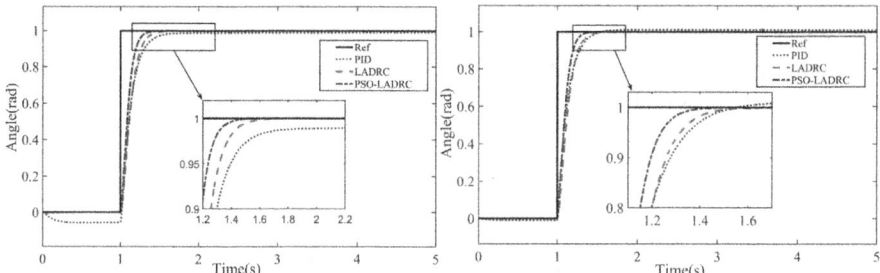

Fig. 5. Simulation graph of the change in the angle of the joints of the manipulator: joint 1 on the left and joint 2 on the right.

The simulation experiment findings demonstrate that the PSO-LADRC controller outperforms the LADRC and PID controllers for controlling behavior of fast, accurate as well as stable tracking of the manipulator.

4.2 PSO-Fuzzy-LADRC Control Performance Test

To validate the effectiveness of PSO-Fuzzy-LADRC controller for controlling the manipulator, it is compared with PSO-PID and PSO-LADRC controllers. The parameters of the Fuzzy-LADRC, LADRC and PID controllers are optimized using the PSO algorithm, and the optimal fitness value is calculated by the fitness function ITAE. Due to the huge workload of this simulation experiment, in order to enhance its optimization efficiency and avoid the diffusion phenomenon, the debugging method above is utilized to carry out simulation experiments on the LADRC and PID controllers first, by debugging the parameters step by step and observing the simulation output results. The LADRC controller parameters range is the same as above. The range of control parameters of the PID controller are $k_p \in [70, 100]$, $k_i \in [5, 15]$, and $k_d \in [10, 15]$.

In order to show that the behavior of PSO-Fuzzy-LADRC is superior to PSO-LADRC and PSO-PID on equal terms, this paper conducts simulation experiments in the following three ways:

Experiment one: The behavior of the PSO-Fuzzy-LADRC controller is examined by applying a desired angle of 1 rad to two joints of the manipulator at a simulation time of 1 s. And it is compared with different controllers under the same conditions. Figure 6 displays the manipulator's single angle variation simulation graph.

Figure 6 (left) shows the simulation curves of joint 1. At the desired angle of 1 rad, PSO-Fuzzy-LADRC attains the steady state without overshooting at 1.3 s, PSO-LADRC attains the steady state without overshooting at 1.5 s, and PSO-PID attains the steady state but does not reach the desired angle at 1.9 s. At 0-1 s, due to the presence of certain weight of connecting rod 1, connecting rod 2, and joint 2, there will be some torque on joint 1, so there will be joint angle regression at the beginning, in which the PSO-PID regressed by 0.05rad and maintained to 0.05 rad, and the PSO-Fuzzy-LADRC and the

PSO-LADRC regressed the maximum of 0.0024 rad and quickly recovered to 0 rad. Figure 6 (right) shows the simulation curves of joint 2. At the desired angle of 1 rad, PSO-Fuzzy-LADRC attains the steady state without overshooting at 1.3 s, PSO-LADRC attains the steady state without overshooting at 1.5 s, and PSO-PID attains the steady state without overshooting at 1.7 s. Similarly, due to the presence of some weight in connecting rod 2, there is some regression of the joint turning angle at 0-1s, but the regression angle is small and negligible.

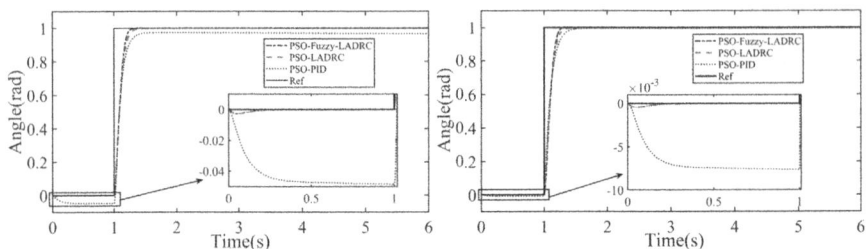

Fig. 6. Simulation graph of single angle change of manipulator: joint 1 on the left and joint 2 on the right.

Experiment two: Based on Experiment one, a desired angle of 2 rad is instantly given to the manipulator at a simulation time of 3s to prove whether the PSO-Fuzzy-LADRC controller can realize fast, accurate and stable control of the manipulator again. Figure 7 displays the simulation graph of the manipulator's dynamic angle change.

Figure 7 (left) shows the simulation curves of joint 1. With a sudden increase of the desired angle by 2 rad at a simulation time of 3 s, the PSO-Fuzzy-LADRC reaches steady state without overshooting at 3.28 s, the PSO-LADRC reaches steady state without overshooting at 3.45 s, and the PSO-PID, although it reaches steady state without overshooting at 3.5 s, exceeds the desired angle and maintained at that angle. Figure 7(right) shows the simulation curves of joint 2. With a sudden increase of the desired angle by 2 rad at a simulation time of 3 s, the PSO-Fuzzy-LADRC reaches steady state without overshooting at 3.3 s, the PSO-LADRC reaches steady state without overshooting at 3.45 s, and the IPSO-PID, although it reaches steady state without overshooting at 3.6 s, does not reach the desired angle and maintained at that angle.

Experiment three: On the basis of experiment one, to validate the designed controller interference immunity, a disturbance with a constant value of $20\,N \cdot m$ and a disturbance with a sinusoidal of $20\sin(\pi t) N \cdot m$ are applied to the manipulator at the simulation time of 3-5 s, respectively. Figure 8 and Fig. 9 display the outcomes of the simulation experiments.

Figure 8 shows the simulation plot with a constant value disturbance applied and Fig. 9 shows the simulation plot with a sinusoidal disturbance applied. It is clear from the figure that both the PSO-Fuzzy-LADRC controller and the LADRC controller have strong anti-disturbance ability regardless of the fixed-value disturbance or sinusoidal

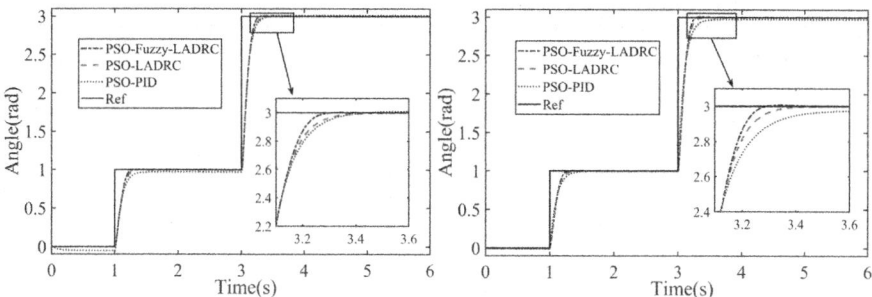

Fig. 7. Simulation graph of dynamic angle change of manipulator: joint 1 on the left and joint 2 on the right.

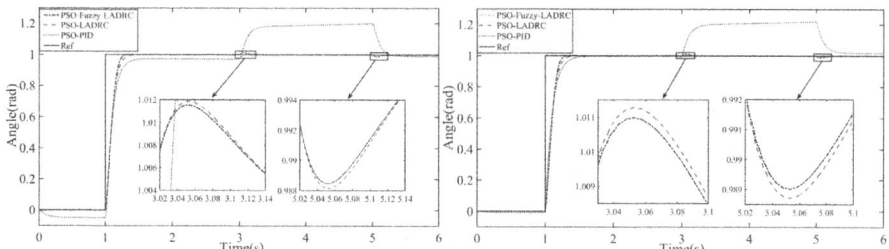

Fig. 8. Apply constant value perturbations to both joint 1 and joint 2.

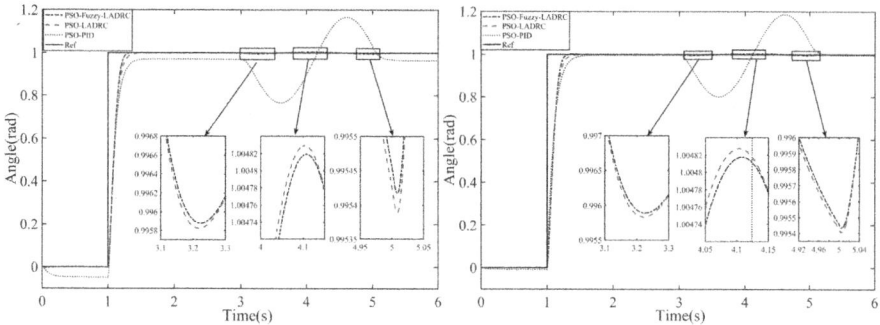

Fig. 9. Apply sinusoidal perturbations to both joint 1 and joint 2.

disturbance, and the zoomed-in figure displays that the controller effect of the PSO-Fuzzy-LADRC controller is superior to the PSO-LADRC controller. However, the PSO-PID controller is less effective than the first two controllers when facing either constant or sinusoidal disturbances.

To visualize the control effect of the three controllers, the ITAE value is used as the performance evaluation index of the controllers, and the ITAE value of each joint controller of the above simulation experiment is recorded as shown in Table 2. And the histogram of each joint controller performance metric is plotted as shown in Fig. 10.

From the three simulation results and ITAE values of each joint controller, the PSO-Fuzzy-LADRC controller designed in this paper can realize fast, accurate and stable angle control of the manipulator, and it has a good capacity to resist interference. The controller designed in this paper performs better compared to PSO-LADRC and PSO-PID controllers.

Table 2. ITAE values for joint 1 and joint 2 controllers

ITAE		PSO-Fuzzy-LADRC	PSO-LADRC	PSO-PID
Joint1	Experiment one	0.101	0.108	0.306
	Experiment two	0.687	0.721	1.335
	Experiment three	0.126	0.133	1.287
Joint2	Experiment one	0.102	0.108	0.234
	Experiment two	0.691	0.723	0.817
	Experiment three	0.127	0.133	1.215

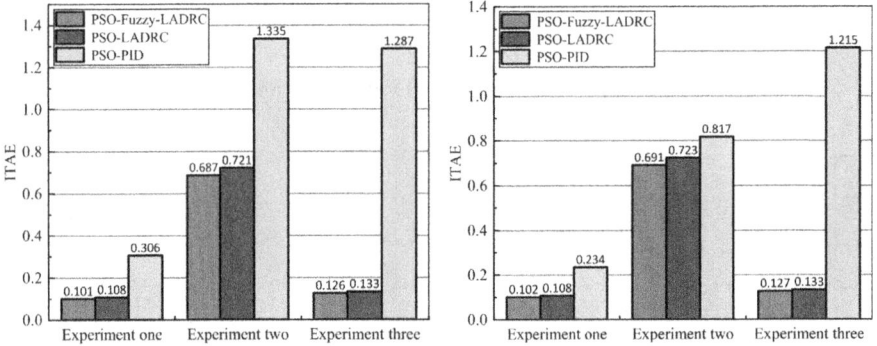

Fig. 10. Histogram of performance indicators for each joint controller: joint 1 on the left and joint 2 on the right.

5 Conclusion

The angular control of a double-jointed manipulator is the main topic of this paper. Since the manipulator is a complex system, and it is very susceptible to uncertainties such as friction and other external disturbances when performing a variable load task, it is difficult to provide fast, accurate and stable angle control in fact. Responding to the above issues, this paper proposes a self-tuning control of the manipulator based on PSO-Fuzzy-LADRC algorithms. Through simulation experiments on the MATLAB /Simulink, the PSO-Fuzzy-LADRC controller is confirmed to have high accuracy and strong robustness in the manipulator control system.

Acknowledgments. This project is supported by the higher education teaching reformation project of Hubei province of China (2022231, 2022216) and department of science and technology of Hubei province of China (2024DJC073).

Disclosure of Interests. The authors have no competing interests to declare that are relevant to the content of this article.

References

1. Rawat, D., Gupta, M.K., Sharma, A.: Intelligent control of robotic manipulators: a comprehensive review. Spat. Inf. Res. **31**(3), 345–357 (2023)
2. Liu, Y., Jiang, D., Yun, J.: Self-tuning control of manipulator positioning based on fuzzy PID and PSO algorithm. Front. Bioeng. Biotechnol. **9**, 817723 (2022)
3. Long, Y., You, J., Huang, Y., Chen, Z.: Robust variable structure PID controller for two-joint flexible manipulator. In: International Conference on Computing, Control and Industrial Engineering, pp. 133–144. Springer Nature Singapore, Singapore (2023)
4. Shang, D.: Dynamic modeling and RBF neural network compensation control for space flexible manipulator with an underactuated hand. Chin. J. Aeronaut. **37**(3), 417–439 (2024)
5. Qiu, B., Guo, J., Mao, M.: A fuzzy-enhanced robust DZNN model for future multi-constrained nonlinear optimization with robotic manipulator control. IEEE Trans. Fuzzy Syst. **32**, 160–173 (2023)
6. Kang, E., Qiao, H., Gao, J.: Neural network-based model predictive tracking control of an uncertain robotic manipulator with input constraints. ISA Trans. **109**, 89–101 (2021)
7. Mary, A.H., Al-Talabi, A., Kara, T.: Adaptive robust tracking control of robotic manipulator based on SMC and fuzzy control strategy. Al-Khwarizmi Eng. J. **20**(1), 63–75 (2024)
8. Li, X., Gao, H., Xiong, L., Zhang, H.: A novel adaptive sliding mode control of robot manipulator based on RBF neural network and exponential convergence observer. Neural Process. Lett. **55**(7), 10037–10052 (2023)
9. Gao, Z.: Scaling and bandwidth-parameterization based controller tuning. In: Proceedings of the 2003 American Control Conference, pp. 4989–4996 (2003)
10. Liu, Y., Li, G., Jiang, D.: Dynamic ensemble multi-strategy based bald eagle search optimization algorithm: a controller parameters tuning approach. Appl. Soft Comput. **148**, 110881 (2023)
11. Hussien, A.G., Amin, M.: Crow search algorithm: theory, recent advances, and applications. IEEE Access **8**, 173548–173565 (2020)
12. Rahayu, E.S.: Particle swarm optimization (PSO) tuning of PID control on DC motor. Int. J. Robot. Control Syst. **2**(2), 435–447 (2022)
13. Nguyen, D.T., Ho, J.R.: A hybrid PSO–GWO fuzzy logic controller with a new fuzzy tuner. Int. J. Fuzzy Syst. **24**(3), 1586–1604 (2022)
14. Jain, M.: An overview of variants and advancements of PSO algorithm. Appl. Sci. **12**(17), 8392 (2022)
15. Ramírez-Ochoa, D.D., Pérez-Domínguez, L.A.: PSO, a swarm intelligence-based evolutionary algorithm as a decision-making strategy: a review. Symmetry **14**(3), 455 (2022)

The Neural Network of College Students Research on the Intelligent Scoring System

Yang Ji(✉) and Huang Shen

Wuhan Vocational College of Software Engineering, Wuhan 430205, China
yj649@sohu.com

Abstract. Using BP neural network function mapping method, combined with the gray system model, the physical health test data of 1500 college students in Wuhan Vocational College of Software Engineering in 2019 were selected to build the intelligent scoring system of college students physical health, and then 350 student test data were selected to test the accuracy of the model. The calculation results show that the model is more accurate than the traditional method of weight coefficient weighted summing, and this method provides a new idea for the physical healthy individual score and group physical condition evaluation of college students.

Keyword: BP neural network · gray system · marking system · college students

1 Introduction

National physical health is one of the basic qualities related to the country as a whole, Over the years by the party and the government attaches great importance to it. The Decision of the CPC Central Committee and The State Council on Deepening Education Reform and Comprehensively Promoting Quality-oriented Education states: "A healthy body is the basic premise for young people to serve the motherland and the people, and is the embodiment of the vigorous vitality of the Chinese nation [1]. "The 2014 national physique monitoring work found that the physical fitness of Chinese college students was declining, and the detection rate of obesity was on the rise [2]. In 2016, the central committee of the communist party of China, the State Council issued the "healthy China 2030" planning outline, "outline" points out: "promoting the construction of healthy China, is to build a well-off society in an all-round way, the important foundation of socialist modernization, is actively involved in global health governance, fulfill the 2030 sustainable development agenda international commitments of major initiatives.[3]" In recent years, colleges and universities have collected a large number of physical health data, college students physical health level and evaluation index between the complex nonlinear relationship, the traditional physical health evaluation method on the basis of the state students each individual body scores, the weighted percentage system way, but the individual scoring criteria and weight coefficient size has been controversial. With the rapid development of artificial intelligence, intelligent scoring can effectively solve these problems. Gray model prediction is the characteristics of single sequence prediction, only in form predicted object sequence model, according to the characteristics of the sequence itself modeling, prediction, and related factors are not directly involved, but many direct, obvious, indirect, hidden, known, unknown factors included in it, as a gray information or gray, find information from their own sequence model, discover and know the inherent law to predict [4].

BP neural network model is a multilevel feedback network [5], which is one of the most widely used neural network models. The main advantage of BP neural network is to deal with the nonlinear problem, which is a special property distributed in the whole network. Compared with the nonlinear regression method, the emergence of neural network avoids the troublesome [6] of selecting which nonlinear function is used. The gray model fluctuates the data due to the special situation in the system, and the prediction error is large. The BP neural network fitting function is fused to train the gray model, so as to correct the model and improve the accuracy of the model [7].

This study will body data monitoring and artificial intelligence analysis technology, comprehensive analysis of college students physical health indicators, using the gray model and BP neural network model, build college students physical health intelligent scoring system, ignore the traditional scoring method of individual index criteria and weight coefficient dispute, is advantageous to college students individual physical score and group physical judgment, for the current college students physical health research has positive significance. This research mainly uses the grey system theory to improve the learning ability of the artificial neural network, and uses it to realize the intelligent scoring of college students physical quality.

2 Model Building

2.1 Establish the Expert Rating Database of College Students Physical Fitness

The construction of college students physical quality scoring database is crucial to this research, and the gray contingency artificial neural network is established on this scoring database. In this study, the physical health test data of 2000 boys of 2019 from Wuhan Software Engineering Vocational College were selected. The indicators included BMI value (BMI = weight (kg)/height 2 (m^2)), vital capacity, 50 m running, standing long jump, sitting forward bending, 1000 m running, and pull-up. 150 invalid data were excluded, and 1500 students were selected to build the model (see Table 1). All the data are not pre-processed (normalized). In order to investigate the scientific nature of the research, we must process the scoring data and build a relatively perfect scoring database. The scoring database is comprehensively scored according to the scoring standards and evaluation methods of the National Physical Health Standards for Students (see Table 1).

Table 1. Expert scoring database of college students physical health level

number	sex	Body mass index	vital capacity	Run 50 m	standing long jump	Sit forward	1000 Meters	pull-up	Expert score
1	1	30.6	3867	8.2	222	14	191	7	65.6
2	1	21.8	5294	8.4	221	22	198	3	67.9
3	1	24.1	4435	7.8	220	14.2	200	5	61.4
4	1	19.8	3850	7.2	225	15	200	0	61.5
5	1	17.6	3599	6.9	235	5.8	201	6	68.6
6	1	21.6	4323	7	218	10.9	201	11	68.6
7	1	19.9	3510	8.5	221	17.6	202	11	68.9
8	1	18.9	3646	7.7	210	11.2	205	5	69.8
9	1	22.1	3517	8.4	189	7.8	206	11	69.9
10	1	22.5	3631	7.7	193	11.4	206	7	70
11	1	17.2	2939	8.6	200	14.4	207	11	70.1
12	1	17.2	3477	7.1	227	9.8	207	11	70.2
13	1	18.3	3142	7.4	233	15.5	207	11	70.2
……	……	……	……	……	……	……	……	……	……
1850	1	20.9	4306	7.1	245	18.2	208	11	70.5

2.2 Integrate the Whitening Contingency and BP Network to Establish the Gray Contingency Neural Network

BP neural network is a multi-layer feed forward network that backdates by error. It is composed of input layer, hidden layer and output layer, and its structure is shown in Fig. 2. After constructing the BP network topology, it is also necessary to train the network on learning, and the learning process consists of forward propagation and feedback correction (Fig. 1).

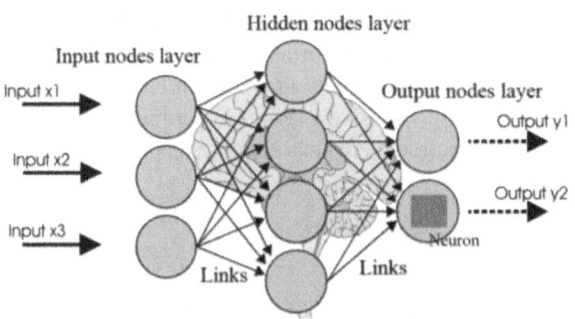

Fig. 1. BP topology of the neural network structure

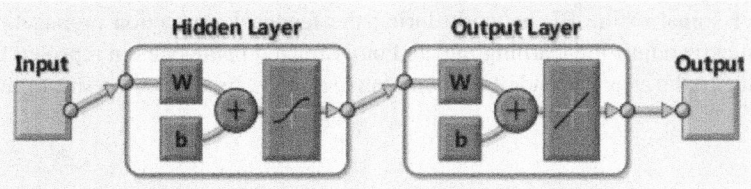

Fig. 2. BP neural network schematic diagram with the topology displayed by MATLAB

In the forward propagation process, the transfer relationship from the input layer to the hidden layer and the hidden layer to the output layer is:

$$y_k = f(\sum_{i=1}^{N} w_{ik} \cdot x_i + \theta_k) \tag{1}$$

$$y_h = f(\sum_{k=1}^{M} w_{kh} \cdot y_k + \theta_h) \tag{2}$$

y is the value of each layer, w is the connection weights of front to back layer of the front layer to the back layer, x_i is the input value of the input layer, θ is the threshold value of each neuron, f is the response function (select the Sigmoid function). Correction to the connection weights w and thresholds θ is crucial during feedback correction propagation. The standard BP neural network weight and threshold modification direction is the gradient direction where the current cumulative error E decreases, and the formula is:

$$w'_{ij} = w_{ij} + \Delta w_{ij} = w_{ij} - a\frac{\partial E}{\partial w_{ij}} = w_{ij} + a \cdot e_j \cdot y_i \tag{3}$$

$$\theta'_j = \theta_j + \Delta \theta_j = w + a \cdot e_j \tag{4}$$

where w'_{ij} is the new w_{ij}, $\frac{\partial E}{\partial w}$ representing the gradient direction of the cumulative error, e is the generalization error, for the prior given learning rate a $(0 < a < 1)$. At this time, the error signal back propagate from the output layer to the input layer and the connection weights and thresholds of each layer are adjusted along the way to reduce the error until the accuracy requirements are met. The feedback weight correction is actually through the repeated training of multiple samples, and then the "fastest gradient descent method" is adopted to make the weight change along the direction of the negative gradient of the error function, and converge to the minimum point or reach the required minimum error value or the maximum number of iterations. The learning rate in BP neural network a is a constant given in advance. As can be seen from (3) and (4), the correction of weight w and threshold θ is "equal step length" for the gradient, which is easy to cause problems such as low learning efficiency, slow convergence speed and local minimum. In view of the deficiency of BP network, "whitening power" can be combined with BP network to establish gray weight neural network to improve the learning efficiency and stability of BP network. The gray weight neural network adds "gray neurons" to the hidden layer and the output layer of the BP network. The function of these gray neurons is to accept the error signal of the BP network during the feedback correction propagation, and automatically adjust the learning rate and other related parameters in repeated training according to the change trend of the cumulative error E. Its structure is shown in Figure Fig. 3.

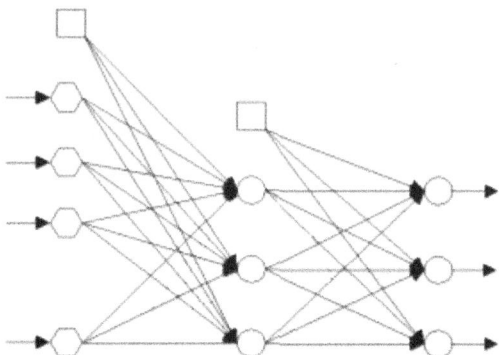

Fig. 3. Schematic of gray variable neural network

If the cumulative error E decreases during repeated training, the current correction direction is correct and can be larger in the next training; if the cumulative error E increases, the current correction direction is not correct and can be smaller in the next training. This becomes a "grey (function) number", with an Grey interval of (0,1). Using the idea of "whitenization weight" in the gray system, we can regard $a^k(E^{k-1})$ as a minus function of E. Here, k refers to the number of training times, so:

$$w_{ij}^{k'} = w_{ij}^k + \Delta w_{ij}^k = w_{ij}^k - a\frac{\partial E^k}{\partial w_{ij}} = w_{ij}^k + a^k(E^{k-1}) \cdot e_j^k \cdot y_i^k \tag{5}$$

$$\theta_j^k = \theta_j^k + \Delta\theta_j^k = \theta_j^k + a^k(E^{k-1}) \cdot e_j^k \tag{6}$$

The learning rate of the gray weight neural network is dynamic, which can be adjusted with the change of E. Judging from (5) and (6), the correction of weight w and threshold θ is no longer "equal step length" for the gradient. When E is large, the neural network can take a little larger steps; when E is smaller. Because the convergence of gray weight neural network is a process of fast and then slow, the disadvantages of BP network can be overcome.

2.3 Build an Intelligent Scoring System of Gray Power Neural Network for College Students Physical Quality

On the basis of the more perfect expert scoring database of college students physical quality, we can use the gray variable weight neural network to build an intelligent scoring system.

Table 2 shows the physical quality and expert scoring data of 1850 male students in Wuhan Software Engineering Vocational College. We select the physical data and expert scoring data of 1500 students as the training sample to establish the gray power neural network, and then use the data of the remaining 350 students as the test sample for verification. It is found that the scoring results of the gray weight neural network are very consistent with the expert scoring results (simulation cumulative error of test sample: 0.9257, simulation cumulative error of training sample: 0.51855) It can be seen that the intelligent scoring system of the gray weight neural network can almost replace the expert scoring.

Table 2. Expert scoring database and intelligent scoring database of college students physical health level

number	sex	Body mass index	vital capacity	Run 50 m	standing long jump	Sit forward	1000 Meters	pull-up	Expert score	Neural network simulation score
1	1	30.6	3867	8.2	222	14	191	7	65.6	65.5
2	1	21.8	5294	8.4	221	22	198	3	67.9	68.0
3	1	24.1	4435	7.8	220	14.2	200	5	61.4	61.5
4	1	19.8	3850	7.2	225	15	200	0	61.5	61.5
5	1	17.6	3599	6.9	235	5.8	201	6	68.6	68.7
6	1	21.6	4323	7	218	10.9	201	11	68.6	69.7
7	1	19.9	3510	8.5	221	17.6	202	11	68.9	70.0
8	1	18.9	3646	7.7	210	11.2	205	5	69.8	68.7
9	1	22.1	3517	8.4	189	7.8	206	11	69.9	70.2
10	1	22.5	3631	7.7	193	11.4	206	7	70	70.3
11	1	17.2	2939	8.6	200	14.4	207	11	70.1	70.8
12	1	17.2	3477	7.1	227	9.8	207	11	70.2	70.0
13	1	18.3	3142	7.4	233	15.5	207	11	70.2	68.9
14	1	20.0	4147	7.6	218	25.2	208	11	70.5	69.3
15	1	20.9	4306	7.1	245	18.2	208	11	70.5	70.0
16	1	20.3	3426	7.3	200	8.4	209	0	70.8	70.6

3 Conclusion

Artificial neural network method has a strong learning ability and the ability to deal with nonlinear problems. This study integrates the grey system model, and provides an effective research tool for college students physical health level intelligent score. This study selected the physical health network scoring system of 1500 male students of 2019 to verify the scoring accuracy of the system with the physical health data of 350 male students of 2019. The study found that the simulation cumulative error of the test sample was 0.9257, and the simulation cumulative error of the training sample was 0.51855, which shows that the constructed scoring system has very high accuracy and can be the scoring basis of college students physical health level, opening up a new idea for the individual scoring of college students physical fitness. The biggest advantage of this method is that it jumps the controversial factor of single score and weight coefficient, and comprehensively analyzes various physical health indicators on the whole by system concept, which is more scientific than the traditional scoring method of single score in the middle line and summing by the weight coefficient. The scoring system has high accuracy, and its extended application can also provide important theoretical guidance conclusions for the comprehensive scoring of the physical health level of college girls.

Acknowledgments. This study was funded by Hubei Education Science Planning project (2024GB445).

Disclosure of Interests. The authors have no competing interests to declare that are relevant to the content of this article.

References

1. The CPC Central Committee and the State Council on deepening education reformand comprehensivelypromoting quality-oriented education [EB/OL]. http://old.moe.gov.cn/publicfiles/business/htmlfiles/moe/moe_177/200407/2478.html. Accessed 13 June1999
2. 2014 National Physical Fitness Monitoring Bulletin [EB/OL]. http://www.sport.gov.cn/n16/n1077/n1227/7328132.html. Accessed 25 Nov 2015
3. The CPC Central Committee and The State Council issued the "Healthy China 2030" plan outline [EB/OL]. http://www.gov.cn/zhengce/2016-10/25/content_5124174.htm. Accessed 25 Oct 2016
4. Chong, Z., Zongping, W.: Comparative study of grey system model and BP neural network model in sports performance prediction. J. Nanjing Inst. Phys. Educ. **20**(6), 134–135 (2006)
5. Haykin, S.: Neural Networks and Learning Machines, 3rd edn., pp. 122–220. Pearson Education, New Jersey (2009)
6. Xia, X., Mao, L., Jin, J.: Correlation between postgraduate physical exercise and comprehensive health in Chinese universities —— empirical study based on BP neural network. Global Educ. Outlook **47**(4), 111–128 (2018)
7. Chang, L., Sun, G., Liu, Y.: Research on sports performance prediction. Value Eng. **34**(20), 191–193 (2015)
8. Yanrong, Z.: Prediction of student PE performance based on improvedgrey neural networks. Electr. Measur. Technol. **42**(22), 86–90 (2019)
9. Chong, Z., Zongping, W.: Comparative study of grey system model and BP neural network model in sports performance prediction. J. Nanjing Inst. Phys. Educ. (Social Science edition) **06**, 134–135 (2006)
10. Deng, K., Xu, L.: A sports performance prediction model based on the grey BP neural network. In: Chinese Society of Sports Science. 2017 National Competitive Sports Science Paper Report paper summary compilation. University of South China, vol. 2 (2017)
11. Ji, X., Yang, Z., Ma, Y., Jia, M., Liu, L.: Progress in AI technology in health promotion, exercise capacity improvement and injury prediction. Sports Sci. Technol. Liter. Bull. (05) (2023)
12. Fan, Y., Zheng, Z., Zheng, W.: Prediction of undergraduate sports performance based on an improved SSA-LS SVM model. Comput. Simul. (01) (2023)

13. Jia, W., Wang, W.: GDP prediction of Hefei metropolitan area based on BP neural network. Proc. Xidian Univ. (Social Science Edition) (01) (2022)
14. Luo, Z.Y.: Student performance prediction based on BP neural network flight. Technol. Inf. (04) (2022)

Uniform Light Field Control System Based on IoT Technology - Solving Eye Discomfort Caused by Changes in Light Intensity

Haobin Yang, Haiying Zhang(✉), and Xunyu Qiao

School of Information Science and Engineering, Linyi University, Linyi, China
zhanghaiying@lyu.edu.cn

Abstract. With the development of computer networks, myopia problems among primary and secondary school students have become more and more serious. The reason is that there is insufficient light supplementation. At the same time, when people enter a new environment, changes in light intensity can cause eye discomfort. Therefore, in response to such problems, this paper uses the Internet of Things technology to connect lights to computers and designs a uniform light field system to solve such problems. External sensors transmit light data to various sensors in the room through nrf24l01. At the same time, the indoor sensors transmit light data to the MCU where each LED is located, forming a three-level network through two-way communication. After each MCU receives the target light intensity, the PID algorithm and stepless dimming are used to make the light transition smoothly, and finally form a uniform light field.

Keywords: two-way communication · three-level networking · PID algorithm · stepless dimming · Internet of Things

1 Introduction

Domestic and foreign research on the myopia problem of primary and secondary school students and the optimization of light environment are gradually increasing. Many scientific research institutions and universities have carried out relevant work, mainly focusing on three aspects: lighting monitoring and control, application of Internet of Things technology, and development of healthy lighting systems. Most of the existing markets use zigbee networking to form lighting supplement systems. This system cannot solve problems such as glare and photophobia caused by light adaptation and dark adaptation of the eyes when the difference between outdoor light intensity and indoor light intensity is large. In response to the above problems, the "uniform light field control system" we designed provides an effective solution. This technology is not only suitable for supplementing light during the day and night, but can also avoid "photophobia" caused by excessive differences in indoor and outdoor brightness. "or "vertigo" and other phenomena, thereby effectively protecting eye health. In this light environment, the ciliary muscles can relax and eye fatigue is significantly reduced, which is especially suitable for school classrooms. In the current lighting design standard, classroom illumination uniformity U_2 is defined as the ratio of minimum illumination E_{min} to average illumination E_{av}: $U_2 = \frac{E_{min}}{E_{av}}$. However, this calculation method fails to fully consider the difficulty of human eyes adapting between maximum illumination E_{max} and minimum illumination. Studies have shown that visual comfort is more susceptible to the maximum and minimum illumination ratio $U_1 = \frac{E_{min}}{E_{max}}$. When $U_2 \geq 0.85$, the classroom light environment is more conducive to visual health. On this basis, it is recommended to further increase the design value of U_1 to $U_1 \geq 0.75$ to reduce the burden on the eyes caused by extreme brightness differences.

The uniform light field control system established in this paper has greatly improved the values of U1 and U2. By collecting the light in places with better lighting, a network is formed to control the lighting in classrooms or places where supplementary lighting is needed. However, the light intensity in different places in the classroom is different. For example, less light supplement is needed near the window than near the window, and it is more dazzling near the large screen than in other places. Therefore, it is necessary to collect the light in one place and control the light through a network to achieve a uniform light field. This process uses an adaptive algorithm to automatically adjust the brightness of the lamp according to environmental information, and uses a stepless dimming algorithm, a PID algorithm, etc. to achieve uniform light adjustment. At the same time, a WeChat applet for room brightness monitoring is established, which is convenient for tracing the source when problems occur and real-time monitoring of the working status of each node. The advantages of this system are accurate algorithm, short light intensity adaptation time, and delicate light intensity changes. At the same time, the nrf24l01 networking cost is low, the power consumption is low, and it conforms to the concept of green industry [1]. The core function of the system is to read multiple probes and automatically adjust the light to achieve uniform light intensity. The innovation is to reduce the difference in light intensity between indoors and outdoors, and control the light intensity of the lamps in real time according to the model we built. At the same time, the stepless dimming is used to solve the flicker problem, making the light intensity change more delicate. The system structure is shown in Fig. 1:

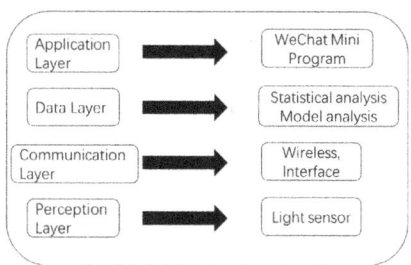

Fig. 1. 1

Perception layer: mainly includes hardware devices such as LED, light sensor, wireless node and single-chip microcomputer, which are responsible for collecting and processing light intensity data and transmitting the data to the data platform.

Communication layer: this layer is responsible for transmitting the light intensity data collected by the hardware layer to the single-chip microcomputer through 2.4G communication, and then transmitting the data to the data platform through network communication.

Data layer: this layer is responsible for receiving and storing the light intensity data transmitted from the hardware layer, and provides data analysis and processing functions, and also interacts with WeChat applet for data. Application layer: this project mainly interacts with the data layer through WeChat applet, and provides a user-friendly interactive interface, allowing users to adjust the lighting mode according to their needs [2].

The specific flow chart and overall design of this system function are shown in Fig. 2:

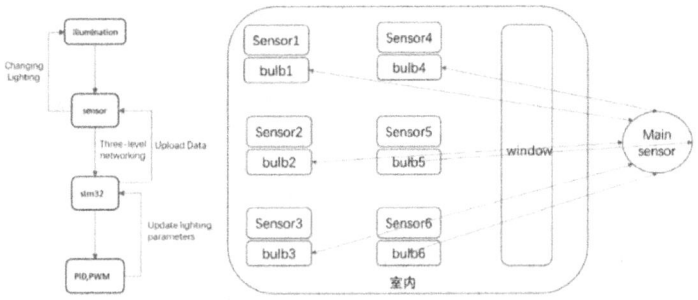

Fig. 2. .

2 Algorithm

2.1 PWM Algorithm

PWM (Pulse Width Modulation) is a technology widely used in LED dimming. It refers to a dimming method that can continuously and uninterruptedly adjust the brightness of the light source. Unlike traditional graded dimming (for example, only three levels of brightness: high, medium, and low), stepless dimming can adjust any brightness within a certain brightness range. It mainly changes the power of the light source by adjusting the current or voltage input to the light source, thereby achieving brightness changes. It controls the brightness of the LED by adjusting the duty cycle of the pulse signal. The duty cycle refers to the proportion of time occupied by the high level in a cycle. For example, if the cycle of a PWM signal is 1 ms and the high level duration is 0.5 ms, then the duty cycle of this PWM signal is 50%. By changing the duty cycle, the brightness of the LED can be accurately controlled, thereby achieving continuous changes in the brightness of the LED. This technology makes the brightness adjustment of the LED smoother and more natural. We comparequantitative dimming with stepless dimming, and the effect is shown in Fig. 3:

Through these two pictures, we can observe the relationship between brightness change and dimming signal in the stepless dimming algorithm, and draw the following conclusions:

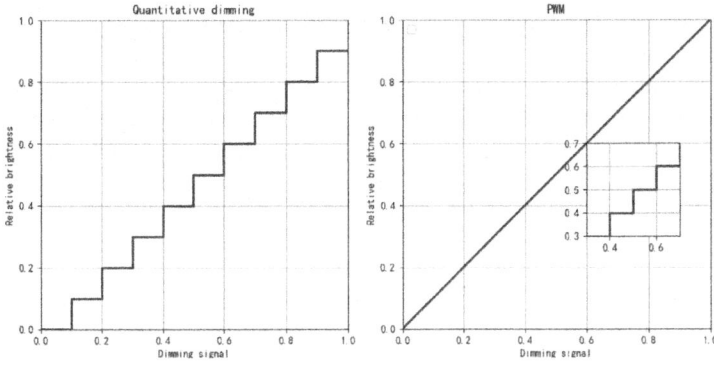

Fig. 3. .

Left picture (step change): The relative brightness does not change continuously, but rises in a discrete step manner. This discreteness will cause obvious brightness jumps in the dimming process, especially in the low brightness range, it is easier to perceive the discontinuous brightness mutation. This jump is more easily perceived by the human eye and makes people feel uncomfortable.

Right picture: The relationship between the dimming signal and the brightness is close to linear, and the brightness changes continuously and smoothly with the signal. The local magnification box shows that there is no step effect in the brightness curve, and the output result is closer to the ideal linear change. The brightness is proportional to the dimming signal, and the user can perceive the effect of brightness adjustment more intuitively.

As can be seen from the figure, the right picture (stepless dimming) is significantly better than the left picture (quantized dimming), especially in terms of brightness smoothness and user experience. Stepless dimming is an ideal goal. Its brightness output is linearly related to the dimming signal, avoiding the jump problem in quantitative dimming, and is the best choice for pursuing a high-quality lighting experience. Although quantitative dimming is simple, it is limited by resolution, and its effect is prone to jumps or unevenness in the low brightness range.

2.2 PID Algorithm

PID control algorithm is a widely used feedback control strategy, often used in industrial control systems to achieve accurate tracking of the expected value (set point). PID is an abbreviation of the first letters of three English words, which stand for proportional, integral and differential. These three control methods have different functions and characteristics. When combined together, they can make the control system have better stability and fast responsiveness. By taking the brightness obtained from the device sensor as the actual value, designing the PID controller and calculating the output PWM duty cycle, the brightness value of the LED is continuously adjusted to achieve the most suitable brightness. In the light intensity control of this system, the actual light intensity is obtained through the light sensor, and it is compared with the set light intensity to obtain the error. The proportional link quickly adjusts the power of the light source according to this error, so that the light intensity approaches the set value as soon as possible. The integral link can accumulate the error of the light intensity for a long time. After the light intensity is stable, it can be used to eliminate small and continuous errors, so that the actual light intensity can reach the set value more accurately. The differential link can adjust the light source power in advance according to the speed of the light intensity change. The relationship between e(t) (error value) and u(t) (output value): [3]

$$u(t) = k_P e(t) + k_1 \int_0^t e(t)dt + k_D \frac{de(t)}{dt}$$

In the formula: KP is the proportional coefficient; KI is the integral coefficient; KD is the differential coefficient, which are all adjustment parameters and need to be adjusted manually.

From this formula, the output value can be obtained as the sum of the proportional gain, integral gain, and differential gain of the error value. The error value e(t) is obtained by subtracting the optimal brightness value obtained by linear fitting from the brightness value obtained from the device sensor. After the proportional gain, integral gain, and differential gain, the output PWM duty cycle value u(t) is obtained. The MCU dims according to this value, and then the brightness value from the device sensor is formed to form a self-feedback system and continuously approach the target value. The system PID working algorithm flow is shown in Fig. 4:

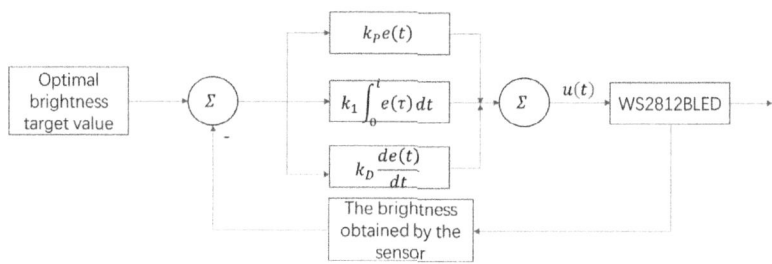

Fig. 4.

PID control and stepless dimming have certain similarities, but there are also differences. Stepless dimming focuses on the feature that the dimming method is continuously variable, mainly by adjusting the voltage, current and other simple and direct methods to change the brightness. The PID algorithm is a complex feedback control algorithm, which needs to adjust the light intensity based on the three factors of the light intensity error (the difference between the set value and the actual value), the integral of the error, and the rate of change of the error. It can more accurately respond to various complex lighting control situations, such as making the lighting system work more stably when there are interference factors. Therefore, we use stepless dimming first and then use the PID algorithm for further precision.

2.3 STM32 Three-Level Networking

2.3.1 nRF24L01 Chip Introduction

Working frequency: supports 2.4GHz ISM band.

Transmission rate: up to 2Mbps data transmission rate.

Power consumption: low power design, suitable for battery-powered applications.

Wireless communication: supports point-to-point communication and multi-point network configuration.

Programmability: can be configured through the SPI interface, including setting the transmission power, data rate, channel selection, etc.

Anti-interference ability: adopts frequency hopping spread spectrum technology (FHSS), which improves the anti-interference ability and transmission distance.

2.3.2 nrf24l01 Networking Ideas

When the host sends commands to multiple nodes, it uses a 1-transmit-N-receive communication mode. In order for the host to successfully send data to multiple slaves, the data width, transceiver address, transceiver channel, and transceiver rate of the host and all slaves must be set to be consistent. After the settings are completed, if the host sends data, theoretically each slave can receive the data. However, in reality, due to interference and other factors, only some slaves can receive the data. When a slave sends data to the host, other slaves may also receive the data sent by the slave. Therefore, when this setting method is used to form a network, the reliability and accuracy of data transmission are poor. In order to improve the performance of data transmission in wireless sensor networks, after research and testing, we propose the following methods to optimize the settings of slaves, thereby improving the reliability and efficiency of data transmission [4].

Method 1: Use different channels. This method is simple and easy to implement and is suitable for situations where the number of slaves is small. However, when the channel interval is close, crosstalk is prone to occur, and there are fewer available channels. For example, set the slaves to different channels, such as slave 1 to channel 0, slave 2 to channel 25, slave 3 to channel 50, and so on.

Method 2: Use different addresses. This method is relatively flexible in address setting and is suitable for situations where there are a large number of slaves. However, the address space is limited and supports a maximum of 256 slaves. For example, set the slaves to different addresses, such as slave 1 address 0x01, slave 2 address 0x02, and slave 3 address 0x03.

Method 3: Use different channel and address combinations The advantage of this method is that the combination methods are diverse, support a large number of slaves, and are suitable for large-scale deployment. However, the configuration complexity is relatively high. Set the slaves to different channels (0~125) and different addresses (0~255), with a total of 32256 combinations.

Regardless of which method is used, it is necessary to ensure that the data width, transceiver address, transceiver channel, and transceiver rate of the host and slave are consistent to ensure the basic conditions for communication. Therefore, in this project, method 1 is used between the first and second levels, and method 2 is used between the second and third levels. Method 3 is used when the site is large. Theoretically, the host can receive data from up to 6 slaves at the same time through the 6 channels in nRF24L01, but in fact, since multiple slaves send data to the host at the same time, the host is prone to data collision, resulting in crashes or data confusion. The idea of solving this problem can still use the method described above, that is, set the channel or address of the host and the slave being communicated to be the same, and then use the polling method, that is, the nRF24L01 networking can be used to realize the one-receive-N-transmit data acquisition function.

2.4 Data Detection Software Design

In order to facilitate data upload and monitoring information visualization after the device is connected, we use the Alibaba Cloud Internet of Things platform. In this project, the light intensity of each node is transmitted to the master node through nrf24l01. The MCU of the master node uploads the data to Alibaba Cloud through esp8266. The WeChat applet subscribes to Alibaba Cloud to obtain data and can search the ID in the WeChat applet to query the light intensity of the corresponding node, so as to analyze the light intensity of each node and comprehensively debug the system. The MQTT protocol is used for communication between Esp8266 and the cloud platform. MQTT is a lightweight message transmission protocol that is particularly suitable for use in resource-constrained environments, such as IoT devices. It adopts a publish/subscribe model to achieve an efficient and reliable message transmission mechanism. Its workflow is shown in Fig. 5:

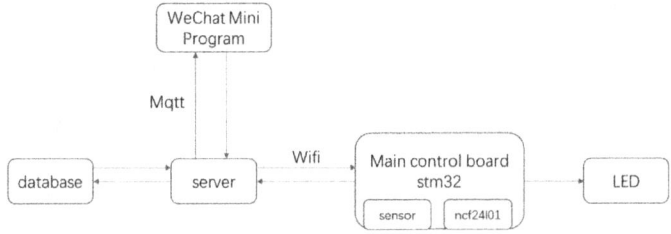

Fig. 5. .

Through the WeChat applet, managers can obtain sensor data in real time to ensure that the data reaches the expected value. By monitoring the system, the algorithm and parameters can be debugged to achieve the expected effect of light changes. The specific page of the applet is as follows in Fig. 6:

3 Experimental Procedures

3.1 DIALUX EVO Introdution

DIALux evo is an international mainstream lighting design software that can not only simulate and calculate indoor, outdoor or street lighting systems according to relevant design standards, but also carry out professional overall planning and design of lighting

Fig. 6.

systems for different types of buildings. Relying on its powerful output function. In order to simulate a more realistic environment, the modeling work of this project was completed by Dialux EVO [5].

3.2 Creat Detailed 3D Models

We chose the classroom as our simulation location, and adjusted the spatial parameters and spatial environment of the 3D model as needed; we made and selected the necessary furniture for the classroom and arranged it in the 3D model, and decorated the 3D model more comprehensively, including the selection of materials such as the ceiling, walls, and floor, so that the virtual model is closer to the real classroom environment. The specific model is shown in Figs. 7 and 8 below.

3.3 Light Source Selection

In DIALUX EVO, it is easy to arrange lights and modify the parameters of lights. For the lighting simulation calculation of indoor chandeliers, we need to select suitable lamps and arrange them according to the corresponding lighting arrangement, and then set the light source parameters and space parameters of the lamps. The indoor chandeliers use T5 fluorescent lamps provided by Sanxiong Aurora in the Dialux EVO lamp library. The lamp model is shown in Fig. 9 below:

Lamp body material: high-quality aluminum alloy Light transmission: polarized lens plus anti-glare grille design Power: 40 W Luminous efficiency: $\geq$ 80 lm/W Luminous flux: 3200 lm Color temperature: 4000 K/5000 K Color rendering: Ra $\geq$ 97, R9 $\geq$

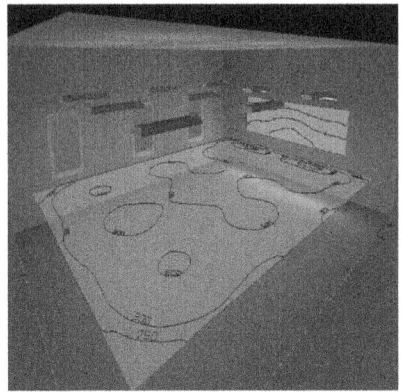
Fig. 7.

Fig. 8.

品德Ⅱ系列LED护眼黑板灯
PAK410073 PAK703000748

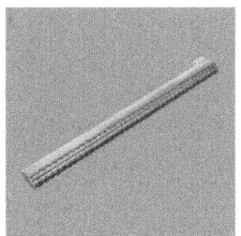
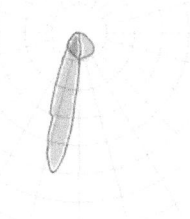

Fig. 9.

95, R15 ≥ 95 Working power supply: 220 V Power factor: ≥0.95 Installation method: boom installation (the matching boom is fixed and telescopic) Power supply: built-in drive Product size: 1205 X 69 X 183 mm Lamp body color: white Others: The product can be used with Sanxiong Aurora wireless intelligent control system.

3.4 Simulate Natural Light

In order to better simulate a more realistic classroom environment, we simulated the experimental environment through DIALux software and set the basic parameters: the wall reflection coefficient is 60%; the window reflection coefficient is 10%, the transmission coefficient is 90%; the floor reflection coefficient is 70%; the desktop reflection coefficient is 40%. The LED lamp power is 30 W; the output luminous flux is 2000 lm, the color temperature is 3000 K, and the test height is 0.75 m. The effect diagram calculated by the software is shown in Fig. 4, and the simulation data calculated by DIALux software is shown in Fig. 10.

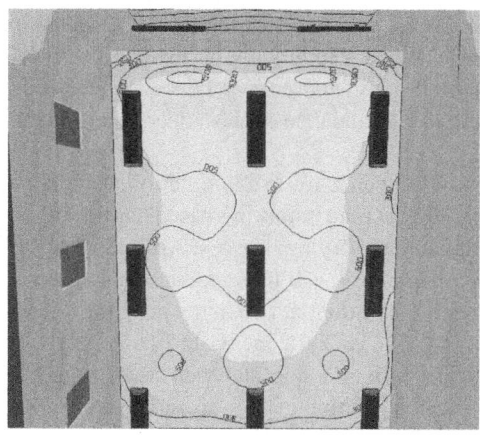

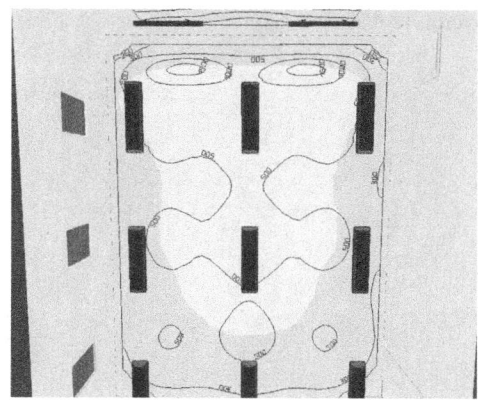

Fig. 10. Pseudo-color images under different lighting conditions

3.5 Test with Different Expected Illumination

We selected scenes such as indoor blinds, and the figure shows the changes in relevant data in multiple areas. The Fig. 11 (a) is the curve of indoor natural light illumination changes during the test. During the period of 8:00–10:00, the indoor natural light illumination value continued to increase over time. During the period of 10:00–11:00, the indoor light illumination value fluctuated to a certain extent due to certain blocking factors (such as building structure, etc.). During the period of 11:00–14:00, the indoor natural light illumination value remained at a relatively high and relatively stable level. After 14:00, the light illumination value gradually decreased.

The figure Fig. 11 (b) shows the indoor temperature fluctuation. It can be seen that the indoor temperature reached a high value around 12:00–14:00, and then gradually decreased. This may be because more solar heat was received around 12:00–14:00, causing the temperature to rise, and then the heat decreased and the temperature dropped accordingly.

The Fig. 11 (c) shows the changes in illumination at Point A and Point B. During 8:00–10:00, the illumination at both points increased. During 10:00–12:00, the illumination at both points fluctuated to varying degrees. Around 12:00–14:00, the illumination at both points reached a high level, and then gradually decreased.

The Fig. 11 (d) also shows the data changes at Point A and Point B. Its trend is similar to the figure on the lower left, increasing from 8:00–10:00, fluctuating from 10:00–12:00, and reaching a high value and then decreasing from 12:00–14:00. This change may be related to the position of the sun and weather conditions. When the sun rises, the illumination and temperature rise, and when the sun is blocked or sets, the illumination and temperature decrease.

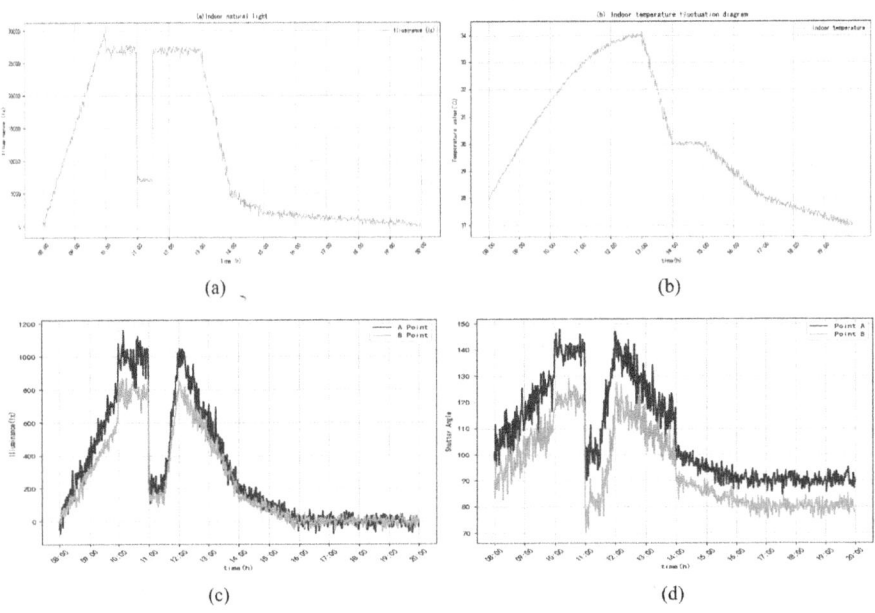

Fig. 11.

Figures 12 and 13 show the illumination values produced by different lights in the model at each detection point in each area. Since each area has different expected illumination values, there must be a buffer area between areas. In this buffer area, the illumination value will decrease from high to low, causing the illumination value of the detection point in the low expected illumination value area close to the high expected illumination value area to be higher than other detection points in the area, such as detection points 4 and 4 in area A. Detection point 2 in area B.

When the influence of natural light is small and uniform illumination is achieved, the actual illumination measurement values at each detection point in each area are shown in Fig. 12(a). The actual illumination values at detection points 1, 2, and 3 are approximately 280 lx, while The illumination value of detection point 4 is about 330 lx. This is because the expected illuminance value in area A is 330 lx and the expected illuminance value in area B is 500 lx. The area with high expected illuminance value has a greater impact on the adjacent area with low expected illuminance value, significantly changing other lights in the area with low expected illuminance value. Distribution of illumination values.

As shown in Fig 13(a), the illumination values of detection points 1, 2, and 3 are in the range of 150 - 175 lx, while the other illumination values of detection point 4 are about 250 lx, which causes other light in area A to The distribution is extremely uneven. Therefore, in order to make the illumination in area A relatively uniform, the illumination value of detection point 4 needs to be about 13% higher than the expected value, while the illumination value of detection points 1, 2, and 3 needs to be about 10% lower than the expected value.

In the same way, as shown in Figs. 12(b) and 13(b), since the expected illumination value in area B is lower than the expected illumination value in area C, and the detection point 2 in area B is more affected by area C, large, so the actual illumination value of detection point 2 in area B is slightly higher than detection points 1, 3, and 4. The expected illumination value in area C is the largest and is less affected by other areas.

Figures 14, 15, and 16 are the illumination uniformity change curves of each test point in each area. The expected illumination value of area A is the smallest. In the test environment with multiple expected illumination values, the lower expected illumination value makes area A more susceptible to the change of illumination in other areas. Test point 4 is close to area B, and the expected illumination value of area B is higher. High expected illumination areas can usually maintain good illumination uniformity within their own range, and at the same time, their influence on the surrounding areas is relatively large. Since test point 4 is close to area B, it is greatly affected by the light of area B, which makes its illumination uniformity reach about 15% at some moments, but it is generally better than area A because area B itself has good illumination stability. Areas B and C have high expected illumination values. High expected illumination values mean that the illumination intensity of these areas themselves is large. In this case, the light distribution inside these areas is relatively uniform, and the degree of interference from light in other areas is relatively small.

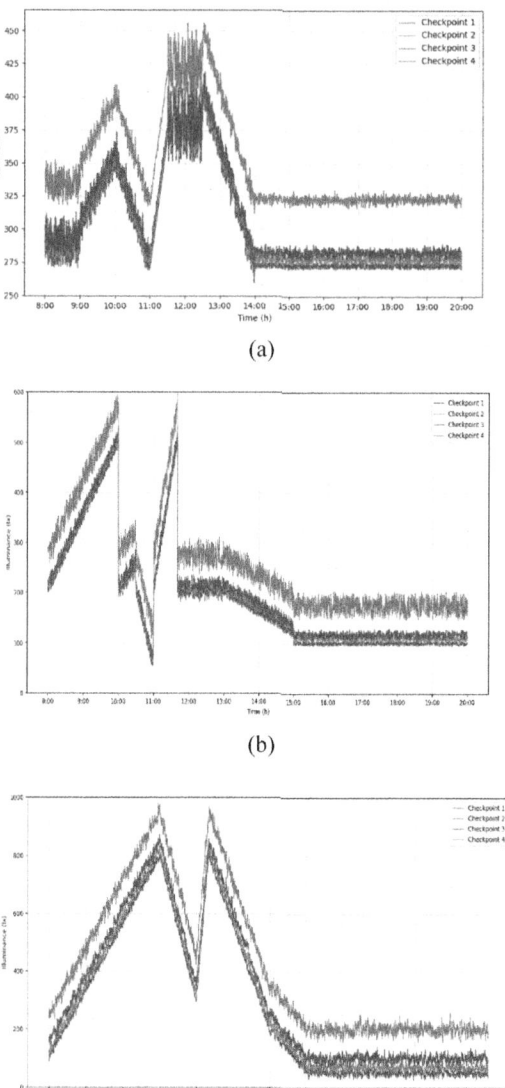

Fig. 12.

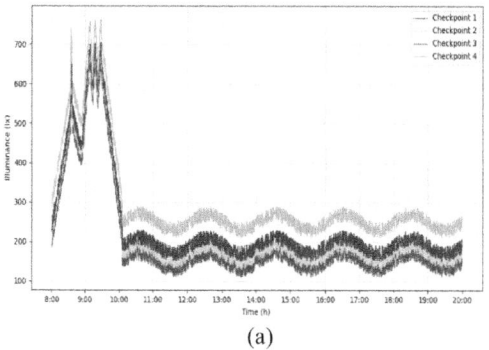

(a)

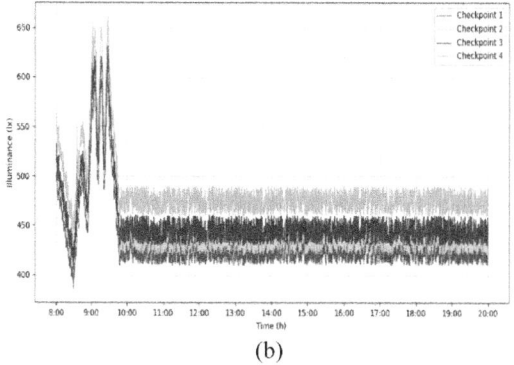

(b)

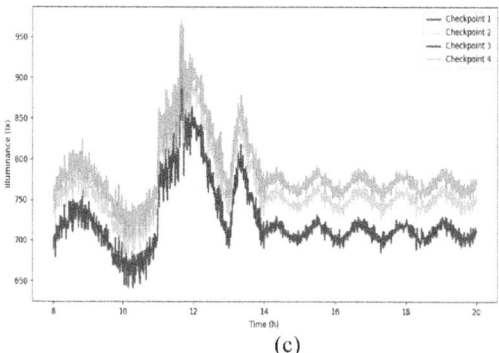

(c)

Fig. 13.

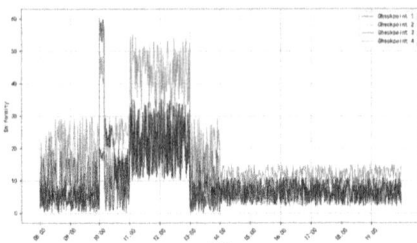

Fig. 14.

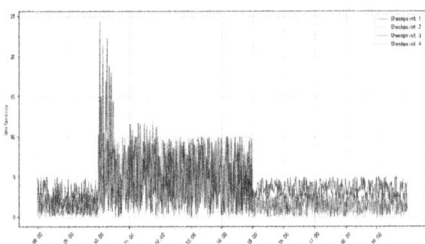

Fig. 15.

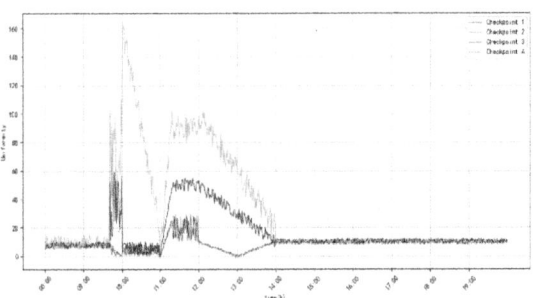

Fig. 16.

4 Experimental Results

4.1 Analyzing Lighting Effects

Before adding the algorithm, the lighting effects is shown in Table 1:

Table 1. .

time\Point	1	2	3	4	5	6
9:00	420	330	512	532	333	701
10:00	442	357	535	538	360	600
11:00	449	332	540	553	326	632
12:00	471	367	526	566	314	755
13:00	480	380	549	580	331	832
14:00	489	360	570	562	582	631
15:00	472	375	549	555	350	513
16:00	444	366	532	530	302	500

After adding the algorithm, the lighting effects is shown in Table 2:

Table 2. .

time\Point	1		2		3		4		5		6	
	Illuminance	Duty ratio	Illuminance	Duty ratio	Illuminance	Duty ratio	Illuminance	Duty ratio	Illuminance	Duty ratio	Illuminances	Duty ratic
9:00	333	65	350	80	345	60	345	65	330	70	300	65
10:00	336	65	363	80	335	60	360	65	3450	70	310	66
11:00	339	65	352	80	351	60	379	65	331	70	320	67
12:00	340	65	361	80	365	60	356	65	351	70	310	68
13:00	345	65	371	80	353	60	346	65	335	70	321	69
14:00	370	65	366	80	364	60	331	65	341	70	325	70
15:00	371	65	372	80	352	60	342	65	339	70	330	71
16:00	357	65	345	80	312	60	315	65	332	70	302	72

Before adding the algorithm, the illumination values of each detection point fluctuated greatly. For example, at 9:00, the illumination value of detection point 6 was 701lx, while the illumination value of detection point 2 was 330lx. After adding the algorithm, the illumination values of each detection point were relatively more balanced, and the fluctuation range was reduced. For example, at 9:00, the illumination value of the detection point was between 300-365lx.

After the algorithm was added, the maximum value of the light intensity difference of each node in the room changed from 200–300 to 20–30, which increased the uniformity tenfold and met the expected requirements. This shows that the algorithm effectively improved the uniformity of indoor lighting.

The experimental results show that the system is real-time, stable and practical.

The system can respond appropriately in real time according to the environmental parameters in the room, and will not cause significant changes in the environmental parameters in the room, thereby ensuring the user experience. During the day, if the natural light is very strong, the system will control the indoor light intensity by adjusting the LED pulse to maintain it at a comfortable level for the human body. At night, when natural light is insufficient, the lighting subsystem will coordinate multiple indoor lamps for uniform lighting to avoid glare and eye fatigue.

In addition, considering that different areas have different requirements for contrast, the system can adjust the lighting to different expected illumination. In the same space, even if the expected illumination in different areas is different, uniform illumination can be achieved in each area. When the expected illumination of each area is the same, its uniformity can be kept within 10%; when the expected illumination of each area is different, except for the uniformity of individual detection points at about 15%, the uniformity of the remaining detection points can also be controlled within 10%. It can be seen that the system has the ability to adjust the light environment in a real learning and office environment.

5 Conclusion and Improvement

The uniform light field control system proposed in this paper cleverly uses DIALUX software for simulation, and combines PID control, PWM dimming, and nRF24L01 wireless communication and other advanced technologies. According to the light intensity received by the sensor, it is uploaded to the computer for algorithm optimization, so as to form a uniform light field in the experiment. This system effectively solves the problem of uneven indoor lighting and provides an effective means to improve the learning and working environment. In addition, the system has the advantages of low cost, easy expansion, and network control of target brightness. In the future, this system can be applied to a wider range of fields, such as medical precision operating rooms, museums, tunnels, scientific research institutes and other areas with strict lighting requirements. In hospitals, clear lighting can reduce surgical risks and improve surgical efficiency, and plays a key role in fine vascular suture or detail repair. Art venues such as museums can enhance the visual color of cultural relics under better lighting conditions, allowing cultural relics to be displayed more clearly to the public. In tunnels, light adaptation and dark adaptation have adverse consequences such as glare and photophobia for drivers, and uniform lighting can greatly reduce this risk. In scientific research institutes, uniform lighting can ensure the accuracy and precision of data, laying a solid foundation for achieving the expected experimental results.

However, this system has some defects that need to be improved, such as the placement of sensors. Sensors in different locations may be affected by certain external factors, resulting in errors in receiving different light intensities, as well as high requirements for the internal networking structure of the lamp and the network of the space in which it is located, which requires real-time online control. In the future, our team will uphold the ability of technological innovation and the spirit of practical exploration, and will be committed to the creation of 3D models for different room types and the encryption security of the Internet of Things. We will clean up the collected data, remove noise points and erroneous data, and calibrate the erroneous data to ensure accurate fusion of data collected in different ways, laying a solid foundation for modeling. A hybrid encryption scheme is adopted, combining the advantages of symmetric encryption and asymmetric encryption algorithms. Asymmetric encryption algorithm is used for key exchange, and symmetric encryption algorithm is used for data encryption and transmission, which ensures data security while speeding up data encryption and decryption.

References

1. Mingxuan, W., et al.: An empirical study on the energy-saving effects of lighting and air conditioning in intelligent building environment control systems. Build. Sci. **40**(08), 199–207 (2024). https://doi.org/10.13614/j.cnki.11-1962/tu.2024.08.24
2. Yang, Y., et al.: Design and implementation of multi-source fuzzy adaptive intelligent lighting network for tunnels. Chin. J. Underground Space Eng. **19**(S2), 955–962+970 (2023)
3. Ding, M., Zheng, T.: Design of supplementary lighting device for LED intelligent dimming in greenhouse. Jiangsu Agric. Sci. **49**(23), 201–206 (2021). https://doi.org/10.15889/j.issn.1002-1302.2021.23.035
4. Zhang, Y., et al.: Research on aircraft status monitoring system based on nRF24L01+. Aeros. Measur. Technol. **41**(04), 60–65+71 (2021)
5. Jiang, X., Zhang, B.: Application of green industrial building lighting design based on DIALUX in the aluminum industry. Light Metals **11**, 53–55 (2022). https://doi.org/10.13662/j.cnki.qjs.2022.11.012

Intelligent Cooperative Control

Attention-Fused Vision Mamba Model for Remote Sensing Image Detection and Segmentation

Zhao Yusen[1], Wu Shan[2], and Tian Liang[3](✉)

[1] Institute of Applied Mathematics, Hebei Academy of Sciences, Shijiazhuang, Hebei, China
[2] School of Mathematical Sciences, Hebei Normal University, Shijiazhuang, Hebei, China
wushan@stu.hebtu.edu.cn
[3] College of Computer and Cyber Security, Hebei Digital Education Collaborative Innovation Center, Hebei Normal University, Shijiazhuang, Hebei, China
tianliang@hebtu.edu.cn

Abstract. Vision Mamba has been growing in popularity across a variety of computer vision applications, including object detection and segmentation in remote sensing images. In order to increase the accuracy of detection and segmentation in remote sensing images, we present a unique model in this study that integrates Transformer and Vision Mamba. We employ the Mamba branch to guarantee high-quality local information extraction because of the vast scale of images and the small size of the objects. Simultaneously, we employ the Transformer branch as a residual component to the Mamba branch, with the aim of enhancing the extraction of global information. To mitigate the complexity of the Transformer branch, we employ axial attention within the self-attention block, which computes self-attention separately along vertical and horizontal directions. By integrating the Transformer and Vision Mamba, we propose the SeaMamba block and construct a novel backbone, testing it on detection and segmentation tasks on remote sensing images as well as image classification using ImageNet-1K. Our experiments demonstrate that this model achieves 83.1% accuracy in classification on the ImageNet-1K dataset. For remote sensing detection, it attains 78.12 mAP on the DOTA dataset with Oriented R-CNN detection head, and 52.32 mIOU on LoveDA dataset with MaskFormer.

Keywords: Vision Mamba · Transformer · Remote Sensing Detection · Remote Sensing Segmentation

---

This research was supported by Doctoral (Postdoctoral) Fund of Hebei Normal University (L2021B39) and Hebei Central Leading Local Science and Technology Development Foundation (236Z0105G).

1 Introduction

As satellite data becomes easier to obtain, there are huge amounts of images need to be processed. The ability of automatically detecting or segmenting high-value object on remote sensing images is becoming more important. Object detection on remote sensing images mainly focuses on cargo ships, airplanes, vehicles, or military targets like fighters, helicopters, naval vessels, submarines. Moreover, we can also detect architectures like airport, harbor etc. on satellite images. Segmentation on remote sensing images aims to label different areas on an image. This capacity is essential in many domains, including agriculture, urban planning, disaster management, and environmental monitoring.

Unlike traditional object detection, there are several characteristics on remote sensing images increase difficulty on detection and segmentation task. Remote sensing images typically have large scale, which can cover extensive geographical areas. Besides, the objects of interest are usually relatively small. Remote sensing images also contain rich geographic information with complex and varied backgrounds where features can differ significantly across regions. So, for detection and segmentation models in remote sensing images, it is essential to extract local features and global context. Local information is for identifying small objects and capturing detailed features within the images, and global information is necessary to understand the broader context and relationships between different regions of the images.

In previous research, numerous attempts have been made to employ convolutional neural networks (CNNs) [9,13,40,44], Transformer models [7,35,39], or hybrid models combining both [34] to accomplish these tasks. CNN model uses convolution to capture features with inductive biases including locality and spatial invariance. Transformer utilizes self-attention to compute attention weights among patches, demonstrating extraordinary performance in various vision tasks. However, CNN is limited by its receptive field which hinders its ability to capture global information. Although Transformer can extract features through global receptive fields, its self-attention mechanism has a quadratic computational cost. Recently, state space model (SSM) [12] have been introduced into computer vision area. Relying on its linear scalability in sequence length and its powerful modeling ability, SSM-based vision Mamba is getting more popular.

Considering the distinctive features of remote sensing images, we propose an innovative backbone architecture that enhances the feature extraction capability by combining the Transformer and Vision Mamba models. The following are this paper's primary contributions: 1. Construct a vision Mamba block: SeaMamba which includes Transformer [27] branch and Mamba branch. Vision Mamba [20] branches are used for modeling both local and global information. Simultaneously, Transformer branches serve as residual connections to enhance global information extraction, which is crucial for segmentation tasks. 2. With Oriented R-CNN [40] and MaskFormer [5], we construct detection model and segmentation model and evaluate them on DOTA [38] and LoveDA [32] dataset. By incorporating these advanced methods, our approach successfully overcomes the difficulties associated with the extensive scale and the small sizes of objects in images. This results in enhanced accuracy and efficiency in both detection and segmentation tasks.

2 Related Work

This section examines studies on object recognition and image segmentation in remote sensing, with a particular emphasis on very high-resolution (VHR) images. VHR images possess high spatial resolution, enabling them to capture detailed surface features such as buildings and vehicles. The primary applications of VHR images include urban planning and management, agricultural monitoring, and military uses. The large data volume and intricate details of VHR images make processing VHR images highly resource-intensive and expensive.

2.1 Object Detection

Object detection on VHR images aims to localize and classify all instances in images. The bounding boxes may be either rectangular or oriented bounding boxes. Previous approaches attempt combining existing two-stage or single-stage detectors based on CNNs and Transformers. Some researches employ DETR [1] detector on VHR images.

In order to improve the recognition of small items in VHR images, [41] presents a backbone called Local Perception Swin-Transformer (LPSW). [46] proposes the GANsformer, which uses a Transformer model to enhance contextual feature extraction. Additionally, it expands the inputs before the backbone by using a generative model. [19] proposes TransConvNet, a novel network design that tackles the issue of rotation invariance in CNNs by combining local and global features. Meanwhile, [17] presents Oriented Rep-Points, a representation system that uses flexible adaptive points and incorporates a quality assessment mechanism to enable robust learning of geometric features in arbitrarily oriented aerial objects. In addition, [47] introduces Point RCNN, a framework that eliminates the need for anchors in rotated object detection, enhancing detection accuracy by avoiding reliance on pre-defined anchor angles. [6] proposes AO2-DETR, which introduces DETR [1] detection framework to accomplish oriented object detection in VHR images. It incorporates a scheme for generating proposals with orientation, a refinement process tailored to rotation-robust features, and a loss function that takes rotational alignment into consideration, collectively boosting detection precision and eliminating unnecessary predictions.

SOAR [28] leverages the SAHI framework on YOLO v9 [30] with Programmable Gradient Information to mitigate information loss in feature extraction, and integrates the Vision Mamba model with a bidirectional State Space Model for improved context modeling. These methods demonstrate significant gains in accuracy and efficiency, offering promising advancements in aerial object recognition.

2.2 Image Segmentation

Image segmentation on remote sensing images is to classify each pixel into semantic categories. [42] presents Efficient-T, a light-weight Transformer model including a mechanism which is capable of enhancing the edge information. [31] proposes CCTNet for crop segmentation. It integrates the strengths of CNN and Transformer, leveraging the CNN's ability to capture fine-grained local features and the Transformer's capacity to extract comprehensive global context. STransFuse [11] applies a self-attention mechanism to merge coarse and fine details from different scales. [34] pairs a U-shaped decoder with Swin Transformer [21] as its backbone to capture long-range interactions. Additionally, it includes an SE block [15] to retain local details.

Samba [49] improves semantic segmentation in remote sensing images by using vanilla Mamba instead of the Self-Attention block, allowing it to better capture global semantic information. [23] uses VSS blocks in an extra branch to support convolution methods, making it easier to include global context in semantic segmentation. RSMamba [3] is designed for both semantic segmentation and change detection tasks. It gathers image context information from all directions using the Mamba-based omni-directional Selective Scan Module. ChangeMamba [2] applies VSS architecture for change detection and building damage assessment, incorporating spatial-temporal relationship modeling to interact with multitemporal features effectively. [50] explores different scanning strategies and proposes 8D Scan block to employ the scan operation in 8 different directions. CM-UNet [18] combines a CNN encoder for local feature extraction with a Mamba decoder featuring the CSMamba block for global information integration. This architecture effectively addresses long-range dependencies and enhances global-local information fusion. PyramidMamba [33] introduces dense spatial pyramid pooling (DSPP) and pyramid fusion Mamba (PFM) to enhance multi-scale feature representation and reduce semantic redundancy. PyramidMamba delivers top-level results on benchmark datasets.

3 Methods

We shall provide our backbone and SeaMamba block which contains Mamba branch and Transformer branch in this section. Mamba branch can process patches to get both local or fine-grained information and global information. To enhance the extraction of global feature which is beneficial for image segmentation task, we employ a Transformer-based residual branch. In addition, axial attention is applied to simplify the self-attention process and lower the complexity of the Transformer branch.

3.1 Mamba Branch

The idea of Mamba [12] originates from the state space model, which has been further developed into the state space sequence model (S4). At time step t, the

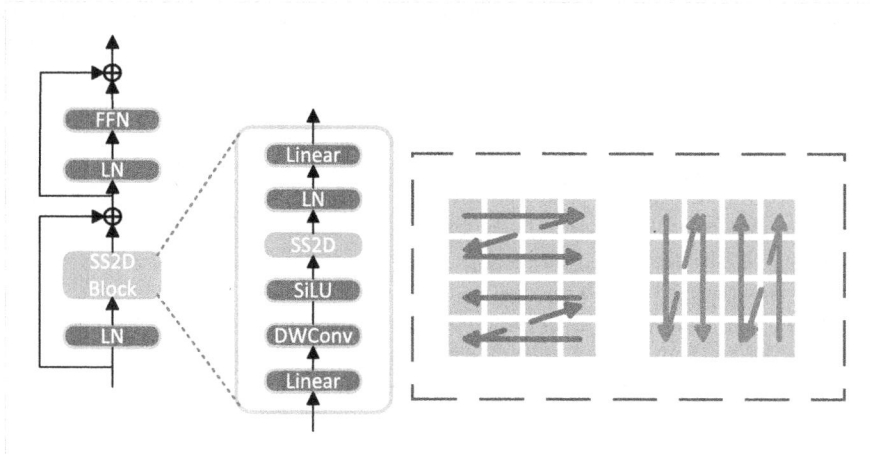

Fig. 1. The pipeline of VSS block. In SS2D module, the input will be scanned in 4 different directions respectively

input $u(t) \in \mathbb{R}$ is processed to produce the output $v(t) \in \mathbb{R}$, using the latent state $z(t) \in \mathbb{R}^M$ in classical SSM.

$$z'(t) = \mathbf{P}z(t) + \mathbf{Q}u(t)$$
$$v(t) = \mathbf{R}z(t) \qquad (1)$$

where $\mathbf{P} \in \mathbb{R}^{M \times M}$, $\mathbf{Q} \in \mathbb{R}^{M \times 1}$, $\mathbf{R} \in \mathbb{R}^{1 \times M}$. In S4, the discretization on continuous parameters $\mathbf{P}$ and $\mathbf{Q}$ is used for dealing with discrete input sequence $\boldsymbol{u}$. Let Θ represent the continuous input $u(t)$'s resolution. The whole discretization by the zero-order hold (ZOH) can be defined as:

$$\bar{\mathbf{P}} = \exp(\Theta \mathbf{P})$$
$$\bar{\mathbf{Q}} = (\Theta \mathbf{P})^{-1}(\exp(\Theta \mathbf{P}) - \mathbf{I}) \cdot \Theta \mathbf{Q} \qquad (2)$$

After converting $\mathbf{P}$ and $\mathbf{Q}$ to $\bar{\mathbf{P}}$ and $\bar{\mathbf{Q}}$, Eq. (1) can be written as:

$$z_t = \bar{\mathbf{P}} z_{t-1} + \bar{\mathbf{Q}} u_t$$
$$v_t = \mathbf{R} z_t \qquad (3)$$

Equation (3) can be processed efficiently by either RNN or CNN style. The CNN formation of Eq. (3) is like:

$$\bar{\mathbf{W}} = (\mathbf{R}\bar{\mathbf{Q}}, \mathbf{R}\bar{\mathbf{P}}\bar{\mathbf{Q}}, \dots, \mathbf{R}\bar{\mathbf{P}}^{L-1}\bar{\mathbf{Q}})$$
$$\boldsymbol{v} = \boldsymbol{u} * \bar{\mathbf{W}} \qquad (4)$$

where L is the sequence length, $\bar{\mathbf{W}} \in \mathbb{R}^L$ can be seen as the SSM convolution kernel.

Mamba proposes a selection mechanism and then constructs the S6 as the core part of Mamba. The selection mechanism makes the parameters of SSM input-dependent. Specifically, make parameters $\mathbf{Q}, \mathbf{R}, \Theta$ calculated by input $\boldsymbol{u}$:

$$\mathbf{Q}, \mathbf{R}, \Theta = Linear(\boldsymbol{u}) \tag{5}$$

The vanilla Mamba block has two main challenges on 2D images. As the first step before Mamba block, we usually split images into patches which can be seen as a sequence Mamba can process. But unlike natural language, 2D images lack a sequential arrangement. In addition, linear scanning strategy is not well-suited for handling spatial information such as images. To overcome these challenges, VMamba [20] introduces Visual State-Space (VSS) blocks. Drawing inspiration from VMamba, this paper adopts the VSS block as the architecture for the Mamba branch, as illustrated in Fig. 1.

The main idea of the VSS is the 2D Selective Scan (SS2D). Input images firstly split into sequences along four different directions. In SS2D, a Cross-Scan operation is employed to scan these patches. A separate S6 block is used for processing each patch sequence in parallel, and then reshapes and merges the result as the output feature-map. By this method, SS2D makes pixels in images effectively integrate together in different directions, and makes the model have global receptive fields in 2D space which allow the model to extract global context information.

3.2 Transformer Branch

Vision Transformer [10] has been widely used in many fields. And models like PVT [36] are trying to reduce the computation complexity in dense prediction tasks. Moreover, several researches have developed many efficient operation to calculate self-attention. For example, linear attention, window-based attention and axial attention etc. In this paper, we will conduct a axial attention to accelerate and simplify the self-attention process.

Inspired by SeaFormer [29], We introduce Sea Attention into our model, which is involving a inverted residual block from MobileNet [24] to extract coarse-grained feature, and an axial attention block to extract global information. This block's construction can be found in Fig. 2. The classical self-attention is defined as:

$$y = softmax(\frac{QK^T}{\sqrt{d_k}})V \tag{6}$$

where Q, K, V are linear projection of input x like $Q = W_q x$, $K = W_k x$, $V = W_v x$. However, calculating self-attention requires quadratic computational cost relative to the input image size, making it inefficient. Based on this situation, axial attention is developed to address this issue. Axial attention splits the 2D attention into two 1D attention operations. This means attention is applied separately along different axes (e.g., rows and columns). Row attention treats the feature map as H rows with W feature vectors each and apply self-attention along the rows. And column attention treats the feature map as W columns with

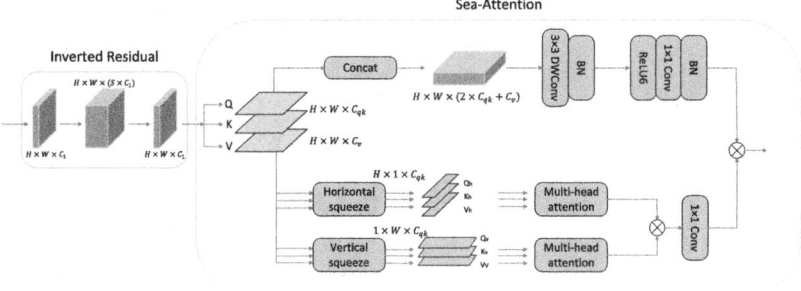

Fig. 2. The whole pipeline of the Transformer branch. On the left side is Inverted Residual block in MobileNet. The channel of the input feature will be transformed into S times bigger than the origin input to enrich the modeling ability. On the right side is Sea-Attention. After the calculating of Q, K and V, we will squeeze the matrix into horizontal and vertical directions, respectively to do the self-attention calculation. The result will multiply a detail enhancement weight gotten by the 3×3 DWConv.

H feature vectors each and apply self-attention along the columns. The complexity decreases from $O((HW)^2)$ to $O(HW(H+W))$, making it better suited for processing large images. To further reduce complexity, SeaFormer proposes SEA Attention, which implements a horizontal squeeze and vertical squeeze strategy. After computing Q, K, and V using the linear projections W_q, W_k, and W_v, We calculate the average of Q, K, and V separately along the horizontal and vertical directions. So in the end, we get $Q_{(h)}$, $K_{(h)}$, $V_{(h)}$ and $Q_{(v)}$, $K_{(v)}$, $V_{(v)}$. The final SEA attention can be defined as:

$$y = softmax(Q_h K_h^T)V_h + softmax(Q_v K_v^T)V_v \qquad (7)$$

By incorporating the squeeze operation, SEA Attention reduces the computational complexity to $O(HW)$.

The squeeze operation retains global information along a single axis but compromises local details. So in SEA Attention, after obtaining Q, K, and V, we concatenate them and use 3×3 DWC to enhance the local details. Then a 1×1 convolution is used to generate detail enhancement weight. Finally, this weight is multiplied by the output of Eq. (7) to obtain the final attention output.

3.3 Architecture of the Backbone

The detail architectures of the SeaMamba block and backbone are described in Fig. 3. We preserve the origin patch embedding and downsampling operation in VMamba. We stack SeaMamba blocks to form a backbone with four stages, capable of outputting multi-scale features. The model structure and parameters follow those of VMamba-tiny. The embedding dimension of Mamba branch is 96. The depth for each stage is $[2, 2, 8, 2]$.

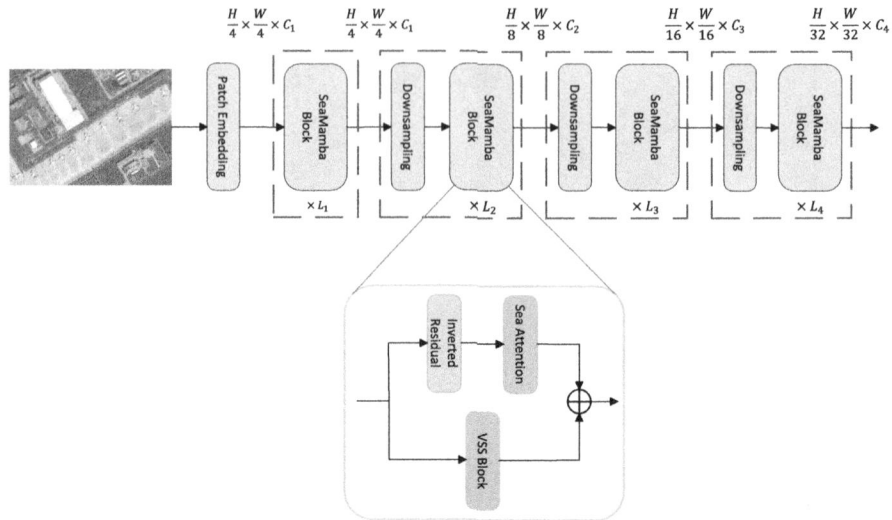

Fig. 3. The whole pipeline of the backbone. The backbone includes 4 stages, each stage has different layers represented by $[L_1, L_2, L_3, L_4]$ in this figure. Due to the patch embedding and downsampling operations, the features will be transformed into half the size of the previous stage. The SeaMamba block has two branches: a Transformer branch on the top and a Mamba branch below.

4 Experiments

This section presents the experiments for our SeaMamba model. We evaluate it through three tasks: image classification, object detection, and image segmentation on VHR images.

4.1 Datasets

We mainly use 3 datasets in the experiments: ImageNet-1K, DOTA, and LoveDA.

- **ImageNet-1K**: The ImageNet-1K dataset [8] is subset of the ImageNet dataset. This dataset contains $1,000$ object classes, $1,281,167$ training images and $50,000$ validation images in total. All images are organized by WordNet hierarchy. The ImageNet-1K dataset is commonly used for pretraining models. In this research, we will pretrain and evaluate the backbone on ImageNet-1K. The pretrained model will be used in successive detection and segmentation tasks.
- **DOTA** is a comprehensive dataset tailored for VHR imagery. It contains objects of different sizes, orientations, and geometric shapes are included, and the images range in size from 800×800 to $20,000 \times 20,000$ pixels. Expert annotators in aerial image interpretation label the instances using arbitrary (8

d.o.f.) quadrilaterals. The DOTA dataset's first release includes 15 frequently viewed categories, 2,806 images, and 188,282 classified instances. Training, validation, and testing sets make up 1/2, 1/6, and 1/3 of the total data, respectively.
- **LoveDA**: LoveDA is a dataset containing 5,987 VHR (0.3 m) remote sensing images taken from the cities of Wuhan, Changzhou, and Nanjing. It focuses on the differences in geographical environments between urban and rural areas. LoveDA is designed to advance tasks in both semantic segmentation and domain adaptation. The dataset presents three significant challenges: handling multi-scale objects, managing complex backgrounds, and dealing with inconsistent class distributions.

4.2 Image Classification

As description in Sect. 3, the architecture of the backbone is followed by VMamba-tiny. We replaced the vanilla VMamba block with our SeaMamba block, Next, we train and test it on ImageNet-1K dataset to evaluate its performance on image classification task. Furthermore, the model we trained in this experiment will be used as the pre-trained model to proceed following experiments on VHR images. The model we construct will be compared with some popular benchmark models. The results are presented in Table 1, where we compare our SeaMamba model with other models based on CNNs and Transformers. The accuracy of SeaMamba is at the same level with popular benchmark models like ResNet, Swin-Transformer, and even can surpass them. The last few columns are the results of SSM-based vision Mamba models. We can see that our SeaMamba can compete to other vision Mamba models with 83.1% classification accuracy.

4.3 Object Detection

Using a backbone pre-trained on ImageNet-1K, we have constructed an object detection model with an Oriented R-CNN detection head and evaluated it on the DOTAv1 dataset. The DOTAv1 dataset is a large-scale collection of VHR images, featuring objects with diverse scales, orientations, and shapes, as well as relatively small sizes. Performing object detection on this dataset thoroughly evaluates the model's capability to capture both global and local features. We use our SeaMamba model and connect it with Oriented R-CNN to finish the detection task. We compare this model with other benchmark models, the result is shown in Table 2. In our experiment, due to the large size of the images, each input image is divided into non-overlapping patches of 224×224, which are then used to train the model. We evaluate the models using mAP, the mean Average Precision, calculated as the average of AP scores for all classes. This considers both accuracy and recall, providing a comprehensive assessment of the model's detection ability.

As the results, our SeaMamba model can reach to 78.12 on mAP, and demonstrates competitive performance against both CNN-based and Transformer-based models, showcasing advantages in several key aspects. This indicates that

Table 1. Classification accuracy of SeaMamba and benchmark models on ImageNet.

Model	Model type	Accuracy (%)
ConvNeXt-T [22]	Conv	82.1
ConvNext-S [22]	Conv	83.1
ResNet50 [14]	Conv	76.1
ResNet101 [14]	Conv	77.4
InceptionNeXt-T [45]	Conv	82.3
DeiT-S [26]	Attn	79.8
DeiT-B [26]	Attn	81.8
Swin-T [21]	Attn	81.3
Swin-S [21]	Attn	83.0
PVTv2-B2 [37]	Attn	82.0
PVTv2-B2-Li [37]	Attn	82.1
Vim-S [48]	Conv+SSM	80.5
VMamba-T [20]	Conv+SSM	82.2
PlainMamba-L3 [43]	Conv+SSM	82.3
LocalVim-S [16]	Conv+SSM	81.2
LocalVMamba-T [16]	Conv+SSM	82.7
Ours	**Attn+SSM**	**83.1**

when used as a backbone, our model benefits from the synergistic interaction between the self-attention and SSM. This combination helps the model improve the capture of both global and local features, especially in large-scale images. The self-attention enables it to identify long-range connections and contextual relationships effectively, while the SSM ensures precise local feature extraction. As a result, our model is capable of handling complex, large-scale images with diverse object orientations and sizes more effectively than conventional approaches. Furthermore, the combination of these components leads to improved robustness in detecting small objects and fine-grained details, making our model well-suited for high-resolution image processing tasks such as remote sensing and aerial imagery analysis. The results substantiate the effectiveness of this hybrid approach in boosting performance across various detection challenges.

Table 2. SeaMamba's detection result on DOTA dataset.

Method	Backbone	mAP (%)
S^2A-Net [13]	ResNet-50	74.12
R^3Det [44]	ResNet-152	73.74
RoI-Transformer [9]	ResNet-101	74.61
Rodformer [7]	ViT-B4	75.60
Oriented RepPoints [17]	Swin-Transformer	77.63
Oriented R-CNN [40]	ResNet-50	75.87
Oriented R-CNN [40]	ResNet-101	76.28
Oriented R-CNN	**SeaMamba**	**78.12**

4.4 Image Segmentation

Furthermore, We evaluate segmentation performance on VHR images with SeaMamba as backbone. We embed our SeaMamba into MaskFormer as backbone to extract the feature. And then the features will be send to Transformer decoder and pixel decoder to accomplish the segmentation task. MaskFormer is a versatile module designed to transform any per-pixel classification model into a mask classification framework. It uses the set prediction approach from DETR, where a Transformer decoder outputs class predictions paired with mask embeddings. Each mask embedding is then merged with per-pixel embeddings from a fully-convolutional network to generate binary mask predictions. We evaluate this model with other VHR image segmentation method and the result is shown in Table 3. Our model reach a 52.32 mIOU on LoveDA dataset, outperforming other CNN-based and Transformer-based benchmark models. Compared to another SSM-based model CM-UNet which has 52.17 mIOU, our model shows better performance on segmentation task. Due to our self-attention and SSM fusion tragedy, the Transformer branch in SeaMamba block can effectively enhance the capability of extracting global information.

Table 3. SeaMamba's detection result on LoveDA dataset.

Method	Backbone	mIOU (%)
DeepLabV3+ [4]	ResNet-50	47.60
Segmenter [25]	ViT-T	47.10
UNetformer [35]	ResNet-18	50.73
DC-Swin [34]	Swin-Transformer	50.60
CM-UNet [18]	ResNet-18	52.17
MaskFormer	**SeaMamba**	**52.32**

5 Conclusion

Our attention-fusion strategy in the Mamba model effectively enhances the ability to extract both global and local features. Thus, SeaMamba model can effectively addresses the challenges posed by large scale and small object sizes in remote sensing images. Vision Mamba splits images into patches, facilitating the efficient extraction of local features. Simultaneously, the self-attention mechanism enables a stronger focus on global features. To further improve computational efficiency, we incorporate an axial attention block, inspired by SeaFormer, to extract global features. The outcomes of the experiment show that the SeaMamba model delivers outstanding performance in image classification, object detection, and image segmentation on VHR images. In future work, we will explore additional strategies for integrating self-attention mechanisms with Mamba model to further enhance model performance.

References

1. Carion, N., Massa, F., Synnaeve, G., Usunier, N., Kirillov, A., Zagoruyko, S.: End-to-end object detection with transformers. In: European Conference on Computer Vision, pp. 213–229. Springer (2020)
2. Chen, H., Song, J., Han, C., Xia, J., Yokoya, N.: ChangeMamba: remote sensing change detection with spatiotemporal state space model. IEEE Trans. Geosci. Remote Sens. **62**, 1–20 (2024)
3. Chen, K., Chen, B., Liu, C., Li, W., Zou, Z., Shi, Z.: RSMamba: remote sensing image classification with state space model. IEEE Geosci. Remote Sens. Lett. (2024)
4. Chen, L.C., Zhu, Y., Papandreou, G., Schroff, F., Adam, H.: Encoder-decoder with Atrous separable convolution for semantic image segmentation. In: Proceedings of the European Conference on Computer Vision (ECCV), pp. 801–818 (2018)
5. Cheng, B., Schwing, A., Kirillov, A.: Per-pixel classification is not all you need for semantic segmentation. In: Advances in Neural Information Processing Systems, vol. 34, pp. 17864–17875 (2021)
6. Dai, L., Liu, H., Tang, H., Wu, Z., Song, P.: AO2-DETR: arbitrary-oriented object detection transformer. IEEE Trans. Circuits Syst. Video Technol. **33**(5), 2342–2356 (2022)
7. Dai, Y., Yu, J., Zhang, D., Hu, T., Zheng, X.: RodFormer: high-precision design for rotating object detection with transformers. Sensors **22**(7), 2633 (2022)
8. Deng, J., Dong, W., Socher, R., Li, L.J., Li, K., Fei-Fei, L.: ImageNet: a large-scale hierarchical image database. In: 2009 IEEE Conference on Computer Vision and Pattern Recognition, pp. 248–255. IEEE (2009)
9. Ding, J., Xue, N., Long, Y., Xia, G.S., Lu, Q.: Learning ROI transformer for oriented object detection in aerial images. In: Proceedings of the IEEE/CVF Conference on Computer Vision and Pattern Recognition, pp. 2849–2858 (2019)
10. Dosovitskiy, A., et al.: An image is worth 16x16 words: transformers for image recognition at scale. ICLR (2021)
11. Gao, L., et al.: STransFuse: fusing Swin transformer and convolutional neural network for remote sensing image semantic segmentation. IEEE J. Sel. Top. Appl. Earth Obser. Remote Sens. **14**, 10990–11003 (2021)
12. Gu, A., Dao, T.: Mamba: linear-time sequence modeling with selective state spaces. arXiv preprint: arXiv:2312.00752 (2023)
13. Han, J., Ding, J., Li, J., Xia, G.S.: Align deep features for oriented object detection. IEEE Trans. Geosci. Remote Sens. **60**, 1–11 (2021)
14. He, K., Zhang, X., Ren, S., Sun, J.: Deep residual learning for image recognition. In: Proceedings of the IEEE Conference on Computer Vision and Pattern Recognition, pp. 770–778 (2016)
15. Hu, J., Shen, L., Sun, G.: Squeeze-and-excitation networks. In: Proceedings of the IEEE Conference on Computer Vision and Pattern Recognition, pp. 7132–7141 (2018)
16. Huang, T., Pei, X., You, S., Wang, F., Qian, C., Xu, C.: LocalMamba: visual state space model with windowed selective scan. arXiv preprint: arXiv:2403.09338 (2024)
17. Li, W., Chen, Y., Hu, K., Zhu, J.: Oriented reppoints for aerial object detection. In: Proceedings of the IEEE/CVF Conference on Computer Vision and Pattern Recognition, pp. 1829–1838 (2022)
18. Liu, M., Dan, J., Lu, Z., Yu, Y., Li, Y., Li, X.: CM-UNet: Hybrid CNN-mamba UNet for remote sensing image semantic segmentation (2024)

19. Liu, X., Ma, S., He, L., Wang, C., Chen, Z.: Hybrid network model: TransConvNet for oriented object detection in remote sensing images. Remote Sens. **14**(9), 2090 (2022)
20. Liu, Y., et al.: VMamba: visual state space model (2024). https://arxiv.org/abs/2401.10166
21. Liu, Z., et al.: Swin transformer: hierarchical vision transformer using shifted windows. In: Proceedings of the IEEE/CVF International Conference on Computer Vision, pp. 10012–10022 (2021)
22. Liu, Z., Mao, H., Wu, C.Y., Feichtenhofer, C., Darrell, T., Xie, S.: A convnet for the 2020s. In: Proceedings of the IEEE/CVF Conference on Computer Vision and Pattern Recognition, pp. 11976–11986 (2022)
23. Ma, X., Zhang, X., Pun, M.O.: RS3Mamba: visual state space model for remote sensing image semantic segmentation. IEEE Geosci. Remote Sens. Lett. (2024)
24. Sandler, M., Howard, A., Zhu, M., Zhmoginov, A., Chen, L.C.: MobileNetV2: inverted residuals and linear bottlenecks. In: Proceedings of the IEEE Conference on Computer Vision and Pattern Recognition, pp. 4510–4520 (2018)
25. Strudel, R., Garcia, R., Laptev, I., Schmid, C.: Segmenter: transformer for semantic segmentation. In: Proceedings of the IEEE/CVF International Conference on Computer Vision, pp. 7262–7272 (2021)
26. Touvron, H., Cord, M., Douze, M., Massa, F., Sablayrolles, A., Jegou, H.: Training data-efficient image transformers & distillation through attention. In: International Conference on Machine Learning, vol. 139, pp. 10347–10357 (2021)
27. Vaswani, A., et al.: Attention is all you need. In: Neural Information Processing Systems (2017). https://api.semanticscholar.org/CorpusID:13756489
28. Verma, T., Singh, J., Bhartari, Y., Jarwal, R., Singh, S., Singh, S.: SOAR: advancements in small body object detection for aerial imagery using state space models and programmable gradients. arXiv preprint: arXiv:2405.01699 (2024)
29. Wan, Q., Huang, Z., Lu, J., Yu, G., Zhang, L.: SeaFormer: squeeze-enhanced axial transformer for mobile semantic segmentation. In: International Conference on Learning Representations (ICLR) (2023)
30. Wang, C.Y., Yeh, I.H., Mark Liao, H.Y.: YOLOV9: learning what you want to learn using programmable gradient information. In: Computer Vision - ECCV 2024: 18th European Conference, Milan, Italy, September 29-October 4, 2024, Proceedings, Part XXXI, pp. 1–21. Springer-Verlag, Berlin, Heidelberg (2024)
31. Wang, H., Chen, X., Zhang, T., Xu, Z., Li, J.: CCTNet: coupled CNN and transformer network for crop segmentation of remote sensing images. Remote Sens. **14**(9), 1956 (2022)
32. Wang, J., Zheng, Z., Ma, A., Lu, X., Zhong, Y.: LoveDA: a remote sensing land-cover dataset for domain adaptive semantic segmentation. arXiv preprint: arXiv:2110.08733 (2021)
33. Wang, L., Li, D., Dong, S., Meng, X., Zhang, X., Hong, D.: PyramidMamba: rethinking pyramid feature fusion with selective space state model for semantic segmentation of remote sensing imagery (2024)
34. Wang, L., Li, R., Duan, C., Zhang, C., Meng, X., Fang, S.: A novel transformer based semantic segmentation scheme for fine-resolution remote sensing images. IEEE Geosci. Remote Sens. Lett. **19**, 1–5 (2022)
35. Wang, L., et al.: UNetFormer: a UNet-like transformer for efficient semantic segmentation of remote sensing urban scene imagery. ISPRS J. Photogramm. Remote. Sens. **190**, 196–214 (2022)

36. Wang, W., et al.: Pyramid vision transformer: a versatile backbone for dense prediction without convolutions. In: Proceedings of the IEEE/CVF International Conference on Computer Vision, pp. 568–578 (2021)
37. Wang, W., et al.: PVT v2: improved baselines with pyramid vision transformer. Comput. Vis. Media **8**(3), 415–424 (2022)
38. Xia, G.S., et al.: DOTA: a large-scale dataset for object detection in aerial images. In: Proceedings of the IEEE Conference on Computer Vision and Pattern Recognition, pp. 3974–3983 (2018)
39. Xie, E., Wang, W., Yu, Z., Anandkumar, A., Alvarez, J.M., Luo, P.: SegFormer: simple and efficient design for semantic segmentation with transformers. In: Advances in Neural Information Processing Systems, vol. 34, pp. 12077–12090 (2021)
40. Xie, X., Cheng, G., Wang, J., Yao, X., Han, J.: Oriented R-CNN for object detection. In: Proceedings of the IEEE/CVF International Conference on Computer Vision (ICCV), pp. 3520–3529 (2021)
41. Xu, X., et al.: An improved Swin transformer-based model for remote sensing object detection and instance segmentation. Remote Sens. **13**(23), 4779 (2021)
42. Xu, Z., Zhang, W., Zhang, T., Yang, Z., Li, J.: Efficient transformer for remote sensing image segmentation. Remote Sens. **13**(18), 3585 (2021)
43. Yang, C., et al.: PlainMamba: improving non-hierarchical mamba in visual recognition. arXiv preprint: arXiv:2403.17695 (2024)
44. Yang, X., Yan, J., Feng, Z., He, T.: R3Det: refined single-stage detector with feature refinement for rotating object. In: Proceedings of the AAAI Conference on Artificial Intelligence, vol. 35, pp. 3163–3171 (2021)
45. Yu, W., Zhou, P., Yan, S., Wang, X.: InceptionNeXt: When inception meets ConvNeXt. In: Proceedings of the IEEE/CVF Conference on Computer Vision and Pattern Recognition, pp. 5672–5683 (2024)
46. Zhang, Y., Liu, X., Wa, S., Chen, S., Ma, Q.: GansFormer: a detection network for aerial images with high performance combining convolutional network and transformer. Remote Sens. **14**(4), 923 (2022)
47. Zhou, Q., Yu, C.: PointRCNN: 3D object proposal generation and detection from point cloud. Remote Sens. **14**(11), 2605 (2022)
48. Zhu, L., Liao, B., Zhang, Q., Wang, X., Liu, W., Wang, X.: Vision Mamba: efficient visual representation learning with bidirectional state space model. arXiv preprint: arXiv:2401.09417 (2024)
49. Zhu, Q., et al.: Samba: semantic segmentation of remotely sensed images with state space model. Heliyon **10**(19) (2024)
50. Zhu, Q., Fang, Y., Cai, Y., Cheng, C., Fan, L.: Rethinking scanning strategies with vision mamba in semantic segmentation of remote sensing imagery: An experimental study. IEEE J. Sel. Top. Appl. Earth Obser. Remote Sens. **PP**, 1–14 (2024)

Control Allocation Based Fault-Tolerant Attitude Control of Hypersonic Reentry Vehicle

Yuan Zhang[1] and Weixin Han[1,2](✉)

[1] School of Automation, Northwestern Polytechnical University, Xi'an, China
yuan_zhang@mail.nwpu.edu.cn, hanweixin2009@163.com
[2] Research and Development Institute of Northwestern Polytechnical University, Shenzhen, China

Abstract. In this paper, a control allocation fault-tolerant control method based on a fault parameter identification for Hypersonic Reentry Vehicle (HRV) is investigated. A backstepping controller is employed to compute the desired torque. A chain distribution strategy facilitates the allocation of this desired torque among the aerodynamic surfaces, with the aerodynamic surface serving as the primary actuator. The least squares method is utilized for parameter identification in scenarios where the aerodynamic surface experiences failure or becomes stuck, resulting in an inability to generate the necessary aerodynamic torque. In addition, an integer programming method is implemented to assign the reaction control system (RCS) to compensate for any remaining torque requirements, thus achieving fault-tolerant control (FTC) to guarantee the tracking performance of the system. Additionally, the stability of the attitude control system is verified theoretically by the Lyapunov approach. Finally, the numerical HRV examples are given to confirmed the effectiveness of the proposed fault-tolerant control scheme.

Keywords: Hypersonic reentry vehicle · Fault-tolerant · Optimal control allocation · Actuator fault

1 Introduction

Hypersonic reentry vehicles (HRVs) [1–3] are increasingly important in the fields of national defense and civil aviation, with more stringent requirements for their flight quality. The reentry process presents a complex flight environment, with dramatic changes in aerodynamic characteristics due to speed variations across different airspaces. This can lead to potential failure of the aerodynamic surfaces. Given the extremely high manufacturing costs and demanding tasks, even minor failures could result in significant economic losses and disasters. Therefore, it is essential to enhance its fault-tolerance control (FTC) ability to ensure timely implementation of effective measures to mitigate faults and prolong service life.

Fault-tolerant controller is a control system that can still achieve the established control target under the failure condition. The prerequisite of FTC is the existence of control redundancy. The source of control redundancy can be divided into physical and analytical redundancy, in which physical redundancy is the most direct and effective fault-tolerant control method. Due to control redundancy, the actual implementation of the actuator to achieve the desired control effect is not unique. Control allocation methods [4–6] utilize the control redundancy to "reallocate" the required torque to the remaining actuators to achieve the expected control effect. When an aerodynamic surface fails, analysis of the redundant capabilities of each channel's control will result in the redistribution of tasks to the intact control redundancy, allowing them to collectively fulfill the responsibilities of the failed rudder and achieve FTC. When implementing FTC through control allocation, there is no need to reconstruct upper-level control laws only requiring the lower-level actuator allocation. Control allocation is combined with other robust control methods for fault-tolerant control generally. The design approach typically involves designing a virtual controller and allocating this virtual controlled quantity to actual actuator systems. This method is characterized by its intuitive efficiency and simplicity in design.

Control allocation (CA) methods are classified into two types depending on whether there is an optimization method or not. The non-optimal CA methods include generalized inverse allocation [7], daisy chaining allocation [8], direct allocation [9], and null-space allocation [10]. This non-optimal allocation method only considers the current performance allocation matrix and does not consider the input-output constraints. Therefore, this method can not make the most of the control redundancy, and the actual control quantity solved may exceed the physical limit. The optimal CA methods can make better use of control redundancy and give overall consideration of other performance indicators. HRV is equipped with the RCS to provide sufficient control redundancy for the coordinated flight of the vehicle [11–13]. When a severe fault occurs that cannot be suppressed only by the aerodynamic surfaces, RCS can replace the faulty component to provide control torque as control redundancy. Therefore, the optimized allocation method is more suitable for solving multiple targets and multi-constraint problems of HRV.

Optimal CA methods include the quadratic programming CA method [14], the CA method based on model prediction [15], the optimization control allocation based on piecewise linearization [16], and the nonlinear optimization CA method [17]. In [8], the multi-objective allocation algorithm is proposed to handle the RCS and aerodynamic surfaces simultaneously. The optimal solution of the multi-objective function is solved to realize the best comprehensive performance. In [12], a multi-objective CA scheme is constructed to address the complex allocation problem of the RCS and aerodynamic surfaces during reentry based on analytic hierarchy process. The aim is to accomplish a comprehensive optimal solution for the optimization function, including minimizing allocation error, maximizing control efficiency, and minimizing fuel consumption. The control allocation problems mentioned above are all deterministic, without considering the uncertain factors caused by the decrease in surface efficiency or the limitation of surface deflection due to unknown faults. The optimization equation and constraints of control allocation are modified by faults, thus it is essential to acquire fault parameters.

Fault diagnosis (FD) technology can be divided into three stages: fault detection, isolation and estimation. Fault estimation is used to obtain the type, position, and signal changes of the system fault. Compared with fault detection and fault isolation techniques, fault estimation technology can provide more comprehensive information about faults, making it more valuable and technically advantageous for practical applications. To realize the fault estimation design of HRV attitude fault-tolerant control, the applications of model-based observer, adaptive robust observer and intelligent method have been widely discussed [18]. In [19], the fault estimation is constructed by two extended state observers to approximate the uncertainties and fault information concurrently. In [20], the fault estimation is achieved by the detection and the identification observers and decision mechanism to accurately estimate actuator fault parameters. However, the diagnostic threshold parameter and how to choose them are not given in the paper. In [21], an intelligent fault estimation which needs to be trained in advance is proposed to diagnosis fault.

Based on the analysis of the above strategies, a control allocation fault-tolerant controller based on fault parameter identification is proposed. The contributions of this paper are shown as follows:

1) A fault estimation using the least squares parameter identification is adopted to get the aerodynamic deflection fault parameter. This scheme can achieve fault parameter estimation, avoid the problem of difficult threshold determination and large amount of data calculation in advance, and reduce the complexity of fault diagnosis.
2) There are only two modes of switching for the RCS, and the distribution result should be an integer. Contrasting to the previous studies [22], the original RCS control distribution nonlinear integer programming problem is altered into a linear integer programming problem, which reduces the computational complexity.

The organizational structure of this paper is as follows. The dynamic model of HRV with faults is described in Sect. 2. The attitude control scheme is designed, and the stability analysis is provided in Sect. 3. The torque control allocation scheme and fault-tolerant torque control allocation are shown in Sect. 4 and Sect. 5. The simulation results are shown in Sect. 6. In Sect. 7, the conclusions are given.

2 Model Dynamics and Problem Formulation

2.1 Reentry Mode

The model of the hypersonic reentry attitude dynamics is composed of attitude angle equation and attitude angle rate equation, which can be described as follow [23]:

$$\dot{\gamma} = R\omega \tag{1}$$

$$J\dot{\omega} = -\Omega J\omega + T_{des} \tag{2}$$

where $\gamma = [\phi, \beta, \alpha]^T$ represents the attitude angle of HRV. The variables ϕ, β and α denote bank angle, sideslip angle and attack angle, respectively, and $\omega = [w_x, w_y, w_z]^T$ represents the attitude angle rate, and w_x, w_y and w_z denote the angular velocities of the rolling, pitching and yawing, respectively, $J \in \Re^{3\times 3}$ and $T_{des} \in \Re^{3\times 1}$ are the inertia and the desired control torque, R and Ω have the following forms

$$R = \begin{bmatrix} \cos\alpha & 0 & \sin\alpha \\ \sin\alpha & 0 & -\cos\alpha \\ 0 & 1 & 0 \end{bmatrix}$$

$$\mathbf{\Omega} = \begin{bmatrix} 0 & -w_z & w_y \\ w_z & 0 & -w_x \\ -w_y & w_x & 0 \end{bmatrix}$$

The expected torque is related to the aerodynamic torque and the RCS torque, and can be described as:

$$\mathbf{T}_{des} = \mathbf{T}_a + \mathbf{T}_{\text{RCS}} \tag{3}$$

where $\mathbf{T}_a \in \Re^{3\times 1}$ is the aerodynamic torque provided by aerodynamic deflection, $\mathbf{T}_{\text{RCS}}$ is the control torque supplied by RCS.

$\mathbf{T}_a$ can be expressed as:

$$\mathbf{T}_a = \mathbf{\Phi}_a \boldsymbol{\delta} \tag{4}$$

where $\mathbf{\Phi}_a$ is the distribution matrix between the torque and the deflection angle, $\boldsymbol{\delta} = [\delta_e, \delta_a, \delta_r]^\mathrm{T}$ is the aerodynamic deflection angle, the variables $\delta_e, \delta_a, \delta_r$ denote the elevator deflection, aileron deflection and rudder deflection, respectively.

The model of aerodynamic deflection can be described as follow [22]:

$$\begin{aligned} \dot{\delta}_i &= \delta_{vi} \\ \dot{\delta}_{vi} &= -k_{2i}\delta_{vi} - k_{1i}\delta_i + k_{1i}\delta_{ci} \end{aligned} \tag{5}$$

where δ_{ci} is the desired deflection generated by control allocation, δ_i is the actuator output deflection, δ_{vi} is the actuator output deflection angle rate, k_{1i} and k_{2i} are the parameters of the model, $k_{1i} \gg 1$ and $k_{1i} \gg k_{2i} > 0$.

The RCS torque can be described as:

$$\mathbf{T}_{RCS} = \mathbf{\Phi}_{i,k} \mathbf{u}_{rcs} \tag{6}$$

where $\mathbf{\Phi}_{i,k} \in \Re^{3\times 10}$ is the distribution matrix between the torque and the RCS jets, $\mathbf{u}_{rcs} = [u_{rcs1}, u_{rcs2}, ..., u_{rcs10}]^\mathrm{T}$ are the output state of RCS.

2.2 Fault Model

The aerodynamic deflection fault model based on the second-order actuator system is shown as:

$$\begin{aligned} \dot{\delta}_i &= l_i \delta_{vi} \\ \dot{\delta}_{vi} &= -k_{2i}\delta_{vi} + k_{1i}l_i(\lambda_i \delta_{ci} - \delta_i) \end{aligned} \tag{7}$$

where $0 < \lambda_i \leq 1$ represents the aerodynamic surface effectiveness damage factor, $l_i \in (0, 1)$ represents the stuck fault sign, when $l_i = 0$ represents the stuck fault occurs, when $l_i = 1$, the stuck fault does not occur.

If the ith aerodynamic surface is stuck, it will stagnate at a fixed position whose rate of change will be zero, which is described as:

$$\begin{cases} \delta_i = \bar{\delta}_i \\ \dot{\delta}_i = 0 \end{cases} \tag{8}$$

where δ_i represents the deflection where the stuck occurs, $\bar{\delta}_i$ represents the surface deflection when the stuck fault occurs.

3 Attitude Control Scheme

3.1 Torque Controller Design

The control torque T directly controls the attitude angle rate ω to achieve the expected attitude angle γ_d tracking. The expected angle rate ω_d is designed to be:

$$\omega_d = R^{-1}(-\kappa_1 \gamma + \kappa_1 \gamma_d + \dot{\gamma}_d) \tag{9}$$

where $\kappa_1 > 0$ is a designer parameter.

From (1) and (8):

$$\dot{\gamma} = -\kappa_1 \gamma + \kappa_1 \gamma_d + \dot{\gamma}_d + R^{-1}(\omega - \omega_d) \tag{10}$$

Remark 1. If ω can accurately track ω_d, γ can be asymptotically approached by γ_d.

In order to make the attitude angular velocity γ tracking γ_d, the expected torque should be designed to:

$$T'_{des} = -\kappa_2 J(\omega - \omega_d) - JR^T(\gamma - \gamma_d) + \Omega J \omega + J\dot{\omega}_d \tag{11}$$

where $\kappa_2 > 0$ is a designer parameter.

3.2 Stability Analysis

Theorem 1. *For system (1) with the controller (10), it is guaranteed that the signals $\tilde{\gamma}$ and $\tilde{\omega}$ in (11) are uniformly ultimately bounded.*

Proof : Choose the Lyapunov function as

$$V = \frac{1}{2}\tilde{\gamma}^T \tilde{\gamma} + \frac{1}{2}\tilde{\omega}^T \tilde{\omega} \tag{12}$$

$\dot{V}$ can be obtained as

$$\begin{aligned}\dot{V} &= \tilde{\gamma}^T(R\omega - \dot{\gamma}_d) + \tilde{\omega}^T(-J^{-1}\Omega J\omega + J^{-1}T_{des} - \dot{\omega}_d) \\ &= \tilde{\gamma}^T(R\omega_d + R\tilde{\omega} - \dot{\gamma}_d) + \tilde{\omega}^T(-J^{-1}\Omega J\omega + J^{-1}T_{des} - \dot{\omega}_d) \end{aligned} \tag{13}$$

The $\dot{V}$ is calculated as

$$\dot{V} = -\kappa_1\tilde{\gamma}^{\mathrm{T}}\tilde{\gamma} + \tilde{\gamma}^{\mathrm{T}}R\tilde{\omega} + \tilde{\omega}^{\mathrm{T}}\left(-J^{-1}\Omega J\omega + J^{-1}T_{des} - \dot{\omega}_d\right) \tag{14}$$

$\dot{V}$ can be written as

$$\dot{V} = -\kappa_1\tilde{\gamma}^{\mathrm{T}}\tilde{\gamma} - \kappa_2\tilde{\omega}^{\mathrm{T}}\tilde{\omega} \tag{15}$$

By choosing $\kappa_1 > 0$ and $\kappa_2 > 0$, it can be obtained as

$$\dot{V} = -\kappa_1\|\tilde{\gamma}\|^2 - \kappa_2\|\tilde{\omega}\|^2 \leq 0 \tag{16}$$

Furthermore, the signals in (8) are uniform ultimate boundedness.
The proof is completed.

4 Torque Control Allocation

When the aerodynamic surface of the hypersonic vehicle fails, the deflection cannot provide the desired aerodynamic torque to stabilize the attitude during reentry. The control allocation will maximize the use of the aerodynamic surface and reasonably optimize the utilization rate of various actuators to produce the desired torque to achieve the control requirements. In this section, a chain distribution strategy is used to distribute the desired control torque between the aerodynamic surface and the RCS system, with the aerodynamic surface as the primary and the RCS as the supplementary. Firstly, Quadratic Programming (QP) is used to maximize the use of the aerodynamic rudder surface to obtain the expected torque T_{des} for maintaining attitude. When the aerodynamic rudder surface fails or reaches saturation, resulting in the aerodynamic surface failing to provide the desired aerodynamic torque, the RCS actuator is launched to compensation. The RCS lighting scheme is determined by integer linear programming.

4.1 Aerodynamic Surface Control Allocation

The control allocation scheme of the aerodynamic surface is based on the QP method to solve the surface deflection. Its primary purpose is to satisfy the minimum differential value between T_{des} and T_a and second to minimize the surface deflection. Therefore, the the control allocation's objective function is described as:

$$\min J_\delta = \frac{1}{2}\left[(1-k_J)(T_{des} - T_a)^{\mathrm{T}}Q_T(T_{des} - T_a) + k_J\delta^{\mathrm{T}}Q_\delta\delta\right]$$

$$s.t \begin{cases} \delta_{\min} \leq \delta \leq \delta_{\max} \\ \dot{\delta}_{\min} \leq \dot{\delta} \leq \dot{\delta}_{\max} \end{cases} \tag{17}$$

where $0 < k_J < 1$ is the target weight for control allocation, $Q_T \in \Re^{3\times 3}$ and $Q_\delta \in \Re^{3\times 3}$ are the weight matrices of δ, $\delta_{\min}$ and $\delta_{\max}$ are the minimum and maximum surface deflection, respectively, $\dot{\delta}_{\min}$ and $\dot{\delta}_{\max}$ are the minimum and maximum surface deflection rate, respectively.

Remark 2 The primary goal of planning is to ensure the minimum torque difference between the expected moment T_{des} and the aerodynamic moment T_a, and then to satisfy the minimum surface deflection. Therefore, the value of k_J in the formula should be as small as possible.

4.2 RCS Control Allocation

When the aerodynamic surface fails to generate aerodynamic torque sufficient to maintain the body attitude, the RCS will be triggered to provide a compensation torque. Therefore, using integer programming to solve the RCS allocation problem, the CA objective function is as follows:

$$\min \sum_{i=1}^{3} Q_{R1} |T_{RCS} - \Phi_{i,k} u_{rcs}| + Q_{R2} u_{rcs} \tag{18}$$

where $T_{RCS} = T_{des} - T_a$ is the expected torque compensated by RCS, $Q_{R1} \in \Re^{3\times 3}$ and $Q_{R2} \in \Re^{3\times 10}$ are the weight matrices, respectively.

The primary objective of the optimization is to satisfy the minimum difference between the required compensatory and the actual torques. The secondary aim is to reduce the number of RCS activations, thereby conserving fuel. The RCS CA converts the continuous command torque vector into an integer torque vector. However, the above nonlinear integer programming problem is difficult to solve directly, so it needs to be transformed into a linear programming problem.

Define $z = T_{RCS} - \Phi_{i,k} u_{rcs}$, the objective function can be written as:

$$\min \sum_{i=1}^{3} Q_{R1} |z| + Q_{R2} u_{rcs} \tag{19}$$

where $z = [z_1, z_2, z_3]^{\mathrm{T}}$.

The new symbols $m_i, n_i \geq 0, i = 1, 2, 3$ are introduced to define $z_i = m_i - n_i$ and $|z_i| = m_i + n_i$. m_i, n_i can be written as:

$$m_i = \frac{z_i + |z_i|}{2}, \quad n_i = \frac{|z_i| - z_i}{2} \tag{20}$$

By substituting the above equation into equation (4)–(22), the mixed integer linear programming problem can be obtained. The optimization objective function can be expressed as follows:

$$\min \sum_{i=1}^{3} Q_{R1}(m+n) + Q_{R2} u_{rcs}$$

$$\text{s.t} \begin{cases} \Phi_{i,k} u_{rcs} + m - n = T_{RCS} \\ m_i, n_i \geq 0, i = 1, 2, 3 \end{cases} \tag{21}$$

The optimization variables by the original u_{rcs} expanding to $\left[u_{rcs}^{\mathrm{T}}, m^{\mathrm{T}}, n^{\mathrm{T}}\right]^{\mathrm{T}}$. Then the optimization solution can be completed.

5 Fault-Tolerant Torque Control Allocation

5.1 Fault Identification

According to the fault model, parameters λ, l can reflect the type of aerodynamic surface fault. Therefore, fault diagnosis can be achieved by identifying the fault model parameters. Define $x_1 = \delta_i$ and $x_2 = \delta_{vi}$, the fault model can be rewritten as

$$\begin{bmatrix} \dot{x}_1 \\ \dot{x}_2 \end{bmatrix} = \begin{bmatrix} 0 & l \\ -k_1 l & -k_2 \end{bmatrix} \begin{bmatrix} x_1 \\ x_2 \end{bmatrix} + \begin{bmatrix} 0 \\ k_1 l \lambda \end{bmatrix} \delta_{ci} \quad (22)$$

The above model can be represented in discrete-from:

$$x_1(k+1) = x_1(k) + lT_s x_2(k) \quad (23)$$

$$x_2(k+1) = x_2(k) - T_s \left(k_1 l x_1(k) + k_2 x_2(k) - k_1 \lambda l \delta_{ci}(k) \right) \quad (24)$$

where T_s is the step.

$x_2(k), x_2(k+1)$ can be expresses as:

$$x_2(k) = \frac{x_1(k+1) - x_1(k)}{lT_s} \quad (25)$$

$$x_2(k+1) = \frac{x_1(k+2) - x_1(k+1)}{lT_s} \quad (26)$$

The model in the form of difference equation can be written as:

$$\frac{x_1(k+2) - x_1(k+1)}{lT_s} = \frac{x_1(k+1) - x_1(k)}{lT_s} - T_s \left(k_1 l x_1(k) + k_2 \frac{x_1(k+1) - x_1(k)}{lT_s} - k_1 \lambda l \delta_c(k) \right) \quad (27)$$

Furthermore, it can be obtained as:

$$\begin{aligned} x_1(k+2) &= 2x_1(k+1) - x_1(k) - l^2 T_s^2 k_1 x_1(k) - T_s k_2 x_1(k+1) + T_s k_2 x_1(k) + k_1 l^2 T_s^2 \lambda \delta_c \\ &= (2 - T_s k_2) x_1(k+1) + \left(-1 - l^2 T_s^2 k_1 + T_s k_2 \right) x_1(k) + k_1 l^2 T_s^2 \lambda \delta_c(k) \end{aligned} \quad (28)$$

Define $\mu = 2 - T_s k_2$ and $y = x_1$, it can be written as:

$$y(k+2) = \mu y(k+1) + a_1 y(k) + a_2 \delta_c(k) \quad (29)$$

where $a_1 = -1 - l^2 T_s^2 k_1 + T_s k_2$, $a_2 = k_1 l^2 T_s^2 \lambda$.

Using the least square method to estimate parameters a_1 and a_2, the fault parameters can be calculated, and they can be represented as:

$$l = \sqrt{\frac{-a_1 - 1 + T_s k_2}{T_s^2 k_1}} \quad (30)$$

$$\lambda = \frac{a_2}{k_1 T_s^2} \quad (31)$$

Remark 3 The above formula has been modified. For the calculation of the parameter λ, l is discarded from the denominator.

5.2 Fault-Tolerant Control Allocation

The fault aerodynamic torque can be described as:

$$T_a = \Phi_a \left(k_\delta \delta + \bar{\delta} \right) \tag{32}$$

Therefore, the objective function of the FTC allocation is as follows:

$$\min J_\delta = \frac{1}{2} \Big[(1 - k_{J1}) \left(T_{\text{des}} - \Phi_a \left(k_\delta \delta + \bar{\delta} \right) \right)^T Q_{T1} \left(T_{\text{des}} - \Phi_a \left(k_\delta \delta + \bar{\delta} \right) \right) \\ + k_{J1} \delta^T Q_{\delta 1} \delta \Big]$$

$$s.t \begin{cases} \delta^i_{\min} - \bar{\delta}^i \leq \delta^i \leq \delta^i_{\max} - \bar{\delta}^i \\ \dot{\delta}^i_{\min} \leq \dot{\delta}^i \leq \dot{\delta}^i_{\max} \end{cases} \tag{33}$$

where $0 < k_{J1} < 1$ is the target weight for control allocation, $Q_{T1} \in \Re^{3 \times 3}$ and $Q_{\delta 1} \in \Re^{3 \times 3}$ are the weight matrices of δ, $\delta^i_{\min}$ and $\delta^i_{\max}$ are the minimum and maximum surface deflection, respectively, $\dot{\delta}^i_{\min}$ and $\dot{\delta}^i_{\max}$ are the minimum and maximum surface deflection rate, respectively.

6 Simulation

In this section, the simulation results of a case study on the X-33 model are discussed and the inertia matrix J is given as $J = \begin{bmatrix} 554486 & 0 & -23002 \\ 0 & 1136949 & 0 \\ -23002 & 0 & 1376852 \end{bmatrix}$. The distribution matrix Φ_a is referred from [24]. The model parameters of aerodynamic deflection are selected as $k_{1i} = 2025$ and $k_{2i} = 72$. The distribution matrix $\Phi_{i,k}$ is set as

$$\Phi_{i,k} = \begin{bmatrix} 0 & -1511 & 8574 & -5098 & 0 & 1515 & -8573 & 5098 & 0 & 0 \\ -367 & 0 & -6982 & 8702 & -367 & 0 & -6981 & 8702 & -367 & -367 \\ 14675 & -11597 & -6918 & -8702 & -14675 & 11597 & 6981 & 8702 & -14675 & 14675 \end{bmatrix}.$$

The controller parameters are selected as $\kappa_1 = 0.6$ and $\kappa_2 = 1.2$, the control allocation parameters are selected as $k_J = 0.1$, $Q_T = \text{diag}\{1, 1, 1\}$, $Q_\delta = \text{diag}\{1, 1, 1\}$, $Q_{R1} = \text{diag}\{1, 1, 1\}$, $Q_{R2} = I^{3 \times 10}$.

The initial state variables and the disturbances torque are selected as $\gamma = [0, 0.1, 0]^T$ deg, $\omega = [2.875, 2.875, 0]^T$ deg/s. The attitude tracking command is defined as
$\gamma_d = [5.73, 0, 2.875]^T$ deg.

The case 1: The rudder deflection occurs a 30% LOE fault in 1s. The proposed is marked as "FTC". And the comparison scheme without RCS is marked as "C". The simulation results are presented in Figs. 1, 2, 3 and 4. In Fig. 1, the attitude angle responses show that the FTC scheme can effectively reduce the deviation of attitude angle and quickly achieve the desired attitude under the unknown rudder deflection LOE fault in 1s. In Figs. 2 and 3, it is obvious that after the fault occurs, the rudder deflection is saturated and the RCS is started for torque compensation. In Figs. 4, the real fault parameter is marked as "R" and the identification result marked as "I". It can be concluded that the fault parameters can be quickly identified after the fault occurs.

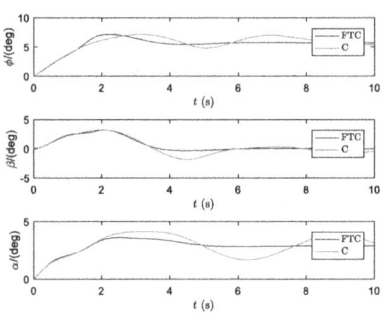

Fig. 1. The Attitude Angles with RCS and without RCS in Case1

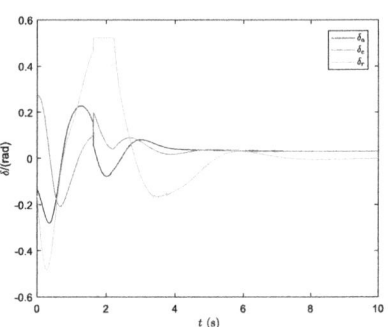

Fig. 2. Deflection in Case1

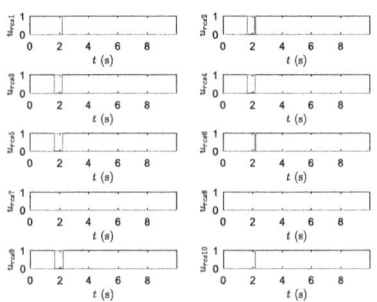

Fig. 3. RCS states in Case1

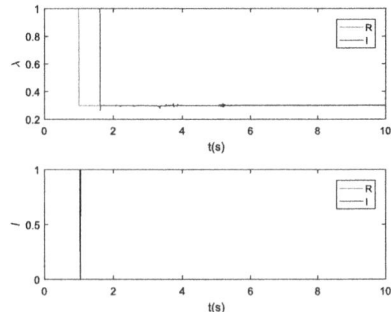

Fig. 4. Identification parameter in Case1

The case 2: The aileron deflection occurs a stuck fault in 3.2s. The simulation results are presented in Figs. 5, 6, 7 and 8. In Fig. 5, the attitude angle responses show that the FTC scheme can effectively reduce the deviation of attitude angle and quickly achieve the desired attitude under the unknown aileron deflection stuck fault in 3.2s. In Figs. 6 and 7, it is obvious that the deflection is stuck and others are adjustable. And the

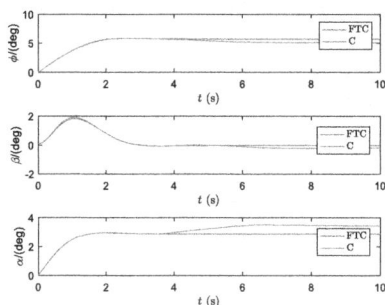

Fig. 5. The Attitude Angles with RCS and without RCS in Case2

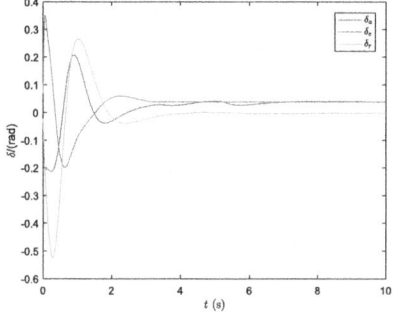

Fig. 6. Deflection in Case2

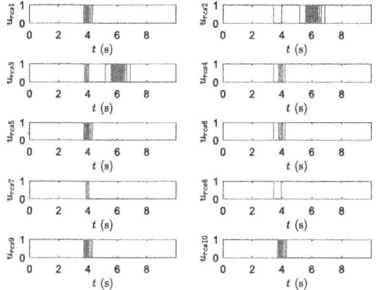

Fig. 7. RCS states in Case2

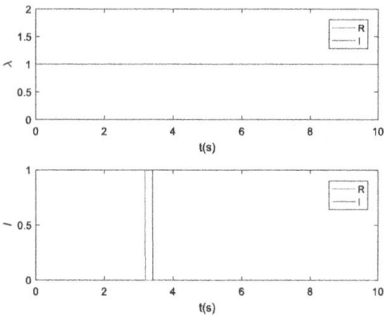

Fig. 8. Identification parameter in Case2

RCS starts quickly to output the compensation torque. In Fig. 8, it can be seen that the damage factor is always equal to 1. This indicates that no LOE fault has occurred. The stuck fault sign becomes 0 after the stuck occurrence, which effectively indicates that the stuck fault occurs. In summary, the FTC scheme can effectively suppress the influence of unknown faults in the reentry stage, improve the system's robustness, and maintain the HRV attitude tracking performance.

7 Conclusion

Considering the aerodynamic surface fault effects during the re-entry of a hypersonic flight vehicle, a CA fault-tolerant controller based on a fault parameter identification is proposed to ensure accurate attitude tracking. The distribution strategy employs the aerodynamic surface as the primary and utilizes the RCS as a secondary to efficiently and precisely deliver the required torque. The least squares can be utilized to identify the parameters of the pneumatic rudder, enabling the determination of torque demands necessitating RCS compensation, thereby fulfilling the objectives of attitude control. The proposed strategy is verified with excellent trajectory tracking performance and fault tolerance capability.

In future work, the HRV's FTC schemes based on the intelligent fault identification will be studied. To solve this problem, an intelligent fault identification is proposed to obtain the fault parameters of different components in real-time. By obtaining multiple types and accurate fault information, the control strategy can achieve better fault-tolerant control performance.

Acknowledgments. This work was partially supported by National Natural Science Foundation of China (Grant No. 62473318, 61933010), Guangdong Basic and Applied Basic Research Foundation 2022A1515240029, Key Research and Development Program of Shaanxi (2021GXLH-01-13), Young Elite Scientists Sponsorship Program by SNAST (20220125).

References

1. Fidan, B., Mirmirani, M., Ioannou, P.: Flight dynamics and control of air-breathing hypersonic vehicles: review and new directions. In: 12th AIAA International Space Planes and Hypersonic Systems and Technologies, pp. 7081–7106, 2003
2. Xu, B., Shou, Y., Shi, Z., Yan, T.: Predefined-time hierarchical coordinated neural control for hypersonic reentry vehicle. IEEE Trans. Neural Netw. Learn. Syst. **34**(11), 8456–8466 (2023)
3. Xu, B., Wang, X., Sun, F., Shi, Z.: Intelligent control of flexible hypersonic flight dynamics with input dead zone using singular perturbation decomposition. IEEE Trans. Neural Netw. Learn. Syst. **34**(9), 5926–5936 (2023)
4. Härkegård, O., Glad, S.T.: Resolving actuator redundancy – optimal control vs. control allocation. Automatica. **41**, 137–144 (2005)
5. Johansen, T.A., Fossen, T.I.: Control allocation – a survey. Automatica **49**(5), 1087–1103 (2013)
6. Blaha, T.M., Smeur, E., Remes, B.: A survey of optimal control allocation for aerial vehicle control. Actuators **12**(7), 1–14 (2023)
7. Guo, Y., Wu, M., Luo, Y.: Optimization of multi-effector aircraft control allocation based on generalized inverse. In: International Conference on Control Automation and Information Sciences, pp. 1–5 (2019)
8. Mu, R., Zhang, X., Wu, P., Chen, J.: RCS and Aero surfaces control allocation research on RLV's Re-Entry phase. Appl. Sci.-Basel **9**(8), 1–12 (2019)
9. Naskar, A.K., Patra, S., Sen, S.: Reconfigurable direct allocation for multiple actuator failures. IEEE Trans. Control Syst. Technol. **23**(1), 397–405 (2015)
10. Xu, D., Yao, T., Ji, N.: Fault tolerant flight-control via control allocation for reusable launch vehicles with aerodynamic control surfaces stuck. In: Control Conference, pp. 6611–6616 (2016)
11. Song, J., Cai, G., Chen, X.: Control allocationcbased command trackingccontrol system for hypersonic re-entry vehicle driven by hybrid effecters. J. Aerosp. Eng. **31**(4), 1–12 (2018)
12. Zhang, X., Mu, R., Chen, J., Wu, P.: Hybrid multi-objective control allocation strategy for reusable launch vehicle in re-entry phase. Aerosp. Sci. Technol. **116**, 1–14 (2021)
13. Shou, Y., Xu, B., Liang, X., Yang, D.: Aerodynamic/reaction-jet compound control of hypersonic reentry vehicle using sliding mode control and neural learning. Aerosp. Sci. Technol. **111**(106564), 1–20 (2021)
14. Zhang, Z., Chen, T., Zheng, L., Luo, Y.: A quadratic programming based neural dynamic controller and its application to UAVs for time-varying tasks. IEEE Trans. Veh. Technol. **70**(7), 6415–6426 (2021)

15. Zhu, W., Wang, Y., Gao, D., Shi, W., Yu, W., Wang, Y.: A thrust allocation strategy for intelligent ships based on model prediction control. Trans. Inst. Meas. Control. **45**(9), 1693–1702 (2023)
16. Bolender, M.A., Doman, D.B.: Nonlinear control allocation using piecewise linear functions. J. Guid. Control. Dyn. **27**(6), 1017–1027 (2004)
17. Su, M., Hu, J., Wang, Y., He, Z., Cong, J., Han, L.: A multi-objective incremental control allocation strategy for tailless aircraft. Int. J. Aeros. Eng. **2022**, 1–18 (2022)
18. Bin, J., Ke, Z., Chun, L., Hao, Y.: Fault diagnosis and accommodation with flight control applications. J. Control Decis. **7**(1), 24–43 (2020)
19. Liu, Y., Wang, Q., Hu, C., Dong, C.: ESO-based fault-tolerant anti-disturbance control for air-breathing hypersonic vehicles with variable geometry inlet. Nonlinear Dyn. **98**(3), 2293–2308 (2019). https://doi.org/10.1007/s11071-019-05329-3
20. Xu, D., Jiang, B., Shi, P.: Robust NSV fault-tolerant control system design against actuator faults and control surface damage under actuator dynamics. IEEE Trans. Industr. Electron. **62**(9), 5919–5928 (2015)
21. Wang, G., Xia, H.: Fault-tolerant learning control of air-breathing hypersonic vehicles with uncertain parameters and actuator faults. Expert Syst. Appl. **238**(121874), 1–12 (2024)
22. He, J., Qi, R., Jiang, B., Zhai, R.: Fault-tolerant control with mixed aerodynamic surfaces and RCS jets for hypersonic reentry vehicles. Chin. J. Aeronaut. **30**(2), 780–795 (2017)
23. Shtessel, Y., McDuffie, J., Jackson, M.: Sliding mode control of the X-33 vehicle in launch and re-entry modes. In: AIAA Guidance, Navigation and Control Conference, Boston, pp. 1352–1362 (1998)
24. Burken, J., Lu, P., Wu, Z.: Reconfigurable flight control designs with application to the X-33 vehicle. In: Proceedings of AIAA Guidance, Navigation, and Control Conference and Exhibit, pp. 951–965 (1999)

Cooperative Multi-target Enclosing Control of Multi-robots Based on Virtual Target Points

Fang Fang Zhang[✉], Zhen Yu Zhu, and Ji Xian Gao

ZhengZhou University, Henan, China
zhangfangfang@zzu.edu.cn

Abstract. This article proposes a controller for the capture problem of multiple moving targets. The controller includes three functions: the repulsion term between robots is used for collision avoidance; the attraction between the robot and the target drives the robot towards the target; the adaptive term is used to estimate the speed of the target. For multiple targets that need to be captured, a strategy of combining virtual target points with minimum enclosing circles is adopted. The center of the smallest circle that can contain all targets is regarded as the virtual target point, which is transformed into the capture of the virtual target point. Combined with a certain collision avoidance distance, the robot's formed enclosing circle can contain the minimum circle where the moving target is located, thereby completing the capture of multiple targets. Finally, the effectiveness of the proposed control strategy was verified through simulation experiments.

Keywords: multi-agent systems · target fencing · cooperative control

1 Introduction

Since the emergence of robots in the 1960s and 1970s [1], robots have been a hot topic of research for scientists. Some progress has been made in mechanical structure, motion control, sensor technology, etc. But single robots often cannot cope with more complex environments. Multi robot systems are equipped with more perception devices, have a wider perception range and flexible topology. Compared with single robots, multi robot systems have the advantages of small size, large quantity, strong flexibility, high robustness, and simple structure [2]. Compared with single robot systems, the application scope of multi robot systems is becoming increasingly wide, playing a huge role in collaborative search in unknown areas [3], mine clearance in military activities [4], target encirclement [5], assembly in industrial production [6], and post disaster rescue [7]. Among them, the collaborative pursuit of targets has been a research hot spot in recent years. How to enable multiple robots to complete the pursuit of multiple targets while considering collision avoidance between robots and between robots and targets is of great significance in daily life, especially in military affairs. Although the research in the field of multi robots started relatively late in China, it has developed rapidly. Currently, China has achieved a series of results in this field. In the early stages of multi robot encirclement research, the main methods used were based on artificial physical models [8], such as methods combining artificial torque [9] and methods based on output feedback controllers [10]. Subsequently, innovation was made on the basis of previous methods, such as abstracting a simplified virtual force model by decomposing capture behavior in complex environments, based on this model, trapping methods were designed [11], and a self-organizing method with loose preference rules was proposed based on decomposed trapping behavior to enable individual robots to form an ideal trapping formation during coordinated self-organizing movements [12]. Wang, et al. considered the situation where the position and number of intruders were unknown, and proposed a dynamic alliance trapping strategy based on an improved contract network protocol for trapping intruders with unknown information [13]. In recent years, with the rapid development of deep learning, more and more deep learning related methods have been introduced into encirclement. For example, Fan et al. [14] modeled a multi robot system using Markov game theory, designed a potential energy function that can control the system to the desired encirclement state and meet obstacle avoidance requirements, and used an improved multi robot reinforcement learning algorithm guided by potential energy models for encirclement. Chen et al. [15] constructed a tracking strategy based on a biologically inspired neural network to dynamically guide all robots in the alliance to track; finally, the formation strategy is adopted to achieve the encirclement of the target. This article proposes a new target encirclement control strategy that introduces an obstacle function to ensure collision avoidance between the robot and the target, and can make multiple targets converge inside the convex hull composed of multiple agents to complete encirclement without specifying the encirclement formation.

2 System Modelling and Problem Description

2.1 System Modelling

Consider a system consisting of individual N intelligences. Assuming that the dynamical properties of each intelligent body are the same, its kinematic equations are as follows:

$$\begin{cases} \dot{p}_i(t) = v_i(t) \\ \dot{v}_i(t) = u_i(t) \end{cases}, i = 1, 2, ..., N \tag{1}$$

where $p_i(t)$ denotes the position of the i intelligence; $v_i(t)$ denotes the velocity of the i intelligence; $u_i(t)$ denotes the acceleration input of the i intelligent body. Consider that the target consists of individual M intelligences. Assuming that the dynamical properties of each intelligent body are the same, its kinematic equations are as follows:

$$\begin{cases} \dot{p}_{oi}(t) = v_{oi}(t) \\ \dot{v}_{oi}(t) = u_{oi}(t) \end{cases}, i = 1, 2, ..., M \tag{2}$$

where $p_{oi}(t)$ denotes the position of the i intelligence; v_{oi} denotes the velocity of the i intelligence; $u_{oi}(t)$ denotes the acceleration input of the i intelligent body.

2.2 Problem Description

Assuming in a two-dimensional space, there are N fencing robots to fence M moving targets($N \geq 3M$). During the encirclement process, the encirclement robot can perceive the position information of the target, but cannot perceive the speed information of the target. When the encirclement robot surrounds all the encirclement targets in the encirclement circle formed by the encirclement robot and the encirclement robot is at the same speed as the target, it is considered a successful encirclement, the conditions for successful encirclement are:

$$\begin{cases} v_i(t) = v_o(t) \\ |\dot{P}_i(t) - \dot{P}_O(t)| \leq 10^{-2} \end{cases} \quad (3)$$

And meanwhile, it is necessary to pay attention to the avoidance of collision between the rounding robot and between the enclosing robot and the target during the process of the rounding process.

3 Main Elements

In the case of multi-target, we consider simplifying the problem of multi-target hunting by multi robots to that of single target hunting by multi robots. Therefore, we propose the concept of virtual target point. The so-called virtual target point is the center of the smallest circle that can contain all target points. As long as the encirclement robot can encircle the center and maintain a certain collision avoidance distance, when the collision avoidance distance is slightly greater than the radius of the smallest circle where all targets are located, the encirclement of multiple targets is completed. In this way, the problem of rounding up multiple points is transformed into the problem of rounding up a point with an area. Since the velocity of multiple targets being rounded up is the same, the virtual velocity of the virtual target point is the velocity of the rounded up target; for the position of the virtual target point, we need to solve by the following method: assume that the location information of the target is respectively $(p_{ox_i}(t), p_{oy_i}(t))$, $i = 1, 2, 3...$. When two targets are rounded up, it is easy to get a circle with two connecting lines as diameters. With the increase of targets, when there are three targets, the corresponding circle can also be obtained according to the principle of determining the unique circle at three points (the collinearity of targets can be transformed into the problem of rounding up two targets), define the following functions:

$$U(x,y) = (p_{oy_2}(t) - p_{oy_1}(t))\left(p^2_{oy_3}(t) - p^2_{oy_1}(t) + p^2_{ox_3}(t) - p^2_{ox_1}(t)\right) + \\ (p_{oy_1}(t) - p_{oy_3}(t))\left(p^2_{oy_2}(t) - p^2_{oy_1}(t) + p^2_{ox_2}(t) - p^2_{ox_1}(t)\right) \quad (4)$$

$$V(x,y) = (p_{ox_3}(t) - p_{ox_1}(t))(p_{oy_2}(t) - p_{oy_1}(t)) - 2(p_{ox_2}(t) - p_{ox_1}(t)) \\ (p_{oy_3}(t) - p_{oy_1}(t)) \quad (5)$$

the centre of the circle coordinates for:

$$\dot{p}_{OX} = \frac{U(x,y)}{V(x,y)} \qquad (6)$$

$$\dot{p}_{OY} = \frac{U(y,x)}{V(y,x)} \qquad (7)$$

the radius of the circle is:

$$R = \sqrt{\left(p^2{}_{OX}(t) - p^2{}_{ox_1}(t)\right) + \left(p^2{}_{OY}(t) - p^2{}_{oy_1}(t)\right)} \qquad (8)$$

When the number of targets is greater than three, we use the following algorithm to find the three target points that will determine the minimum circle (Fig. 1):

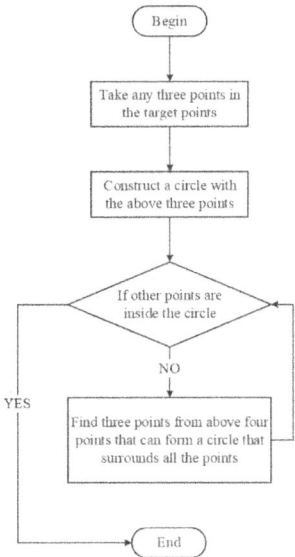

Fig. 1. Flowchart for determining the three target points that make up a circle when the number of targets is greater than three

When the three target points are determined the radius and centre coordinates of the smallest circle can be obtained by using the method (4)–(8) above. Since then, we can get the position information of the virtual target point as well as the velocity information by corresponding calculation in the process of rounding up. Based on the above problems, although the velocity of the virtual target is the same as the target's velocity. But given that the speed of the target is unknown to the roundup robot, the speed of the virtual target is likewise unknown. Therefore, a velocity observer v_i is designed for this. The position of the virtual target $\dot{p}_0$ can be calculated based on the position of the target during the roundup process, so the following controller can be created:

$$\begin{aligned} u_i^{I_1} &= \phi_i + k_1(\dot{P}_o - p_i) + v_i \\ \dot{v}_i &= \phi_i + k_2(\dot{P}_o - p_i) \\ \phi_i &= \sum_{j \in NB_i} \alpha(||p_{ij}||) \frac{p_{ij}}{||p_{ij}||} \end{aligned} \qquad (9)$$

where $k_1 > 0, k_2 > 0$, and the collision avoidance function is as follows:

$$\alpha(||p_{ij}||) = \frac{1}{||p_{ij}|| - d} - \frac{1}{\mu - d}, ||p_{ij}|| \in (d, \mu] \qquad (10)$$

where the exclusion term ϕ_i is used to avoid collisions between an intelligence and its neighbouring intelligences; the position information of the virtual target $\dot{p}_0$ can be obtained by the previously described calculations; attractors $k_1(\dot{P}_o - p_i)$ drive individual intelligences towards a virtual goal; the collision avoidance distance d needs to be greater than the radius of the smallest circle containing all the targets; μ is a parameter greater than the collision avoidance distance d. In order to better understand the different forces generated by the controller, Fig. 2 provides a graphical illustration of three intelligent agents and their targets.

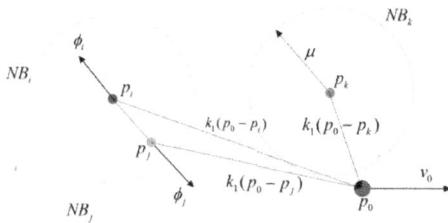

Fig. 2. Schematic diagram of the attraction and repulsion effects between three intelligent agents and the target

The strategy of the entire algorithm is as follows: during the hunting process, the center and radius of the smallest circle containing the target point are calculated in real time based on the position information of the hunting target. The position information of the center of the circle is the position information of the virtual target point, and the speed is the same as that of the target. Then the hunting robot hunts the target when the collision avoidance distance is greater than the radius of the center of the circle. When the virtual target point is enclosed in the formed enclosure and moves at the same speed as the virtual target point, since the collision avoidance distance is greater than the radius of the circle, a smaller enclosure centered at the virtual target point and with a radius slightly larger than the radius of the circle is formed inside the hunting robot. This enclosure contains all the circles where the target points are located, which completes the hunting of the target.

4 Experimental Simulation

In this section, six robots round up three robots as an example to verify the effectiveness of the proposed method. Given the initial positions of the six robots are: $p_i(0) = 10 + 7[\cos(i-1)/n*2\pi, \sin(i-1)/n*2\pi]^T, i = 1,2,3$, $p_i(0) = 10 + 9[\cos(i-1)/n*2\pi, \sin(i-1)/n*2\pi]^T, i = 4,5,6$, the initial positions of the three targets are : $p_1(0) = [25, 11]$, $p_2(0) = [24, 9.5]$, $p_3(0) = [25.5, 9.5]$, the velocity is $v_0 = [3, 1.5]$, the parameters of the controller is $k_1 = 5, k_2 = 4, \mu = 15$, next we give the simulation of the system under the action of the controller (9).

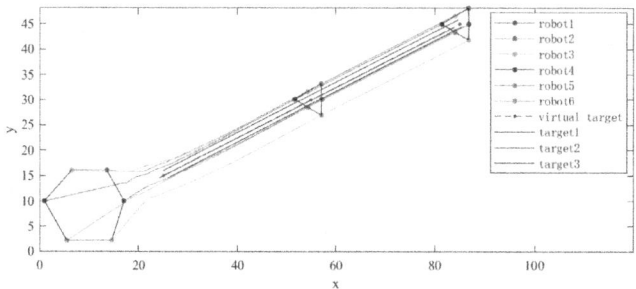

Fig. 3. Motion trajectories of multi-mobile robots in roundup control

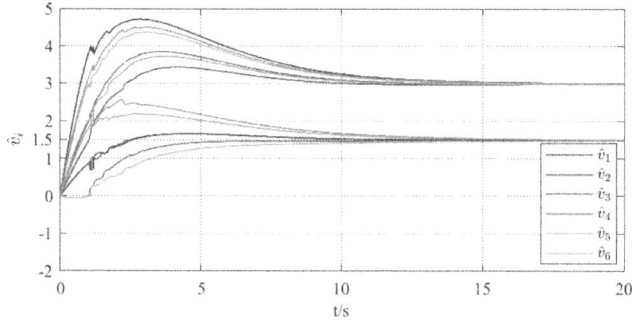

Fig. 4. Speed of fence robots

The trajectories of the robot and the targets during the rounding process are shown in Fig. 3, and the black polygons are the convex envelopes formed by the rounding robot, which eventually completed the rounding of all targets. Figure 4 shows the speed change curve of the trapping robot, and it can be seen that the robot finally reaches the same speed with the target. Figure 5 draws the minimum distance between the enclosing robot and the virtual target, and it can be seen that the distance is always greater than the collision avoidance distance. Figure 6 draws the position error of the virtual target, and it can be seen that it finally converges to the centre of the convex packet to complete the enclosure.

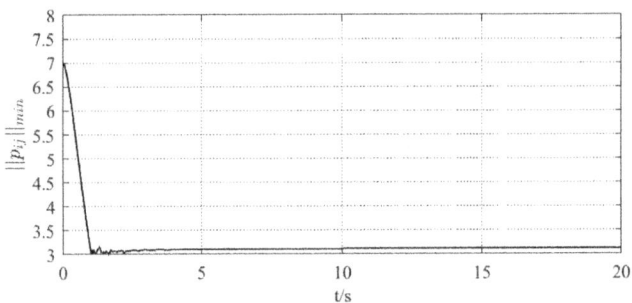

Fig. 5. Minimum distance between the roundup robot and the virtual target

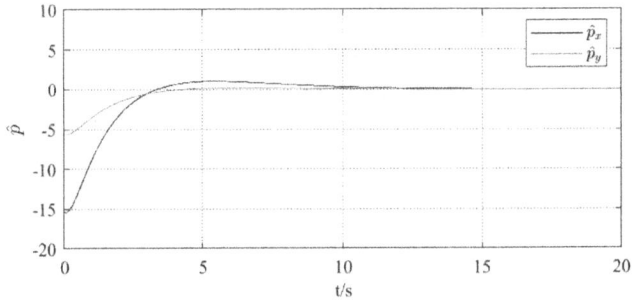

Fig. 6. Position error of virtual targets

In addition, we conducted multiple experiments with different quantities of targets, all of which are able to meet the requirements for trapping. The average trapping time for the ten experiments is shown in the table below (Table 1).

Table 1. Average capture time for different target numbers.

Number of targets	Number of experiments	average time/s
2	10	16.4
3	10	21.2
4	10	24.6

5 Conclusion

This paper investigates the hunting control problem of multiple mobile robot systems for multiple moving targets. During the encirclement process, the pursuer does not need to set up a formation in advance. Multiple targets are transformed into a virtual target point through the method of the smallest circle containing the targets. Simulation results verify that the controller can effectively estimate the unknown velocities of multiple moving targets, avoid obstacles between robots, and complete hunting control of multiple targets. In the future, we can consider the situation that there are still or even moving obstacles in the environment and the hunting strategy when the hunting robot breaks down in the process of hunting.

References

1. Finžgar, M., Podržaj, P.: Machine-vision-based human-oriented mobile robots: a review. J. Mech. Engineering/Strojniški Vestnik **63**(5) (2017)
2. Husni, N. L., Handayani, A. S., Nurmaini, S., Yani, I.: Cooperative searching strategy for swarm robot. In: 2017 International Conference on Electrical Engineering and Computer Science (ICECOS), pp. 92–97. IEEE (2017)
3. Queralta, J.P., et al.: Collaborative multi-robot search and rescue: planning, coordination, perception, and active vision. IEEE Access **8**, 191617–191643 (2020)
4. Habib, M.K., Baudoin, Y.: Mechanical mine clearance: development, applicability and difficulties. In: Using Robots in Hazardous Environments, pp. 299–326. Elsevier (2011)
5. Kurdi, H., AlDaood, M.F., Al-Megren, S., Aloboud, E., Aldawood, A.S., Youcef-Toumi, K.: Adaptive task allocation for multi-UAV systems based on bacteria foraging behaviour. Appl. Soft Comput. **83**, 105643 (2019)
6. Rizk, Y., Awad, M., Tunstel, E.W.: Cooperative heterogeneous multi-robot systems: a survey. ACM Comput. Surv. (CSUR) **52**(2), 1–31 (2019)
7. Lee, W., Lee, Y., Park, G., Hong, S., Kang, Y.: A whole-body rescue motion control with task-priority strategy for a rescue robot. Auton. Robot. **41**, 243–258 (2017)
8. Hong, Y., Zhai, C.: Dynamic coordination and distributed control design of multi-agent systems. Control Theory Appl. **28**(10), 1506–1512 (2011)
9. Zhen, Q., Wan, L., Li, Y., Jiang, D.: Formation control of a multi-AUVs system based on virtual structure and artificial potential field on SE (3). Ocean Eng. **253**, 111148 (2022)
10. Knorn, S., Chen, Z., Middleton, R.H.: Overview: collective control of multiagent systems. IEEE Trans. Control Netw. Syst. **3**(4), 334–347 (2015)
11. Lianghong, W.: Hunting in unknown environments with dynamic deforming obstacles by swarm robots. Int. J. Control Autom. **8**(11), 385–406 (2015)
12. Huang, T.Y., Xue-Bo, C., Wang-Bao, X.U., Zi-Wei, Z., Zhi-Yong, R.: A self-organizing cooperative hunting by swarm robotic systems based on loose-preference rule. Acta Automatica Sinica **39**(1), 57–68 (2013)
13. Wang, Y., Dong, L., Sun, C.: Cooperative control for multi-player pursuit-evasion games with reinforcement learning. Neurocomputing **412**, 101–114 (2020)
14. Zhilin, F., Yang, H., Yilin, H.: Reinforcement learning-based target rounding control for multi-intelligent body systems. J. Aeronaut. **44**(S1), 236–245 (2023)
15. Chen, Z., Zou, A.: Research on multi robot cooperative hunting based on biological heuristic neural network. Electr. Measur. Technol. **44**(10), 82–90 (2021)

Event-Triggered Secure Consensus Control of Nonlinear Multiagent Systems with hybrid DoS attacks

Yingxin Zhang$^{(\boxtimes)}$, Huixia Cui, and Senping Jia

Faculty of Robot Science and Engineering, Northeastern University,
110819 Shenyang, Liaoning, China
2202076@stu.neu.edu.cn

Abstract. This paper studies secure consensus control for leader-following multi-agent systems (MASs) facing random hybrid denial-of-service (DoS) attacks and an event-triggered strategy (ETS). Unlike previous work, it categorizes communication topologies into cases with and without spanning trees, considering more realistic scenarios. Topologies with spanning trees under weak DoS attacks are modeled as a semi-Markovian jump process with an uncertain transition matrix, while those without spanning trees under strong attacks are treated as separate modes. Using the integral mean value theorem, sampling time delay is incorporated into the Lyapunov-Krasovskii functional (LKF). Sufficient conditions for secure consensus are derived via Lyapunov stability theory, and a corollary addresses partial spanning tree cases. A simulation example validates the methods.

Keywords: Event-triggered Secure Consensus · Random hybrid DoS attacks · Semi-Markovian switching topologies · Generally uncertain transition matrix

1 Introduction

Over the past decade, coordination control of MASs has attracted significant attention and been applied in areas like unmanned vehicles, industrial IoT, and wireless sensor networks. This includes consensus, formation control, and flocking, with consensus control focusing on how agents reach agreements using local information and distributed protocols.

Information exchange among agents depends on distributed communication networks, which are vulnerable to cyber-attacks that disrupt connectivity and affect system consensus. DoS attacks are especially destructive, blocking transmission channels and causing data loss [1–3]. These attacks can be periodic, intermittent, or random [4]. Real-world attackers often use random intervals to increase success and conceal their actions, highlighting the need to study DoS attacks within a random framework. Many existing studies assume that the probabilities of DoS attacks are known in advance, which is unrealistic due to the hidden nature of attacks and the difficulty in accurately determining attack probabilities. To address this, we use a generally uncertain transition matrix to model the jump process of switching topologies. Additionally, most consensus conditions for leader-following MASs assume that each attacked topology contains a directed spanning tree rooted at the leader [5,6]. However, in practice, hybrid DoS attacks can prevent such spanning trees from existing. Therefore, this paper investigates cases with both spanning tree and non-spanning tree topologies under DoS attacks.

This article addresses the consensus control problem for nonlinear MASs under DoS attacks, the key contributions are as follows. Unlike previous work focusing only on topologies with spanning trees under DoS attacks [7,8], this paper considers comprehensive topologies with and without spanning trees caused by weak and strong hybrid attacks, enabling application to more complex scenarios. The sampling time delay is incorporated into the LKF using the integral mean value theorem, allowing for more sampling information and reducing the conservativeness of the secure consensus conditions. The generally uncertain semi-Markovian switching topologies are introduced, extending beyond traditional Markovian models [9,10], to account for weak attack effects.

2 Preliminaries

2.1 Notation

$\mathcal{R}^p$ denotes the set of p-dimensional real vectors, $\mathcal{R}^{p \times q}$ represents the set of $p \times q$ real matrices. The symbol $\|\cdot\|$ denotes the Euclidean norm, and $\otimes$ stands for the Kronecker product. $\mathfrak{L}_2([0, +\infty), \mathcal{R}^p)$ denotes the space of p-dimensional square summable vector functions. $\text{Sym}\{Z\} = Z + Z^T$. Furthermore, $\lambda_{\min}(Z)$ and $\lambda_{\max}(Z)$ represent the minimum and maximum eigenvalues of the matrix Z. I_p stands for the p-dimensional identity matrix. The symbol $\mathcal{E}$ denotes the mathematical expectation operator. The expression $e_\xi = [O_{Nn \times (\xi-1)Nn}, I_{Nn}, O_{Nn \times (7-\xi)Nn}]$ defines e_ξ for $\xi = 1, 2, \cdots, 7$, where O indicates the zero matrix.

2.2 Algebraic Graph Theory

The directed graph $\mathbb{G} = (\mathbb{V}, \mathbb{E}, \mathbb{A})$ models the connections among N followers, where $\mathbb{V} = \{v_1, v_2, \ldots, v_N\}$ is the set of nodes representing followers, and $\mathbb{E} \subseteq \mathbb{V} \times \mathbb{V}$ is the set of directed edges. The neighbor index set of node v_k is defined as $\mathbb{N}_k = \{l | (v_k, v_l) \in \mathbb{E}\}$. The adjacency matrix $\mathbb{A} = (a_{kl})_{N \times N}$ captures the weights of the edges, where $a_{kl} > 0$ if $(v_k, v_l) \in \mathbb{E}$, and $a_{kl} = 0$ otherwise. The Laplacian matrix $\mathbb{L} = (l_{kl})_{N \times N}$ is defined by $l_{kk} = \sum_{l \in \mathbb{N}_k} a_{kl}$ and $l_{kl} = -a_{kl}$ for $k \neq l$.

2.3 Assumption

Assumption 1. *[11] All agents synchronize their sampling periods using the same clock, and the sampling period is denoted by h.*
Assumption 2. *The communication topology switches only at sampling instants.*

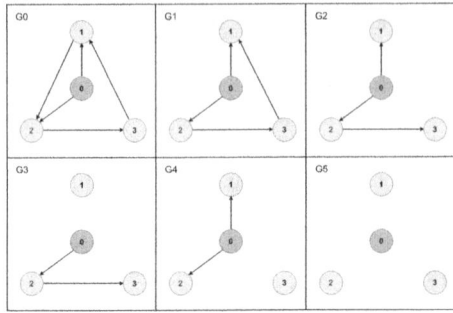

Fig. 1. A topology without attacks and five topologies under attacks.

2.4 Properties of the Switching Topologies

To simplify analysis, communication topologies in MASs are categorized as follows. SPT case, which the topology contains a directed spanning tree with all agents, rooted at the leader. PSPT case, which the topology contains a directed spanning tree with a subset of agents, rooted at the leader. NSPT case, which the topology lacks a directed spanning tree rooted at the leader, including fully blocked communication, termed the WORST case. The set of SPT is denoted as $\tilde{\mathbb{G}}_t$, and the collective set of PSPT and NSPT is represented by $\tilde{\mathbb{G}}_{ut}$.

Figure 1 illustrates the attack-free topology G0 and five topologies under DoS attacks for a MAS, with the leader labeled as 0 and followers as $1, 2, 3$. Topologies G1 and G2 result from predictable weak DoS attacks, switching stochastically among G0, G1, and G2 as a semi-Markovian jump process. Unpredictable strong DoS attacks lead to topologies G3 to G5, analyzed as additional modes.

For any $t > s > t_0$, let $\mathcal{T}_t(t, s)$ be the total time of $\tilde{\mathbb{G}}_t$ active interval on $[s, t)$, and $\mathcal{T}_{ut}(t, s)$ be the total time of $\tilde{\mathbb{G}}_{ut}$ active interval on $[s, t)$, furthermore, the $\mathcal{T}_{pt}$ represents the total time of PSPT time in $\mathcal{T}_{ut}$. The cumulative number of occurrences when $\tilde{\mathbb{G}}_{ut}$ is active is denoted as $\mathcal{N}_{ut}(t, s)$. The set of followers not included in the spanning tree is $\mathcal{A}_{ut}$, and the set of followers within the spanning tree is $\mathcal{A}_t$.

$\{r_t, t \geq 0\}$ is a right-continuous semi-Markovian process. Its value is the finite state space $\mathcal{S} = \{1, 2, \cdots, \mathcal{M}\}$ with $\Pi = (\pi_{ij})$ $(i, j \in \mathcal{S})$.

$$Pr\{r(t+\Delta) = j \mid r(t) = i\} = \begin{cases} \pi_{ij}(\Delta)\Delta + o(\Delta), & i \neq j \\ 1 + \pi_{ii}(\Delta)\Delta + o(\Delta), & i = j \end{cases}$$

where $\Delta > 0$, and $\lim_{\Delta \to 0}(o(\Delta)/\Delta) = 0$. $\pi_{ij}(\Delta) \geq 0$ $(i \neq j)$ is the transition rate from mode i at time t to mode j at time $t + \Delta$, and $\pi_{ii}(\Delta) = -\sum_{j=1, j \neq i}^{\mathcal{M}} \pi_{ij}(\Delta)$.

Note that in this paper, the transition rates (TRs) are assumed to be generally uncertain, encompassing both completely unknown case and uncertain case. For the uncertain case, there are $\underline{\pi}_{ij} \leq \pi_{ij}(\Delta) \leq \overline{\pi}_{ij}$, rewrite as $\pi_{ij}(\Delta) = \tilde{\pi}_{ij} + \delta\tilde{\pi}_{ij}(\Delta)$ with $\tilde{\pi}_{ij} = \frac{1}{2}(\underline{\pi}_{ij} + \overline{\pi}_{ij})$, $\|\delta\tilde{\pi}_{ij}(\Delta)\| \leq \lambda_{ij}$ and $\lambda_{ij} = \frac{1}{2}(\overline{\pi}_{ij} - \underline{\pi}_{ij})$, where $\delta \in \mathcal{R}$, $\tilde{\pi}_{ij}$ and λ_{ij} are known values. "?" denotes the completely unknown case. Therefore, the transition rate matrix (TRM) Π is represented as follows:

$$\Pi = \begin{pmatrix} \tilde{\pi}_{11} + \delta\tilde{\pi}_{11}(\Delta) & ? & \tilde{\pi}_{13} + \delta\tilde{\pi}_{13}(\Delta) & \cdots & ? \\ \tilde{\pi}_{21} + \delta\tilde{\pi}_{21}(\Delta) & ? & ? & \cdots & ? \\ \vdots & \vdots & \vdots & \ddots & \vdots \\ \tilde{\pi}_{\mathcal{M}1} + \delta\tilde{\pi}_{\mathcal{M}1}(\Delta) & ? & ? & \cdots & ? \end{pmatrix} \quad (1)$$

For $\forall\, i \in \mathcal{S}$, we denote $\mathcal{S}^i = \mathcal{S}_k^i \bigcup \mathcal{S}_{uk}^i$, where $\mathcal{S}_k^i = \{j : \pi_{ij}(\Delta) \neq \text{"?"}, j \in \mathcal{S}\}$ and $\mathcal{S}_{uk}^i = \{j : \pi_{ij}(\Delta) = \text{"?"}, j \in \mathcal{S}\}$. $\mathcal{S}_t = \mathcal{S}$ is the set of all possible modes corresponding to the $\tilde{\mathbb{G}}_t$, while the $\mathcal{S}_{ut} = \{j : j > \mathcal{M}\}$ corresponding to the $\tilde{\mathbb{G}}_{ut}$. Additionally, denote the set of all PSPT cases as $\mathcal{S}_{pt}$.

2.5 Problem Formulation

The following nonlinear MASs with N following agents are considered in this paper. The dynamics of the kth agent are given as

$$\begin{cases} \dot{x}_k(t) = Ax_k(t) + Bf(x_k(t)) + Cu_k(t), \\ \psi_k(t) = Dx_k(t), k = 0, 1, 2, \cdots, N. \end{cases} \quad (2)$$

where $x_k(t) \in \mathcal{R}^n$, $u_k(t) \in \mathcal{R}^n$, and $\psi_k(t) \in \mathcal{R}^n$ represent the state, control input, and output signal of the k-th agent, respectively. The leader agent is indexed as $k = 0$, and its control input is defined as $u_0(t) = 0$. The matrices $A \in \mathcal{R}^{n \times n}$, $B \in \mathcal{R}^{n \times n}$, $C \in \mathcal{R}^{n \times 1}$, and $D \in \mathcal{R}^{n \times n}$ are constant. The function $f(x_k(t)) : \mathcal{R}^n \times [0, +\infty) \to \mathcal{R}^n$ is a nonlinear, continuously differentiable function that satisfies the Assumption 1 in [12].

The initial conditions of system (2) are given as follows

$$x_k(s) = \phi_{k0}(s) \in \mathcal{C}([-\varphi, 0], \mathcal{R}^n), \quad k = 0, 1, 2, \ldots, N,$$

where $\mathcal{C}([-\varphi, 0], \mathcal{R}^n)$ denotes the set of continuous functions from the interval $[-\varphi, 0]$ to $\mathcal{R}^n$.

To save network communication resources, a distributed ETS is equipped with each agent. If the ETS is satisfied and $t_p^k + \beta h \in (\mathcal{T}_t \cup \mathcal{T}_{pt})$, the agent $k \in \mathcal{A}_t$ would send its sampled-data to its neighbors. Define the next transmission instant t_{p+1}^k for the kth agent

$$\begin{aligned} t_{p+1}^k = t_p^k + \min_{\beta \geq 1}\{\beta h | y_k^T(t_p^k + \beta h)\Omega_i y_k(t_p^k + \beta h) \\ > \theta_k w_k^T(t_p^k + \beta h)\Omega_i w_k(t_p^k + \beta h)\}, \end{aligned} \quad (3)$$

where $\theta_k > 0$ represents the threshold parameter. The event-triggered matrix Ω_i is defined such that $\Omega_i \in \mathcal{R}^{n \times n}$ and $\Omega_i > 0$. The t_p^k indicates the pth event-triggered instant for agent k. The current sampling instant is given by $t_p^k + \beta h$, with $y_k(t_p^k + \beta h) = x_k(t_p^k) - x_k(t_p^k + \beta h)$ and $w_k(t_p^k + \beta h) = \sum_{l \in \mathbb{N}_k} a_{kl}[x_k(t_p^k) - x_l(t_{\hat{p}}^l)] + b_k[x_k(t_p^k) - x_0(t_p^k + \beta h)]$, where $t_{\hat{p}}^l = \max\{t \mid t \in \{t_p^l, p = 1, 2, 3, \cdots\}, t \leq t_p^l + \beta h\}$. Zeno behavior does not occur because the minimum event interval time $t_{p+1}^k - t_p^k \geq h$.

In the absence of PSPT case, consider the MASs consensus protocol, for $k = 1, 2, \cdots, N$,

$$u_k(t) = -K(r_t)\left\{\sum_{l \in \mathbb{N}_k} a_{kl}(r_t)[x_k(t_p^k) - x_l(t_{\hat{p}}^l)] + b_k(r_t)[x_k(t_p^k) - x_0(qh)]\right\}, \quad (4)$$

where $t \in [t_p^k, t_{p+1}^k) \cap [qh, (q+1)h)$, and q is an integer. The matrix $K(r_t)$ represents the feedback gain. The term $a_{kl}(r_t)$ is an element of the adjacency matrix $\mathbb{A}(r_t)$. For agent $k \in \mathcal{A}_{ut}$, we set $a_{kl}(r_t) = 0$. The term $b_k(r_t)$ is an element of the leader adjacency matrix, where $b_k(r_t) > 0$ indicates that the k-th agent can receive information from the leader, otherwise $b_k(r_t) = 0$. Define $y_k(qh) = x_k(t_p^k) - x_k(qh)$, $\beta h = qh - x_p^k$, and $v_k(qh) = x_k(qh) - x_0(qh)$. For $t \in [qh, (q+1)h)$, the consensus protocol can be expressed as follows

$$u_k(t) = -K(r_t)\{\sum_{l \in \mathbb{N}_k} a_{kl}(r_t)[y_k(qh) - y_l(qh) + v_k(qh)$$
$$- v_l(qh)] + b_k(r_t)[y_k(qh) + v_k(qh)]\}, \quad k = 1, 2, \cdots, N. \quad (5)$$

According to Eq. (2) and Eq. (5), one gets the following error dynamics system

$$\begin{cases} \dot{v}_k(t) = Av_k(t) + Bg(v_k(t)) - CK(r_t)\{\sum_{l \in \mathbb{N}_k} a_{kl}(r_t)[y_k(qh) - y_l(qh) \\ \quad + v_k(qh) - v_l(qh)] + b_k(r_t)[y_k(qh) + v_k(qh)]\}, \\ h_k(t) = Dv_k(t), t \in [qh, (q+1)h), \quad k = 1, 2, \cdots, N. \end{cases} \quad (6)$$

where $v_k(t) = x_k(t) - x_0(t)$, $\tilde{f}_{nl}(v_k(t)) = f(x_k(t)) - f(x_0(t))$, $h_k(t) = \psi_k(t) - \psi_0(t)$.

Assume $v(t) = [v_1^T(t), v_2^T(t), \cdots, v_N^T(t)]^T$, $y(t) = [y_1^T(t), y_2^T(t), \cdots, y_N^T(t)]^T$, $h(t) = [h_1^T(t), h_2^T(t), \cdots, h_N^T(t)]^T$, $w(t) = [w_1^T(t), w_2^T(t), \cdots, w_N^T(t)]^T$, $\tilde{F}_{nl}(v(t)) = [\tilde{f}_{nl}^T(v_1(t)), \tilde{f}_{nl}^T(v_2(t)), \cdots, \tilde{f}_{nl}^T(v_N(t))]^T$, $G_i = \mathbb{L}_i + \mathbb{B}_i$, $\mathbb{B}_i = \text{diag}\{b_{1,i}, b_{2,i}, \cdots, b_{N,i}\}$. Each value of r_t is represented by i ($i \in (\mathcal{S}_t \cup \mathcal{S}_{ut})$). According to Eq. (6), one gets

$$\begin{cases} \dot{v}(t) = (I_N \otimes A)v(t) + (I_N \otimes B)\tilde{F}_{nl}(v(t)) \\ \qquad = -(G_i \otimes CK_i)(v(t - d(t)) + y(t - d(t))), \\ h(t) = (I_N \otimes D)v(t), t \in [qh, (q+1)h), \end{cases} \quad (7)$$

where $t - d(t) = qh$, $0 \leq d(t) = t - qh < h$, and $d(t)$ is a piecewise linear function with $\dot{d}(t) = 1$ for $t \neq qh$. The initial state condition for $v(t)$ is specified as $v(0) = [v_1^T(0), v_2^T(0), \cdots, v_N^T(0)]^T$.

Definition 1. *MAS is said to have average non-tree ratio $\varpi \in (0,1)$, if there exists $\mathcal{T}_0 \geq 0$ such that*

$$\mathcal{T}_{ut}(t,s) \leq \varpi(t-s) + \mathcal{T}_0, \text{ for } t > s \geq t_0, \tag{8}$$

where $\mathcal{T}_0$ is the elasticity number. From Eq. (8), it indicates $\mathcal{T}_t(t,s) = t - s - \mathcal{T}_{ut}(t,s) \geq (1-\varpi)(t-s) - \mathcal{T}_0$.

3 Main results

In this section, we assume that DoS attacks result in two types of communication topologies, the SPT and the NSPT. The SPT cases are modeled as a generally uncertain semi-Markovian jump process, while the NSPT cases are conservatively analyzed as the WORST case.

Theorem 1. *Under Assumptions 1 and 2, scale factor $m > 0, \alpha_k > 0, \alpha_{uk} > 0, \alpha_t = \min\{\alpha_k, \alpha_{uk}\}, \alpha_{ut} > 0$, the secure consensus control of nonlinear MASs (2) under DoS attack is achieved if there are invertible symmetric matrix $F \in \mathcal{R}^{n \times n}$, symmetric positive definite matrices $\tilde{P}_i \in \mathcal{R}^{n \times n}, \tilde{\Omega}_i \in \mathcal{R}^{n \times n}, \tilde{M} \in \mathcal{R}^{n \times n}, \tilde{T}_{ij} \in \mathcal{R}^{nN \times nN}, \tilde{Z}_{ij} \in \mathcal{R}^{n \times n}$ and positive diagonal matrices $\tilde{R} \in \mathcal{R}^{n \times n}$, such that for any $i \in \mathcal{S}$, the following LMIs are satisfied.*

$$\alpha_t - \varpi(\alpha_t + \alpha_{ut}) < 0. \tag{9}$$

If $i \in \mathcal{S}_t, i \in \mathcal{S}_k^i, \forall l \in \mathcal{S}_{uk}^i, \mathcal{S}_k^i = \{k_1^i, \ldots, k_{mi1}^i\}$,

$$\begin{bmatrix} \tilde{\Xi}_i + \alpha_k \tilde{\nu}_i + \tilde{\eta}_{k,i,l} & \tilde{\Gamma}_{k,i}^{21} \\ * & \tilde{\Gamma}_{k,i}^{22} \end{bmatrix} < 0. \tag{10}$$

If $i \in \mathcal{S}_t, i \in \mathcal{S}_{uk}^i, \forall l \in \mathcal{S}_{uk}^i, l \neq i, \mathcal{S}_k^i = \{k_1^i, \ldots, k_{mi2}^i\}$,

$$\tilde{P}_i - \tilde{P}_l \geq 0, \begin{bmatrix} \tilde{\Xi}_i + \alpha_{uk} \tilde{\nu}_i + \tilde{\eta}_{uk,i,l} & \tilde{\Gamma}_{uk,i}^{21} \\ * & \tilde{\Gamma}_{uk,i}^{22} \end{bmatrix} < 0. \tag{11}$$

If $i \in \mathcal{S}_{ut}$,

$$\tilde{\Xi}_i - \alpha_{ut} \tilde{\nu}_i < 0. \tag{12}$$

Where

$$\Xi_i = \tilde{\Xi}_i + \sum_{j=1}^{\mathcal{M}} \pi_{ij}(\Delta) e_1^T (I_N \otimes \tilde{P}_j) e_1, \tilde{\Gamma}_k^{21} = e_1^T \left[\tilde{P}_{k_1^i} - \tilde{P}_l, \ldots, \tilde{P}_{k_{m_{i1}}^i} - \tilde{P}_l \right],$$

$$\tilde{\eta}_{k,i,l} = e_1^T \left[\sum_{j\in\mathcal{S}_k^i} \left(\frac{\lambda_{ij}^2}{4}\tilde{T}_{ij} + \tilde{\pi}_{ij}\left(\tilde{P}_j - \tilde{P}_l\right)\right)\right] e_1, \tilde{\Gamma}_{uk}^{21} = e_1^T \left[\tilde{P}_{k_1^i} - \tilde{P}_l, \ldots,\right.$$
$$\left.\tilde{P}_{k_{mi2}^i} - \tilde{P}_l\right],$$
$$\eta_{uk,i,l} = e_1^T \left[\sum_{j\in\mathcal{S}_k^i} \left(\frac{\lambda_{ij}^2}{4}\tilde{Z}_{ij} + \tilde{\pi}_{ij}\left(\tilde{P}_j - \tilde{P}_l\right)\right)\right] e_1, \tilde{\Gamma}_k^{22} = \mathrm{diag}\left\{-\tilde{T}_{ik_1^i}, \ldots,\right.$$
$$\left.-\tilde{T}_{ik_{mi1}^i}\right\},$$
$$\tilde{\Gamma}_{uk}^{22} = \mathrm{diag}\left\{-\tilde{Z}_{ik_1^i}, \ldots, -\tilde{Z}_{ik_{i2}^i}^i\right\}, \Upsilon = I_{8N} \otimes \tilde{F},$$

$$\varepsilon(t) = \left[v^T(t), \tilde{F}_{nl}^T(v(t)), \dot{v}^T(t), v^T(qh), y^T(qh), v^T(\xi), \tilde{F}_{nl}^T(v(\xi)), \dot{v}^T(\xi)\right]^T,$$
$$\varepsilon_{12}(t) = \left[v^T(t), \tilde{F}_{nl}^T(v(t))\right]^T, \varepsilon_{67}(\xi) = \left[v^T(\xi), \tilde{F}_{nl}^T(v(\xi))\right]^T, \tilde{\varepsilon}(t) = \Upsilon\varepsilon(t),$$

$$E_1 = \mathrm{diag}\{e_1^+ e_1^-, \cdots, e_N^+ e_N^-\}, E_2 = \mathrm{diag}\{\frac{e_1^+ + e_1^-}{2}, \cdots, \frac{e_N^+ + e_N^-}{2}\}, K_i = \tilde{K}_i \tilde{F}.$$

If $i \in \mathcal{S}_t$,

$$\tilde{\Xi}_i = \mathrm{Sym}\{e_1^T(I_N \otimes \tilde{P}_i)e_3\} + h^2 e_3^T(I_N \otimes \tilde{M})e_3 - \frac{\pi^2}{4}(e_1 - e_4)^T(I_N \otimes \tilde{M})(e_1 - e_4)$$
$$+ \mathrm{Sym}\{(e_1 + me_3)^T[-(I_N \otimes F)e_3 + (I_N \otimes AF)e_1 + (I_N \otimes BF)e_2$$
$$- (G_i \otimes C\tilde{K}_i)(e_4 + e_5)]\} - e_1^T(I_N \otimes \tilde{R}E_1)e_1 + \mathrm{Sym}\{e_1^T(I_N \otimes \tilde{R}E_2)e_2\}$$
$$- e_2^T(I_N \otimes \tilde{R})e_2 - e_5^T(I_N \otimes \tilde{\Omega})e_5 + (e_4 + e_5)^T(G_i^T \Theta G_i \otimes \tilde{\Omega}_i)(e_4 + e_5),$$

$$\tilde{\nu}_i = e_1^T(I_N \otimes \tilde{P}_i)e_1 + h^3 e_8^T(I_N \otimes \tilde{M})e_8 - \frac{\pi^2}{4}h(e_6 - e_4)^T(I_N \otimes \tilde{M})(e_6 - e_4)$$
$$+ \mathrm{Sym}\{(e_6 + me_8)^T[-(I_N \otimes F)e_8 + (I_N \otimes AF)e_6 + (I_N \otimes BF)e_7$$
$$- (G_i \otimes C\tilde{K}_i)(e_4 + e_5)]\} - e_6^T(I_N \otimes \tilde{R}E_1)e_6 + \mathrm{Sym}\{e_6^T(I_N \otimes \tilde{R}E_2)e_7\}$$
$$- e_7^T(I_N \otimes \tilde{R})e_7.$$

If $i \in \mathcal{S}_{ut}$,

$$\tilde{\Xi}_i = \mathrm{Sym}\{e_1^T(I_N \otimes \tilde{P}_i)e_3\} + h^2 e_3^T(I_N \otimes \tilde{M})e_3 + \mathrm{Sym}\{(e_1 + me_3)^T[-(I_N \otimes F)e_3$$
$$+ (I_N \otimes AF)e_1 + (I_N \otimes BF)e_2 - (G_i \otimes C\tilde{K}_i)(e_4 + e_5)]\}$$
$$- e_1^T(I_N \otimes \tilde{R}E_1)e_1 + \mathrm{Sym}\{e_1^T(I_N \otimes \tilde{R}E_2)e_2\} + e_2^T(I_N \otimes \tilde{R})e_2,$$

$$\tilde{\nu}_i = e_1^T(I_N \otimes \tilde{P}_i)e_1.$$

Proof. Choose the LKF for nonlinear MASs with generally uncertain switching topologies as follows.

$$V(t,i) = \begin{cases} V_1(t,i) + V_2(t,i), & \text{if } i \in \mathcal{S}_t, \\ V_1(t,i), & \text{if } i \in \mathcal{S}_{ut}, \end{cases} \quad (13)$$

$$V_1(t,i) = v^T(t)(I_N \otimes P_i)v(t), \quad (14)$$

$$V_2(t,i) = h^2 \int_{qh}^{t} \dot{v}^T(s)(I_N \otimes M)\dot{v}(s)ds \quad (15)$$

$$-\frac{\pi^2}{4}\int_{qh}^{t}(v(s)-v(qh))^T(I_N \otimes M)(v(s)-v(qh))ds. \quad (16)$$

where $t \in [qh, (q+1)h)$, $\tilde{F} = F^{-1}$, $P_i = \tilde{F}^T\tilde{P}_i\tilde{F}$, $\Omega_i = \tilde{F}^T\tilde{\Omega}_i\tilde{F}$, $M = \tilde{F}^T\tilde{M}\tilde{F}$, $R = \tilde{F}^T\tilde{R}\tilde{F}$.

From Lemmma 2 in [13] one gets $V_2(t,i) \geq 0$. $\mathcal{L}$ is defined as the weak infinitesimal operator. Then, calculating the time derivative of $V(t,i)$, one has

$$\mathcal{L}V_1(t,i) = 2v^T(t)(I_N \otimes P_i)\dot{v}(t) + \sum_{j=1}^{\mathcal{M}}\pi_{ij}v^T(t)(I_N \otimes P_j)v(t), \quad (17)$$

$$\mathcal{L}V_2(t,i) = h^2\dot{v}^T(t)(I_N \otimes M)\dot{v}(t) - \frac{\pi^2}{4}(v(t)-v(qh))^T(I_N \otimes M)(v(t)-v(qh)). \quad (18)$$

From Assumption 4, one gets

$$\varepsilon_{12}^T(t)\begin{bmatrix} -I_N \otimes RE_1 & I_N \otimes RE_2 \\ * & -I_N \otimes R \end{bmatrix}\varepsilon_{12}(t) \geq 0, \quad (19)$$

For invertible symmetric matrix $\tilde{F}$ and Eq. (7), one gets

$$2[v^T(t)(I_N \otimes \tilde{F}) + m\dot{v}^T(t)(I_N \otimes \tilde{F})][-\dot{v}(t) + (I_N \otimes A)v(t) + (I_N \otimes B)\tilde{F}_{nl}(v(t)) - (G_i \otimes CK_i)(v(t-d(t)) - y(t-d(t)))] = 0. \quad (20)$$

Assuming $K_i = \tilde{K}_i\tilde{F}$, one gets

$$\text{Sym}\{[v^T(t) + m\dot{v}^T(t)][-(I_N \otimes \tilde{F})\dot{v}(t) + (I_N \otimes \tilde{F}A)v(t) + (I_N \otimes \tilde{F}B)\tilde{F}_{nl}(v(t)) - (G_i \otimes \tilde{F}C\tilde{K}_i\tilde{F})(v(t-d(t)) - y(t-d(t)))]\} = 0. \quad (21)$$

From Eq. (3), we get the following formula

$$\sum_{k=1}^{N}y_k(qh)\Omega_i y_k(qh) = y^T(qh)(I_N \otimes \Omega_i)y(qh)$$

$$\leq w^T(qh)(\Theta \otimes \Omega_i)w(qh) = \sum_{k=1}^{N}\theta_k w_k^T(qh)\Omega_i w_k(qh), \quad (22)$$

where $\Theta = \text{diag}\{\theta_1, \theta_2, \ldots, \theta_N\}$, $w(qh) = (G_i \otimes I_n)v(qh) + (G_i \otimes I_n)y(qh)$, $y(qh) = [y_1^T(qh), y_2^T(qh), \cdots, y_N^T(qh)]^T$, $w(qh) = [w_1^T(qh), w_2^T(qh), \cdots, w_N^T(qh)]^T$. One gets

$$y^T(t-d(t))(I_N \otimes \Omega_i)y(t-d(t)) \leq [v^T(t-d(t)) \\ + y^T(t-d(t))](G_i^T \Theta G_i \otimes \Omega_i)[v(t-d(t)) + y(t-d(t))] \quad (23)$$

mean value theorem, $\xi \in [qh, (q+1)h)$, one obtains

$$V_2(t, i) \quad (24)$$

$$\leq h^3 \dot{v}^T(\xi)(I_N \otimes M)\dot{v}(\xi) - \frac{\pi^2}{4}h(v(\xi) - v(qh))^T(I_N \otimes M)(v(\xi) - v(qh)) \quad (25)$$

for $t = \xi$, one gets corresponding Eq. (19) and (21), then we have

$$V(t, i) \leq \tilde{\varepsilon}^T(t)\tilde{\nu}_i\tilde{\varepsilon}(t) \quad (26)$$

When $t \in [qh, (q+1)h) \subseteq \mathcal{T}_t$, $i \in \mathcal{S}_t$. Then combining Eq. (13), (17), (18), (19), (21), (23) and (26), we can infer that $\mathcal{E}\{\mathcal{L}V(t,i)\} \leq \mathcal{E}\{\tilde{\varepsilon}^T(t)\Xi_i\tilde{\varepsilon}(t)\} \leq -\alpha_t \mathcal{E}\{\tilde{\varepsilon}^T(t)\tilde{\nu}_i\tilde{\varepsilon}(t)\} \leq -\alpha_t \mathcal{E}\{V(t,i)\} < 0$, if $\mathcal{E}\{\tilde{\varepsilon}^T(t)(\Xi_i + \alpha_t\tilde{\nu}_i)\tilde{\varepsilon}(t)\} \leq 0$ holds, by using the comparison principle, one gets

$$\mathcal{E}\{V(t,i)\} \leq e^{-\alpha_t(t-qh)}\mathcal{E}\{V(qh,i)\} \quad (27)$$

When $t \in [qh, (q+1)h) \subseteq \mathcal{T}_{ut}, i \in \mathcal{S}_{ut}$. Then combining Eq. (13), (17), (19), (21), (26), we can infer that $\mathcal{E}\{\mathcal{L}V(t,i)\} \leq \mathcal{E}\{\tilde{\varepsilon}^T(t)\tilde{\Xi}_i\tilde{\varepsilon}(t)\} \leq \alpha_{ut}\mathcal{E}\{\tilde{\varepsilon}^T(t)\tilde{\nu}_i\tilde{\varepsilon}(t)\} = \alpha_{ut}\mathcal{E}\{V(t,i)\}$, if $\mathcal{E}\{\tilde{\varepsilon}^T(t)(\tilde{\Xi}_i - \alpha_{ut}\tilde{\nu}_i)\tilde{\varepsilon}(t)\} \leq 0$ holds, by using the comparison principle, one gets

$$\mathcal{E}\{V(t,i)\} \leq e^{\alpha_{ut}(t-qh)}\mathcal{E}\{V(qh,i)\} \quad (28)$$

To facilitate analysis, denote $\mathcal{E}\{V(t)\} = \mathcal{E}\{V(t,i)\}$, then for $t \in [qh, (q+1)h)$, together with Eq. (27) and (28), one can get

$$\mathcal{E}\{V(t)\} \leq \left(\prod_{n=1}^{q}\lambda_n\right)e^{-\alpha_t \mathcal{T}_t + \alpha_{ut}\mathcal{T}_{ut}}\mathcal{E}\{V(0)\}, \quad (29)$$

where $\lambda_n > 0, n = 1, 2, \ldots, q$, if no mode switching occurs at the sampling point qh, the $\lambda_n = 1$. If $nh \in \mathcal{T}_t$, then $\alpha_n = -\alpha_t$, else $\alpha_n = \alpha_{ut}$.

From Eq. (1), one gets $-\alpha_t \mathcal{T}_t + \alpha_{ut}\mathcal{T}_{ut} \leq -\alpha_t t + (\alpha_t + \alpha_{ut})\varpi t + \mathcal{T}_0$, then we have

$$\mathcal{E}\|V(t)\| \leq \tilde{\omega}e^{-\tilde{\eta}t}\mathcal{E}\|V(0)\|, \quad (30)$$

where $\tilde{\omega} = \left(\prod_{n=1}^{q}\lambda_n\right)e^{\mathcal{T}_0}$, and from Eq. (9), $\tilde{\eta} = \alpha_t - \varpi(\alpha_t + \alpha_{ut}) < 0$.

From Eq. (30), one get

$$\mathcal{E}\|v(t)\|^2 \leq \omega e^{-\eta t}\mathcal{E}\|v(0)\|^2 \quad (31)$$

where $\omega = \tilde{\omega}\left[\dfrac{\min_{i \in S}\{\lambda_{max}(P_i)\}}{\max_{i \in S}\{\lambda_{min}(P_i)\}}\right], \eta = \tilde{\eta}$.

That is to say, the mean-square leader-following consensus of MASs under DoS attacks can be achieved. Next, we divide the proof into two cases to discuss the generally uncertain TRs in the $\bar{\varepsilon}^T(t)\Xi_i\bar{\varepsilon}(t)$. Denote $\rho_k^i = \sum_{j \in \mathcal{S}_k^i} \pi_{ij}(\Delta)$, $\tilde{P}_i^I = I_N \otimes \tilde{P}_i$.

Case 1. $i \in \mathcal{S}_t, i \in \mathcal{S}_k^i$. In this case, we have

$$\sum_{j \in \mathcal{S}} \pi_{ij}(\Delta)\tilde{P}_j^I = \sum_{j \in \mathcal{S}_k^i} \pi_{ij}(\Delta)\tilde{P}_j^I - \rho_k^i \sum_{j \in \mathcal{S}_{uk}^i} \frac{\pi_{ij}(\Delta)}{-\rho_k^i}\tilde{P}_j^I \tag{32}$$

Then we obtain

$$\Xi_i + \alpha_k \tilde{\nu}_i = -\sum_{l \in \mathcal{S}_{uk}^i} \frac{\pi_{il}(\Delta)}{\rho_k^i}\left[\tilde{\Xi}_i + \alpha_k \tilde{\nu}_i + e_1^T\big(\sum_{j \in \mathcal{S}_k^i} \pi_{ij}(\Delta)(\tilde{P}_j^I - \tilde{P}_l^I)\big)e_1\right],$$

and thus, $\Xi_i + \alpha_k \tilde{\nu}_i < 0$ holds if for any $l \in \mathcal{S}_{uk}^i$, which is

$$\tilde{\Xi}_i + \alpha_k \tilde{\nu}_i + e_1^T \left(\sum_{j \in \mathcal{S}_k^i} \pi_{ij}(\Delta)\left(\tilde{P}_j^I - \tilde{P}_l^I\right)\right)e_1 < 0. \tag{33}$$

Note that $\sum_{j \in \mathcal{S}_k^i} \pi_{ij}(\Delta)\left(\tilde{P}_j^I - \tilde{P}_l^I\right) = \sum_{j \in \mathcal{S}_k^i} (\tilde{\pi}_{ij} + \delta\tilde{\pi}_{ij}(\Delta))\left(\tilde{P}_j^I - \tilde{P}_l^I\right)$. Then, from Lemma 3 in [14], we know that for any $\tilde{T}_{ij} > 0, \sum_{j \in \mathcal{S}_k^i} \delta\tilde{\pi}_{ij}(\Delta)$ $\left(\tilde{P}_j^I - \tilde{P}_l^I\right) \leq \sum_{j \in \mathcal{S}_k^i} \left[(\lambda_{ij}^2/4)\tilde{T}_{ij} + \left(\tilde{P}_j^I - \tilde{P}_l^I\right)\tilde{T}_{ij}^{-1}\left(\tilde{P}_j^I - \tilde{P}_l^I\right)\right]$. Hence, with the aid of Schur complement, $\Xi_i + \alpha_k \tilde{\nu}_i < 0$ is satisfied if for any $l \in \mathcal{S}_{uk}^i$, Eq. (10) holds.

Case 2. $i \in \mathcal{S}_t, i \in \mathcal{S}_{uk}^i$. We rewrite $\sum_{j \in \mathcal{S}} \pi_{ij}(\Delta)\tilde{P}_j^I$ as

$$\sum_{j \in \mathcal{S}} \pi_{ij}(\Delta)\tilde{P}_j^I = \sum_{j \in \mathcal{S}_k^i} \pi_{ij}(\Delta)\tilde{P}_j^I + \pi_{ii}(\Delta)\tilde{P}_i^I - (\pi_{ii}(\Delta)$$
$$+\rho_k^i)\sum_{j \in \mathcal{S}_{uk}^i, j \neq i} \frac{\pi_{ij}(\Delta)\tilde{P}_j^I}{-\pi_{ii}(\Delta) - \rho_k^i}. \tag{34}$$

If $\tilde{P}_i - \tilde{P}_l \geq 0$, then similar to Case 1 above, we can derive that for any $l \in \mathcal{S}_{uk}^i, l \neq i$, Eq. (11) holds.

Finally, from Case 1 and Case 2 above, we have $\mathcal{E}\{\bar{\varepsilon}^T(t)(\Xi_i + \alpha_t \tilde{\nu}_i)\bar{\varepsilon}(t)\} \leq 0$, and from Eq. (12), we have $\mathcal{E}\{\bar{\varepsilon}^T(t)(\tilde{\Xi}_i - \alpha_{ut}\tilde{\nu}_i)\bar{\varepsilon}(t)\} \leq 0$ holds, the system (2) is leader-following consensus with generally uncertain TRs according to the Definition 1 in [15]. It completes the proof.

Corollary 1. *Extending the preconditions of Theorem 1, let us assume that the inclusion of $\mathcal{S}_{ut}$ encompasses both PSPT and NSPT cases. And the PSPT case control gains K_i can be determined by solving the following LMIs.*

If $i \in \mathcal{S}_{pt}$, set row k and column k of the $\mathbb{L}_i$ to zero, $k \in \mathcal{A}_{ut}$. $\alpha_{i,pt} > 0$.

$$\tilde{\Xi}_{i,pt} < 0, \tag{35}$$

where

$$\tilde{\Xi}_{i,pt} = \mathrm{Sym}\{e_1^T(I_{N,i} \otimes \tilde{P}_i)e_3\} + h^2 e_3^T(I_{N,i} \otimes \tilde{M})e_3 - \frac{\pi^2}{4}(e_1 - e_4)^T(I_{N,i} \otimes \tilde{M})(e_1 - e_4)$$
$$+ \mathrm{Sym}\{(e_1 + me_3)^T[-(I_{N,i} \otimes F)e_3 + (I_{N,i} \otimes AF)e_1 + (I_{N,i} \otimes BF)e_2$$
$$- (G_i \otimes C\tilde{K}_i)(e_4 + e_5)]\} - e_1^T(I_{N,i} \otimes \tilde{R}E_1)e_1 + \mathrm{Sym}\{e_1^T(I_{N,i} \otimes \tilde{R}E_2)e_2\}$$
$$- e_2^T(I_{N,i} \otimes \tilde{R})e_2 - e_5^T(I_{N,i} \otimes \tilde{\Omega})e_5 + (e_4 + e_5)^T(G_i^T \Theta G_i \otimes \tilde{\Omega}_i)(e_4 + e_5),$$

$I_{N,i} = \mathrm{diag}\{\mathcal{I}_1, \mathcal{I}_2, \ldots, \mathcal{I}_N\}$, if $k \in \mathcal{A}_t$, $\mathcal{I}_k = 1$, else $\mathcal{I}_k = 0$. $K_i = \tilde{K}_i \tilde{F}$.

Proof. Choose the LKF as $V(t,i) = V_1(t,i) + V_2(t,i)$, where $V_1(t,i)$ and $V_2(t,i)$ are the same as defined in Theorem 1. By applying the same processing method and then referring to Eq. 35, it can be observed that the derivative of the LKF is less than zero. This implies that the states of the followers in the PSPT mode are converging toward the states of the leader.

4 Simulation

Consider the consensus control of the nonlinear MASs (2) with $h = 0.01\,\mathrm{s}$,
$f(x(t)) = \begin{bmatrix} tanh(x_1(t)) \\ tanh(x_2(t)) \end{bmatrix}$, $A = \begin{pmatrix} -0.5 & 1 \\ -1.5 & -1 \end{pmatrix}$, $B = \begin{pmatrix} 1 & 0 \\ 0 & 1 \end{pmatrix}$, $C = \begin{pmatrix} 1 \\ 1 \end{pmatrix}$, $D = \begin{pmatrix} 1 & 0 \\ 0 & 1 \end{pmatrix}$, $E_1 = \begin{pmatrix} 0 & 0 \\ 0 & 0 \end{pmatrix}$, $E_2 = \begin{pmatrix} 0.5 & 0 \\ 0 & 0.5 \end{pmatrix}$.

According to Fig. 1, the Laplacian matrices $\mathbb{L}_i$ and the leader adjacency matrices β_i ($i = 0, 1, 2, 3, 4, 5$) could be obtained. The transition matrix is given as follows $\tilde{\pi}_{11} = -4.6$, $\tilde{\pi}_{12} = 1.95$, $\tilde{\pi}_{23} = 3$, $\tilde{\pi}_{33} = -6$, $\lambda_{11} = 3.3$, $\lambda_{12} = 1.45$, $\lambda_{23} = 2$, $\lambda_{33} = 4$.

Choose $m = 0.2$, $\alpha_{uk} = 0.0499$, $\alpha_k = 0.049$, $\alpha_{ut} = 1.2$, and the event-triggered parameters are given as $\theta_1 = 0.002$, $\theta_2 = 0.002$, $\theta_3 = 0.005$. We get the following consensus feedback matrices by adopting Matlab YALMIP toolbox and MOSEK solver [16,17]. $K_0 = (2.9178\ 1.0999)$, $K_1 = (3.1682\ 1.1948)$, $K_2 = (3.7480\ 1.4001)$, $K_3 = (5.5493\ 1.5549)$, $K_4 = (23.8899\ 5.1047)$, $K_5 = (0\ 0)$, meanwhile the event-triggered parameter matrices $\tilde{\Omega}_0 = \begin{pmatrix} 0.1312 & 0.2047 \\ 0.2047 & 0.3200 \end{pmatrix}$, $\tilde{\Omega}_1 = \begin{pmatrix} 0.1314 & 0.2062 \\ 0.2062 & 0.3251 \end{pmatrix}$, $\tilde{\Omega}_2 = \begin{pmatrix} 0.1401 & 0.2138 \\ 0.2138 & 0.3416 \end{pmatrix}$, $\tilde{\Omega}_3 = \begin{pmatrix} 0.5579 & 0.2944 \\ 0.2944 & 0.7225 \end{pmatrix}$, $\tilde{\Omega}_4 = \begin{pmatrix} 0.9470 & 0.3628 \\ 0.3628 & 0.9601 \end{pmatrix}$, and the average non-tree ratio $\varpi < 0.0392$. It is worth pointing out that the control gain and event-triggered matrices for modes 3 and

4 were only applied to the followers that form the partial spanning tree, the mode 5 is not required ETS.

The initial conditions of the leader and the followers are given as $x_0(t) = \begin{bmatrix} 2 & -2 \end{bmatrix}^T$, $x_1(t) = \begin{bmatrix} 4 & 4 \end{bmatrix}^T$, $x_2(t) = \begin{bmatrix} -3 & 3 \end{bmatrix}^T$, $x_3(t) = \begin{bmatrix} 1 & -4 \end{bmatrix}^T$.

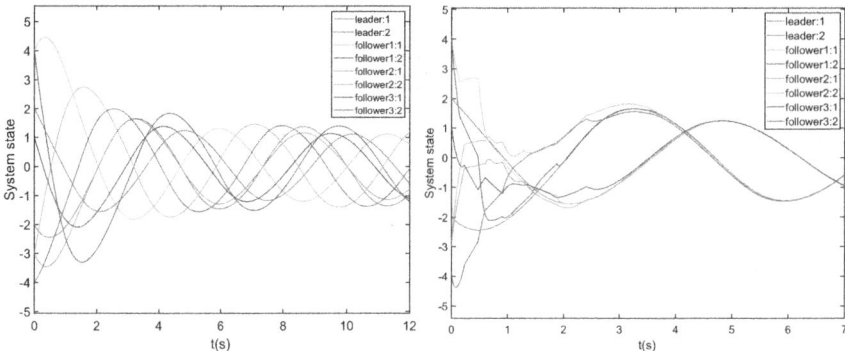

Fig. 2. Comparison of system states without and with controllers.

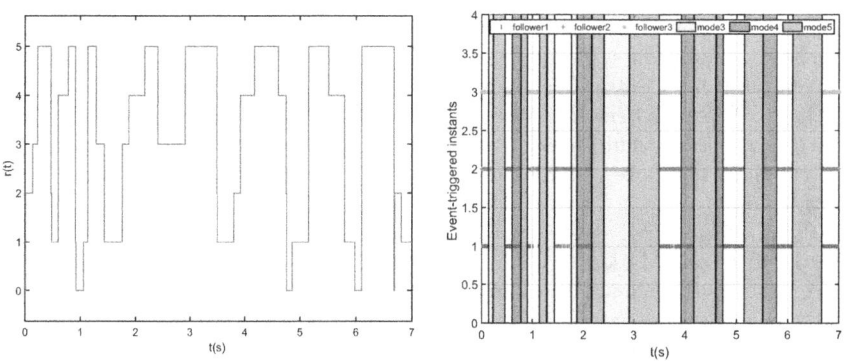

Fig. 3. Switching signal and event triggered instants

The error states without and with controllers are displayed in Fig. 2. Concurrently, the switching signal and the event-triggered instants are illustrated in Fig. 3. It is evident that secure consensus control can be effectively achieved in leader-following nonlinear MASs (2).

References

1. Feng, S., Ishii, H.: Dynamic quantized consensus of general linear multiagent systems under denial-of-service attacks. IEEE Trans. Control Netw. Syst. **9**(2), 562–574 (2022)
2. Chen, P., Liu, S., Chen, B., Yu, L.: Multi-agent reinforcement learning for decentralized resilient secondary control of energy storage systems against DoS attacks. IEEE Trans. Smart Grid **13**(3), 1739–1750 (2022)
3. Amini, A., Asif, A., Mohammadi, A.: RQ-CEASE: a resilient quantized collaborative event-triggered average-consensus sampled-data framework under denial of service attack. IEEE Trans. Syst. Man Cybern. Syst. **51**(11), 7027–7039 (2021)
4. Wang, J., Deng, X., Guo, J., Zeng, Z.: Resilient consensus control for multi-agent systems: a comparative survey. Sensors **23**(6), 2904 (2023)
5. Zhao, M., Peng, C., Han, Q.-L., Zhang, X.-M.: Cluster consensus of multiagent systems with weighted antagonistic interactions. IEEE Trans. Cybern. **51**(11), 5609–5618 (2021)
6. Zuo, R., Li, Y., Lv, M., Liu, Z.: Distributed asynchronous consensus control of nonlinear multi-agent systems under directed switching topologies. Automatica **152**, 110952 (2023)
7. Wang, X., Park, J.H., Yang, H.: An improved protocol to consensus of delayed MASs with UNMS and aperiodic DoS cyber-attacks. IEEE Trans. Netw. Sci. Eng. **8**(3), 2506–2516 (2021)
8. Xiao, Y., Che, W.-W.: Neural-networks-based event-triggered consensus tracking control for nonlinear MASs with DoS attacks. Neurocomputing **501**, 451–462 (2022)
9. Wang, J., Wen, G., Duan, Z.: Distributed antiwindup consensus control of heterogeneous multiagent systems over markovian randomly switching Topologies. IEEE Trans. Autom. Control **67**(11), 6310–6317 (2022)
10. Li, B., Wen, G., Peng, Z., Wen, S., Huang, T.: Time-varying formation control of general linear multi-agent systems under markovian switching topologies and communication noises. IEEE Trans. Circuits Syst. II Express Briefs **68**(4), 1303–1307 (2021)
11. Dai, J., Guo, G.: Event-triggered leader-following consensus for multi-agent systems with semi-Markov switching topologies. Inf. Sci. **459**, 290–301 (2018)
12. Ye, Z., Ji, H., Zhang, H.: Passivity analysis of Markovian switching complex dynamic networks with multiple time-varying delays and stochastic perturbations. Chaos Solitons Fractals **83**, 147–157 (2016)
13. Liu, K., Fridman, E.: Wirtinger's inequality and Lyapunov-based sampled-data stabilization. Automatica **48**(1), 102–108 (2012)
14. Hu, S., Yue, D.: Event-triggered control design of linear networked systems with quantizations. ISA Trans. **51**(1), 153–162 (2012)
15. Wang, H., Xue, B., Xue, A.: Leader-following consensus control for semi-markov jump multi-agent systems: an adaptive event-triggered scheme. J. Franklin Inst. **358**(1), 428–447 (2021)
16. Costa da Silva Campos, V., Nguyen, A.-T., Martínez Palhares, R.: Adaptive gain-scheduling control for continuous-time systems with polytopic uncertainties: an LMI-based approach. Automatica **133**, 109856 (2021)
17. MOSEK. ApS. The MOSEK Optimization Toolbox for MATLAB Manual. Version 10.0. (2022)

FastLSLO: An Efficient LiDAR Odometry Based on Improved Lie Group B-Splines

Xinyang Tang[1,2], Wei Yuan[3], Chenxi Yang[1,2], Chunxiang Wang[1,2], Bing Wang[1,2], and Ming Yang[1(✉)]

[1] Department of Automation, Shanghai Jiao Tong University,
Shanghai 200240, CN, China
mingyang@sjtu.edu.cn
[2] Key Laboratory of System Control and Information Processing,
Ministry of Education of China, Shanghai 200240, CN, China
[3] Global Institute of Future Technology, Shanghai Jiao Tong University,
Shanghai 200240, China

Abstract. Motion distortion correction is critical in LiDAR odometry. Traditional scan-based methods alleviate inner-scan motion based on constant-velocity assumptions or Inertial Measurement Units (IMUs). However, these methods treat such correction as a preprocessing, and the inevitable correction errors cannot be eliminated in subsequent stages. On the other hand, The continuous trajectory representation based on B-spline can effectively address such problems, but it typically suffers from low computational efficiency, requiring sacrificing point cloud density to achieve better real-time performance. In this study, we introduce an efficient method for implementing B-spline curves and trajectory optimization. It can achieve real-time performance without the need for feature extraction or significant downsampling of point clouds, which contributes to obtaining higher odometry accuracy and local map density. Furthermore, our method can be easily integrated into various other B-spline-based methods to help improve their computational efficiency.

Keywords: SLAM · Pointcloud · LiDAR · B-Spline · Continuous-time trajectory.

1 Introduction

Odometry is a crucial function for autonomous mobile robots, providing essential positioning information for robot navigation. However, traditional integrated navigation systems are constrained by their application environments and may face challenges such as GPS signal loss in indoor environments. Simultaneous Localization and Mapping (SLAM) technology, estimating the pose of a robot by aligning the online observation with pre-prepared environment map, offers a robust solution to these issues. Among the perception sensors, Light Detection And Ranging (LiDAR) is widely used in SLAM thanks to its robustness and rich information.

LiDAR acquires the three-dimensional structure of the environment asynchronously, owing to the incorporation of built-in rotators operating in one or more dimensions. Therefore, conventional LiDAR methods like [9,18,21,22] accumulate data over a period of time into a single "scan" before processing. However, due to the sensor's own motion, the points within each scan cannot be readily considered synchronized representations of the environment. This phenomenon is referred to as point cloud motion distortion. Failure to address motion distortion can adversely affect odometry accuracy.

Some methods [12,18,22] rely on the assumption of constant velocity to correct motion distortion, but these approaches do not consider abrupt motion changes. Another category of methods [15,20,21] utilizes IMUs (Inertial Measurement Units) for integration, calculating sensor motion within a scan to correct motion distortion. However, these methods introduce complexity into the system and are constrained by the accuracy and data frequency of the IMU. Furthermore, both of these approaches treat distortion correction as a preprocessing step, which prevents the elimination of distortion errors in subsequent stages, limiting the system's accuracy.

Recently, there has been an increase in methods [6,10] that utilize continuous-time trajectories to determine the pose for each LiDAR point at each moment. These methods hold promise for addressing the motion distortion problem. However, some approaches [5,8,13] rely on linear interpolation, which restricts their ability to accurately represent high-frequency motion. Meanwhile, other methods [1,14] utilize spline curves, but they suffer from computational complexity and lack real-time capabilities.

In our study, we present a novel method for representing continuous trajectories based on improved B-spline curves. This method uniformly processes inner-scan and inter-scan points in LiDAR odometry processing, as depicted in Fig. 1. Through an approximate computational approach, significant improvements in the computational efficiency of B-splines are achieved, which holds crucial significance for the deployment of B-spline-based odometry methods. Experimental results demonstrate that the proposed approach enhances computational efficiency while maintaining accuracy.

The main contributions of this paper are as follows:

- A novel approximation method for B-splines on Lie groups, which enhances computational efficiency.
- A sliding window optimization method for the improved B-spline curve in this paper.
- Our method can be easily integrated, which facilitates the industrial application of B-spline approaches.

2 Related Work

LiDAR odometry and map has undergone extensive research and typically follows a well-defined pipeline. In this pipeline, raw point clouds are accumulated

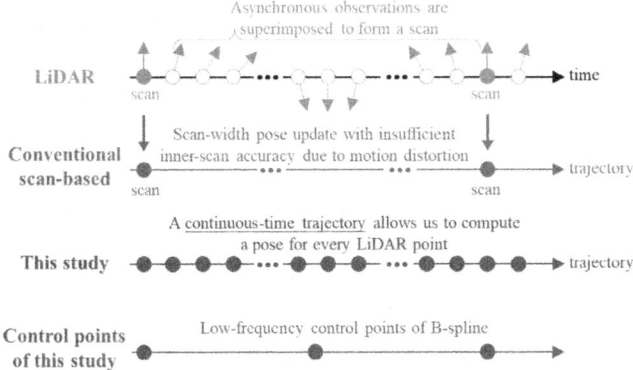

Fig. 1. A comparison of the methods in this paper with the conventional scan-based method. Conventional methods accumulate LiDAR points over a period of time into a single scan and compute a single pose for it. Our approach can compute a pose for each individual LiDAR point.

into scans, which are then aligned with historical scans or map to estimate the current pose. This estimated pose is subsequently used to update the map. Throughout this process, addressing point cloud motion distortion is crucial. LOAM [22], a landmark work in LiDAR odometry and mapping, operates under the the assumption that the sensor moves at a constant velocity. For each scan, the point cloud is corrected for motion distortion using the velocity estimated from the previous scan. Such method for correcting motion distortion is also adopted by many subsequent approaches like [18]. Alternatively, in [2,15,20,21], Inertial Measurement Unit (IMU) is employed for point cloud correction. By integrating the IMU data starting from the first point, it becomes possible to compute the corrected pose for each point. However, these methods all share a common characteristic: they treat distortion correction as a preprocessing step and maintain the correction results throughout subsequent Iterative Closest Point (ICP) matching and state estimation processes. Methods based on the constant velocity assumption perform poorly in cases of abrupt acceleration, deceleration, or sharp turns, while methods relying on IMU integration face constraints due to sensor accuracy, and the IMU data feedback frequency is much lower than the LiDAR point frequency.

Some methods dynamically perform point cloud distortion correction to address motion distortion more effectively during the ICP matching process. For example, VICP [8,19] approximates the inner-scan motion as constant velocity and repeatedly applies distortion correction during the iterative pose estimation process. Before each iteration of optimization begins, the point cloud is corrected using the results from the previous iteration. Works like Elastic LiDAR Fusion [13] and CT-ICP [5] based on linear interpolation achieve continuous pose representation. However, these methods are limited by the constant velocity assumption and may lose high-frequency motion information. Point-LIO [7] breaks away from the concept of scan by directly processing individual LiDAR points and immediately updating the Kalman filter for each point. However, this approach may lead to LiDAR points joining the map without sufficient constraints, reducing map accuracy.

On the other hand, a class of methods based on spline curves represents poses as continuous curves, thus better capturing high-frequency motion information. [16,17] proposed a fast computation and differentiation method for B-spline curves on Lie groups, providing a solid theoretical foundation for pose representation based on B-spline curves. In this method, B-spline curves on the Lie group are represented incrementally. A B-spline curve of order N can be computed by multiplying N Lie group elements. [11] implemented event camera odometry based on B-spline curves, which is a sensor with similar characteristics to LiDAR, both of which asynchronously collect environmental information. [1] defined six poses for each scan and represented the trajectory using B-spline curves, and [10] implemented a LiDAR-Inertial odometry with B-spline curves, but these methods are not real-time. [14] improved real-time performance with multi-resolution B-spline curves, but this approach smoothed the trajectory and reduced its ability to represent high-frequency motion.

This paper uses B-spline curves as trajectory representation and sliding window method for optimization. But compared to [16], our proposed method does not directly optimize control points in Lie group. Instead, it optimizes their Lie algebra. B-spline interpolation is performed on the tangent space and then transformed back to Lie group by exp mapping. To address potential singular point issues, control points of each segment B-spline are assigned a distinct local coordinate system.

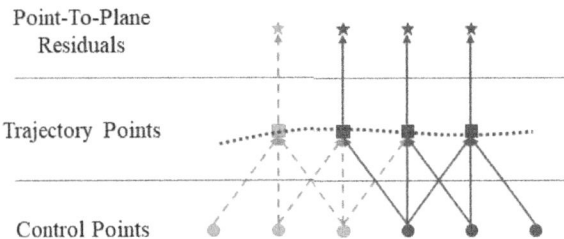

Fig. 2. The blue control points are optimized within the sliding window, while the gray points, outside the sliding window, are set to fixed values. At least one control point, which remains within the sliding window, will contribute to the calculation of the residual terms (blue pentagrams).

3 Methodology

This work aims to utilize B-spline curves as a continuous pose representation method to address the issue of point cloud motion distortion without relying on IMU data. Additionally, it improves B-spline curves on Lie group to enhance computational performance while maintaining accuracy and meeting the real-time requirements.

3.1 Overview

For each scan $P = \{p_i | p_i \in \mathbb{R}^3\}$, we first preprocess it by removing points that are too close to the sensor and then applying voxel filter to down-sample point cloud $P_{preprocessed} \subseteq P$. Then we calculate the residual function for each point and optimize the trajectory control points using a sliding window method (which will be introduced in the following sections). As the sliding window moves, the LiDAR points that are removed from the window are added to the local map. The local map is represented using a hash voxel map [2], which offers efficient nearest-neighbor search capabilities.

Fig. 3. The cyberrock experimental plantform.

3.2 Trajectory Representation

A uniform b-spline of order k in a vector space with control points $p_i (0 \leq i \leq N)$ has the form:

$$p(t) = \sum_{i=0}^{N} B_{i,k}(t) p_i \quad (1)$$

In which, the coefficient $B_{i,j}(t)$ are given by the De Boor-Cox recurrence relation [3,4].

$$B_{i,0}(t) = \begin{cases} 1, & t_i \leq t < t_{i+1} \\ 0, & otherwise \end{cases} \quad (2)$$

$$B_{i,j}(t) = \frac{t - t_i}{j \Delta t} B_{i,j-1}(t) + \frac{t_{i+j+1} - t}{j \Delta t} B_{i+1,j-1}(t) \quad (3)$$

For computational simplicity, we define the normalized time $u(t) := (t - t_0)/\Delta t - i$ for the segment $[t_i, t_{i+1}]$. The following formulations will use this time representation.

At time $(t_i \leq t < t_{i+1})$, the uniform b-spline can also be rewritten in cumulative form:

$$p_i(u) = p_i + \sum_{j=1}^{k-1} \lambda_j(u) * d_j^i \tag{4}$$

$$d_j^i = p_{i+j} - p_{i+j-1} \tag{5}$$

In which, $\lambda_j(u)$ is the coefficient for the cumulative form.

Based on Eq. 4, [16] gives the cumulative form of B-spline for order k in a Lie Group with control points $X_i \in \mathcal{L}$:

$$X_i(u) = X_i * \prod_{j=1}^{k-1} Exp(\lambda_j(u) * d_j^i) \tag{6}$$

$$d_j^i = Log(X_{i+j-1}^{-1} X_{i+j}) \in \mathbb{R}^d \tag{7}$$

In which, $Exp : \mathbb{R}^d \to \mathcal{L}$ and $Log : \mathcal{L} \to \mathbb{R}^d$. An implementation following Eq. 6 needs to perform (k–1) matrix multiply, exponential mapping and logarithmic mapping.

To reduce computational complexity, we can compute B-splines in the Lie algebra space and then map the results back to the Lie group space:

$$X_i(u) = Exp(\sum_{j=0}^{k-1} B_{i+j,k}(u) * d_j^i) \tag{8}$$

$$d_j^i = Log(X_{i+j}) \tag{9}$$

However, when control points are far from the origin, singular point issues may arise, so we define a local coordinate system for each segment of B-spline $(t_i \leq t < t_{i+1})$. The origin of the local coordinate system is defined as the starting point of the B-spline segment:

$$X_i(u) = X_{origin_i} * Exp(\sum_{j=0}^{k-1} B_{i+j,k}(u) * d_j^i) \tag{10}$$

$$= X_{origin_i} * Exp(f_i(u)) \tag{11}$$

$$d_j^i = Log(X_{origin_i}^{-1} * X_{i+j}) \tag{12}$$

The derivatives of control points are given by:

$$\frac{\partial X_i(u)}{\partial d_j^i} = X_{origin_i} * J_{i,r} * B_{i+j,k}(u) \tag{13}$$

In which, $J_{i,r}$ is the right Jacobian of $Exp(f_i(u))$.

The time derivatives are given by:

$$\dot{X}_i(u) = X_{origin_i} * (J_{i,r} * \dot{f}_i(u))^\wedge \tag{14}$$

An implementation following these formulas only perform single matrix multiply and single exponential mapping, which are the most computational expensive parts. Compared to the previous method (Eq. 6), the computational complexity of B-spline order in this method (Eq. 10) has decreased from O(n) to O(1). This is why our method can enhance computational efficiency.

These formulas above only consider a single segment $(t_{i+1} \le t < t_{i+2})$. For the next segment of b-spline $(t_{i+1} \le t < t_{i+2})$, we can get the following constrain:

$$X_{origin_{i+1}} = X_{origin_i} * Exp(f_i(1)) \quad (15)$$
$$d_j^{i+1} = d_{j+1}^i - f_i(1) \quad (16)$$

This method can only provide an approximate first-order continuity:

$$\dot{X}_i(1^-) = X_i(1) * (J_{i,r}(1) * \dot{f}_i(1))^\wedge \quad (17)$$
$$\approx X_i(1) * \dot{f}_i(1)^\wedge \quad (18)$$

$$\dot{X}_{i+1}(0^+) = X_{i+1}(0) * (J_{i+1,r}(0) * \dot{f}_{i+1}(0))^\wedge \quad (19)$$
$$= X_{i+1}(0) * \dot{f}_{i+1}(0)^\wedge \quad (20)$$
$$\approx \dot{X}(1^-) \quad (21)$$

Note that $\dot{f}_i(1)$ equals $\dot{f}_{i+1}(0)$ and $X_i(1)$ equals $X_{i+1}(0)$ by definition.

Despite the approximate first-order continuity, experiments have demonstrated that this method can improve computational efficiency without sacrificing accuracy.

3.3 Trajectory Optimization

In this paper, we use the point-to-plane ICP matching method. The residual function can be given as follows:

$$e_n = N^T * X_i(u_n) * P_n \quad (22)$$

P_n is a sampled LiDAR point and $N^{(i)}$ is the corresponding plane fitted from the nearest five neighbor points. u_n is the normalized timestamp for P_n. X_i is the B-spline segment where Pn resides.

The minimization problem can be expressed as follows:

$$\operatorname{argmin}_{d_j^i} \sum_n ||\rho(e_n)||^2 \quad (23)$$
$$\rho(s) = log(1+s) \quad (24)$$

Note that we optimize d_j^i in Eq. 10 instead of calculating them through Eq. 12. Constrains between different segments of B-spline are given by Eq. 15. Although $f_i(1)$ in Eq.15 is a function of d_j^i, we treat it as a constant in each optimization iteration. This way, d_j^i in different segments differ only by a constant.

The minimization problem can be optimized by the sliding window method illustrated in Fig. 2. Control points outside the sliding window are marginalized. New control point is initialized to be the same as the last control point.

During each optimization iteration, we randomly select a subset of the input point cloud $P_{iteration} \subseteq P_{downsampled}$ for computation, similar to what is done in stochastic gradient descent, and use the entire input point cloud $P_{downsampled}$ in the final iteration.

4 Experiments

This section presents a series of qualitative and quantitative experimental results to evaluate the effectiveness of this work. The experimental data were collected using our Cyber-Rock platform, whose configuration is depicted in Fig. 3. In this platform, the Pandar-40P LiDAR serves as the input sensor for the method described in this paper, while RTK-GPS provides high-precision positioning results as reference ground truth. Additionally, the vehicle is equipped with an inertial navigation system that includes a 100 Hz IMU. Although the LiDAR odometry in this paper does not utilize this sensor, the comparative method

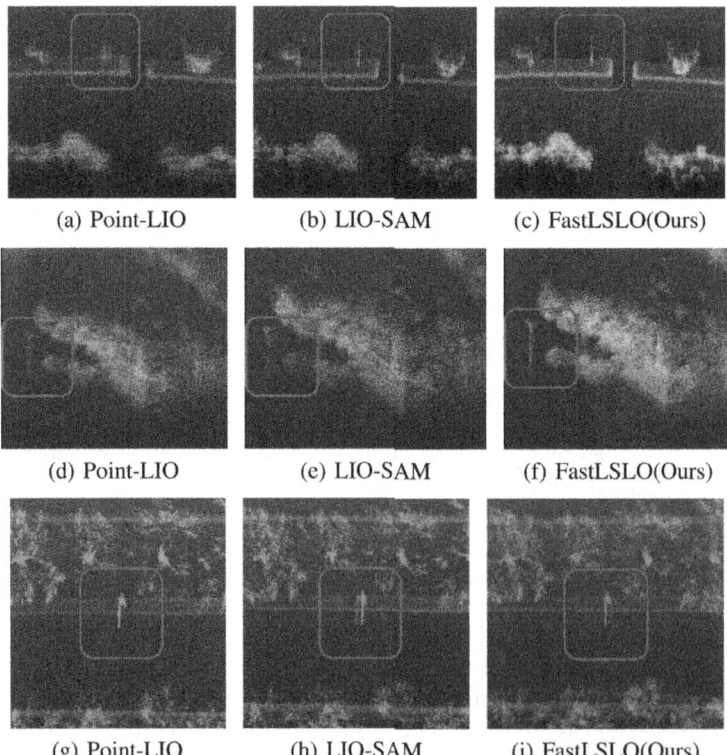

Fig. 4. Mapping detail shows how each approach performs in surrounding environments.

based on LiDAR-Inertial odoemtry will use it as an input. The data used in the experiments include scenarios involving rapid acceleration, rapid deceleration, sharp turns, and bumpy road sections, to assess the system's adaptability to intense motion. It also includes scenarios such as loops and parking lots to evaluate the system's positional drift accuracy. In the experiments, we selected B-spline curves with a degree of k = 3 and a sliding window size of 5. The experiments were conducted on a desktop computer with an i5-12490F CPU and 16 GB of RAM, using Ubuntu-20.04 and ROS as the development environment.

4.1 Intense Motion Experiments

To demonstrate the effect of our approach under conditions of intense motion, we deliberately performed frequent braking and rapid acceleration during the data

Table 1. Odometry Experiments

Method	APE(m)			RPE(m)		
	Mean	RMSE	Std	Mean	RMSE	Std
scenario with max acceleration = $5.53\,\text{m/s}^2$						
Kiss-ICP	0.099	0.110	0.047	0.104	0.107	0.023
Point-LIO	0.123	0.143	0.053	0.103	0.105	0.020
LIO-SAM	0.076	0.084	0.035	0.127	0.130	0.026
F-LOAM	0.111	0.121	0.047	**0.102**	0.105	0.019
FastLSLO(Ours)	**0.073**	**0.081**	**0.033**	0.103	**0.105**	**0.019**
max acceleration = $2.43\,\text{m/s}^2$ max rotation = $31.53\,\text{degree/s}$						
Kiss-ICP	0.255	0.274	**0.100**	0.072	0.095	0.062
Point-LIO	0.267	0.295	0.125	0.062	0.082	0.053
LIO-SAM	0.313	0.337	0.122	**0.044**	**0.048**	0.021
F-LOAM	0.271	0.297	0.121	0.062	0.082	0.053
FastLSLO(Ours)	**0.242**	**0.270**	0.119	0.049	0.053	**0.020**
max speed = 33.2 km/h, length = 450 m						
Kiss-ICP	0.550	0.614	0.274	0.156	0.174	0.077
Point-LIO	0.548	0.586	0.207	0.069	0.084	0.047
LIO-SAM	0.568	0.608	0.215	**0.060**	**0.072**	0.039
F-LOAM	0.569	0.602	0.196	0.069	0.076	0.031
FastLSLO(Ours)	**0.463**	**0.501**	**0.190**	0.068	0.074	**0.030**
max speed = 40.5 km/h, length = 1.2 km						
Kiss-ICP	1.772	2.044	1.019	0.115	0.122	0.045
Point-LIO	1.178	1.234	**0.365**	0.106	0.115	0.045
LIO-SAM	1.396	1.507	0.567	**0.092**	**0.106**	0.052
F-LOAM	1.404	1.568	0.698	0.106	0.115	0.046
FastLSLO(Ours)	**1.063**	**1.158**	0.458	0.103	0.114	**0.043**

acquisition process. We analyzed the linear acceleration data from the IMU and extracted segments of data with accelerations exceeding $3\,\mathrm{m/s}^2$. Each segment covered approximately 100 m of travel. Subsequently, we assessed different methods for Absolute Pose Error (APE) and Relative Pose Error (RPE) for each of these segments. Figure 4 illustrates the qualitative results of our intense motion experiments. Our approach preserves more details of surrounding environments compared to other competing methods.

4.2 Odometry Experiments

This section presents the experimental results of our approach and compares them with other methods using the untrimmed complete dataset. The dataset consists of multiple driving sequences within a campus environment, with each sequence covering approximately 1 km in distance. Moreover, the starting and ending positions of each sequence remain largely consistent, providing an assessment of the drift accuracy of the methods. In the experimental process, the loopclosure and GPS functionalities of LIO-SAM were disabled to obtain a relatively fair comparison result. Table 1 presents the quantitative results. Fig. 5 illustrates the results of various methods in the Z-axis. Our method achieved the best performance among all methods. OOur method achieves relatively better results.

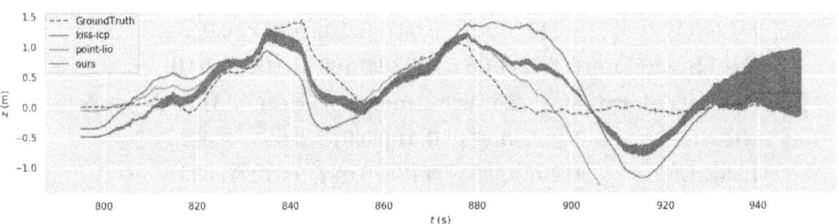

Fig. 5. The z-axis results on the experimental dataset. KISS-ICP, based on the constant velocity assumption, exhibited pose oscillations. Point-LIO exhibited the highest drift in the Z-axis in the final segment. Our method yielded relatively accurate results.

4.3 Computational Efficiency Experiments

In this experimental section, a separate comparison was conducted between the improved B-spline curve proposed in this paper and the traditional B-spline on Lie group implemented by [16]. The evaluation focused on the speed of single-pose computation and derivative calculation for the two methods at different orders. The time consumption of other steps in the ICP process, such as nearest neighbor search in the point cloud, was ignored. The derivative calculation was performed using the Ceres-Solver library for automatic differentiation. Figure 6 indicates that our method significantly enhances the computational efficiency of B-spline.

Table 2. Computational Efficiency Experiments

Methods	Translation APE RMSE(m)	Rotation APE RMSE(deg)	Frame Process Time(ms)	Trajectory Solve Time(ms)
clins before replace (order=3)	**0.563463**	0.955933	70.973	47.612
clins after replace (order=3)	0.638160	**0.871429**	**40.924**	**25.246**
clins before replace (order=4)	0.665574	**0.878723**	91.416	62.477
clins after replace (order=4)	**0.558381**	1.159868	**47.370**	**30.827**

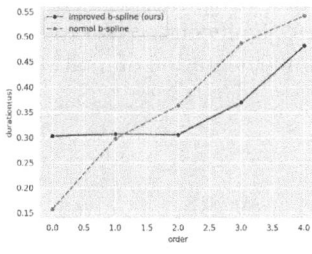
(a) Spline computation duration

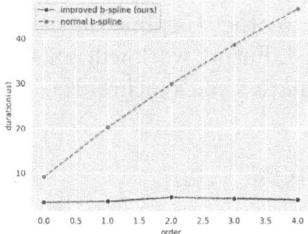
(b) Spline derivative duration

Fig. 6. Comparing the computational efficiency between normal B-spline on Lie group and the proposed improved B-spline on Lie algebra. It must be emphasized that this graph depicts the time taken for a single B-spline computation. During a trajectory optimization process, such computations may need to be performed tens of thousands of times.

To better illustrate the effectiveness of our approach in improving B-spline curves on Lie groups, we replaced the B-spline representation used in the clins [10] framework with the method described in our paper. Specifically, clins employed the trajectory representation based on $SO(3) + t$, and we replaced it with the trajectory representation based on $SE(3)$ using our proposed implementation. Additionally, to showcase the accuracy of odometry, we disabled the loopclosure functionality in clins. The Table 2 presents the comparison of odometry accuracy and speed metrics before and after the replacement, considering different spline orders. Notably, while maintaining similar accuracy, the trajectory optimization speed improved by 52%. It's essential to emphasize that during the trajectory optimization phase, besides spline computation time, other computational tasks are involved. Despite this, we still achieved a significant speed enhancement, indicating that spline calculations dominate the overall computational workload in trajectory optimization.

5 Conclusion

This paper presents a novel LiDAR odometry method based on an improved B-spline curve. The method demonstrates improved adaptability to point cloud motion distortion without relying on an IMU, while maintaining acceptable real-time performance. Our method represents trajectories as continuous curves in continuous time using B-spline curve, establishes point-to-plane constraints to build residual function, and optimizes continuous trajectories using a sliding window approach. By introducing an approximate computation and optimization method for B-spline curve on Lie group, the computational complexity has been reduced while ensuring that accuracy is not compromised. The improved B-spline curve proposed in this paper can be easily integrated into other methods, resulting in enhanced computational efficiency. The experiments demonstrate that our proposed method achieves better accuracy when the vehicle exhibits intense motion, and the generated maps exhibit finer details. By integrating our method into other B-spline-based approaches, significant acceleration effects can be achieved. Future work will consider the implementation of multi-sensor fusion and automatic time synchronization.

References

1. Alismail, H., Baker, L.D., Browning, B.: Continuous trajectory estimation for 3d slam from actuated lidar. In: 2014 IEEE International Conference on Robotics and Automation (ICRA), pp. 6096–6101. IEEE (2014)
2. Bai, C., Xiao, T., Chen, Y., Wang, H., Zhang, F., Gao, X.: Faster-lio: lightweight tightly coupled lidar-inertial odometry using parallel sparse incremental voxels. IEEE Robot. Autom. Lett **7**(2), 4861–4868 (2022)
3. Cox, M.G.: The numerical evaluation of b-splines. IMA J. Appl. Math. **10**(2), 134–149 (1972)
4. De Boor, C.: On calculating with b-splines. J. Approx. Theory **6**(1), 50–62 (1972)
5. Dellenbach, P., Deschaud, J.E., Jacquet, B., Goulette, F.: Ct-icp: real-time elastic lidar odometry with loop closure. In: 2022 International Conference on Robotics and Automation (ICRA), pp. 5580–5586. IEEE (2022)
6. Dubé, R., Sommer, H., Gawel, A., Bosse, M., Siegwart, R.: Non-uniform sampling strategies for continuous correction based trajectory estimation. In: 2016 IEEE International Conference on Robotics and Automation (ICRA), pp. 4792–4798. IEEE (2016)
7. He, D., Xu, W., Chen, N., Kong, F., Yuan, C., Zhang, F.: Point-lio: robust high-bandwidth light detection and ranging inertial odometry. Adv. Intell. Syst. **5**(7), 2200459 (2023)
8. Hong, S., Ko, H., Kim, J.: Vicp: velocity updating iterative closest point algorithm. In: 2010 IEEE International Conference on Robotics and Automation, pp. 1893–1898. IEEE (2010)
9. Lin, J., Zhang, F.: Loam livox: a fast, robust, high-precision lidar odometry and mapping package for lidars of small fov. In: 2020 IEEE International Conference on Robotics and Automation (ICRA), pp. 3126–3131. IEEE (2020)
10. Lv, J., Hu, K., Xu, J., Liu, Y., Ma, X., Zuo, X.: Clins: continuous-time trajectory estimation for lidar-inertial system. In: 2021 IEEE/RSJ International Conference on Intelligent Robots and Systems (IROS), pp. 6657–6663. IEEE (2021)
11. Mueggler, E., Gallego, G., Rebecq, H., Scaramuzza, D.: Continuous-time visual-inertial odometry for event cameras. IEEE Trans. Rob. **34**(6), 1425–1440 (2018)
12. Pan, Y., Xiao, P., He, Y., Shao, Z., Li, Z.: Mulls: versatile lidar slam via multi-metric linear least square. In: 2021 IEEE International Conference on Robotics and Automation (ICRA), pp. 11633–11640. IEEE (2021)

13. Park, C., Moghadam, P., Kim, S., Elfes, A., Fookes, C., Sridharan, S.: Elastic lidar fusion: dense map-centric continuous-time slam. In: 2018 IEEE International Conference on Robotics and Automation (ICRA), pp. 1206–1213. IEEE (2018)
14. Quenzel, J., Behnke, S.: Real-time multi-adaptive-resolution-surfel 6d lidar odometry using continuous-time trajectory optimization. In: 2021 IEEE/RSJ International Conference on Intelligent Robots and Systems (IROS), pp. 5499–5506. IEEE (2021)
15. Shan, T., Englot, B., Meyers, D., Wang, W., Ratti, C., Rus, D.: Lio-sam: tightly-coupled lidar inertial odometry via smoothing and mapping. In: 2020 IEEE/RSJ international conference on intelligent robots and systems (IROS), pp. 5135–5142. IEEE (2020)
16. Sommer, C., Usenko, V., Schubert, D., Demmel, N., Cremers, D.: Efficient derivative computation for cumulative b-splines on lie groups. In: Proceedings of the IEEE/CVF Conference on Computer Vision and Pattern Recognition, pp. 11148–11156 (2020)
17. Sommer, H., Forbes, J.R., Siegwart, R., Furgale, P.: Continuous-time estimation of attitude using b-splines on lie groups. J. Guid. Control. Dyn. **39**(2), 242–261 (2016)
18. Vizzo, I., Guadagnino, T., Mersch, B., Wiesmann, L., Behley, J., Stachniss, C.: Kiss-icp: In defense of point-to-point icp-simple, accurate, and robust registration if done the right way. IEEE Robot. Autom. Lett. **8**(2), 1029–1036 (2023)
19. Wang, H., Wang, C., Chen, C.L., Xie, L.: F-loam: fast lidar odometry and mapping. In: 2021 IEEE/RSJ International Conference on Intelligent Robots and Systems (IROS), pp. 4390–4396. IEEE (2021)
20. Xu, W., Cai, Y., He, D., Lin, J., Zhang, F.: Fast-lio2: fast direct lidar-inertial odometry. IEEE Trans. Rob. **38**(4), 2053–2073 (2022)
21. Xu, W., Zhang, F.: Fast-lio: a fast, robust lidar-inertial odometry package by tightly-coupled iterated kalman filter. IEEE Robot. Autom. Lett. **6**(2), 3317–3324 (2021)
22. Zhang, J., Singh, S.: Loam: lidar odometry and mapping in real-time. In: Robotics: Science and systems. vol. 2, pp. 1–9. Berkeley, CA (2014)

Improved BP Neural Network Based Deck Motion Prediction and Landing Control for Carrier-Based Aircrafts

Zhaoxing Li[✉] and Xingzhao Zhang

School of Automation, Northwestern Polytechnical University, Xi'an 710072, China
lzx_nwpu@163.com

Abstract. Considering that carrier-based aircrafts are affected by deck motion disturbances during landing, a prediction method based on the improved BP neural network (NN) using Grey Wolf (GW) algorithm is proposed. The model of deck motion is developed in this paper. Considering the irregular characteristics of deck motion, the BP NN is used to train it and predict the future deck motion information in advance. Aiming at the defects of BP such as slow convergence speed and converging local optimum, the GW algorithm is designed to optimize its initial threshold and weight to achieve quickly global convergence. The predicted information is used as the correction term of landing guidance command and the back-stepping method is used for control. Simulation results show that the proposed method can estimate and predict deck motion more effectively and enhance the control accuracy of landing trajectory tracking.

Keywords: Carrier-based aircraft · Deck motion prediction · BP neural network

1 Introduction

Carrier-based aircraft is a special type of aircraft that relies on the carrier as platform, and its precise, safe, and stable landing directly affects the overall combat capability of the carrier. Compared with landing control, carrier-based aircraft need to control its altitude and lateral deviation to track and maintain the desired landing trajectory during landing process. Due to various factors such as sea breeze and waves, carrier will experience various forms of deck motion such as roll, bow, pitch, and heave [1]. As a result, the expected landing trajectory will also constantly change, increasing the difficulty of landing. The deck motion seriously affects the safety and success rate of the landing process, so it is necessary to predict and introduce the prediction results into the guidance control to offset the height and lateral deviation.

Deck motion prediction (DMP) is based on historical data combined with the current fluctuation to predict future results [2]. Reference [3] shows that if the data for the next 2–4 s are accurately predicted, the landing trajectory can be adjusted timely to improve

---

Sponsored by Innovation Foundation for Doctor Dissertation of Northwestern Polytechnical University(Grant CX2024071).

tracking control accuracy. At present, scholars have achieved certain results in predicting it through multiple methods such as statistics [4, 7], convolutional neural networks [5], and Kalman filtering [6]. In practical engineering, it is difficult to obtain accurate predictions due to the complexity of wave motions and the differences in ships. The BP neural network with iterative correction is an effective method for solving accurate DMP due to its powerful ability to approximate any nonlinear function. However, BP neural networks also have shortcomings such as being prone to local optima and high computational complexity, which cause great challenges to the real-time and accuracy [8].

Based on above-mentioned analysis, a DMP method based on the improved BP NN of GW algorithm for quickly, accurately prediction is proposed which utilizes the back-stepping method to control the landing process. Simulation results show that the proposed algorithm can improve the accuracy of predictions and landing control. The models of deck motion is illustrated in Sect. 2. Section 3 presents the steps for improved BP prediction of deck motion based on Grey Wolf algorithm. The proposed method are tested in Sect. 4 while Sect. 5 summarizes this whole article.

2 Problem Formulation

At present, there are extensively studied models describing deck motion. However, different sea conditions and ship types in engineering practice makes it difficult to unify. Two kinds of models to describe deck motion are mainly studied: the stochastic model based on power spectrum and the combined model based on sine wave superposition. In this paper, the deck motion model under general sea conditions given in [10] is analyzed.

The movement of carrier caused by the irregular fluctuation of sea waves during navigation causes the ideal landing point to change constantly, which affects the accuracy of landing position. Define Δx_s, Δy_s, Δz_s, θ_s, ϕ_s and ψ_s as deck motion variables, where Δx_s is surge, Δy_s is sway, Δz_s is heave, θ_s is pitch, ϕ_s is roll and ψ_s is yaw respectively, as show in Fig. 1.

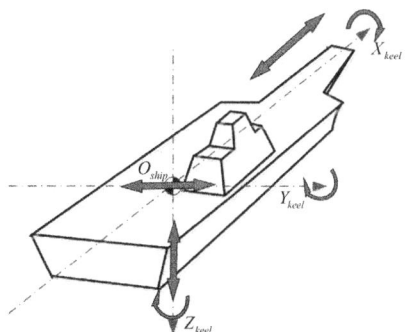

Fig. 1. Deck motion in three axes

The equation is introduced and the deck motion are described by the transfer function of linear motion and angular motion, with the expressed as follows

$$G_T(s) = \frac{b_3 s^2 + b_2 s + b_1}{s^4 + a_4 s^3 + a_3 s^2 + a_2 s + a_1}$$
$$G_A(s) = \frac{o_3 s^2 + o_2 s + o_1}{s^4 + h_4 s^3 + h_3 s^2 + h_2 s + h_1}$$
(1)

where a_i, b_j, h_i and $o_j (i = 1, 2, 3, 4; j = 1, 2, 3)$ are passing function parameter values, respectively.

3 Deck Motion Prediction

Remark 1: As a feedforward neural network composed of multiple layers, BP neural network can discover the characteristics and rules of the dataset through training, thus solving the prediction problem of time-series data. However, BP neural network also has shortcomings such as falling into local optima, low efficiency and et al. This section considers the shortcomings of BP and optimizes it through Grey Wolf algorithm to improve its defects and deficiencies.

3.1 BP Neural Network

The structure of BP NN mainly includes input, output and hidden layers. Each layer is connected by neurons. Currently, the most common structure is the hidden layer BP NN as depicted in Fig. 2.

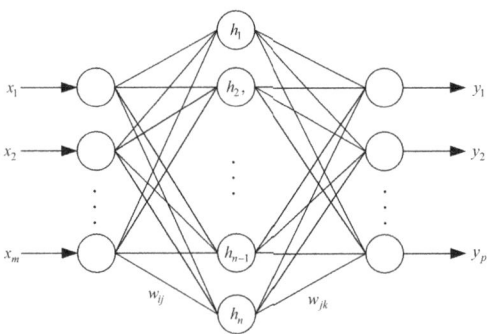

Fig. 2. The structure of BP NN

Define $x_1, x_2, \cdots, x_m$ is input signal, $h_1, h_2, \cdots, h_n$ is hidden layer processing structure, $y_1, y_2, \cdots, y_p$ is output signal, $\tilde{\theta}$ is threshold, w_{ij} and w_{jk} are the weight values corresponding to each neural unit, respectively. Then the signals in different layers are calculated as

$$h_j = f(\sum_{i=0}^{m} w_{ij}x_i - \tilde{\theta}_j) \; j = 1, 2, \cdots, n$$
$$y_k = f(\sum_{i=0}^{n} w_{jk}h_j - \tilde{\theta}_k) \; k = 1, 2, \cdots, p \qquad (2)$$

where $f(*)$ is activation functions between layers and can be expressed as

$$f(x) = \frac{1}{1+e^{-x}} \qquad (3)$$

The deviation of δ_k is calculated as

$$\delta_k = y_k - \bar{y}_k, \; (k = 1, 2, \cdots, p) \qquad (4)$$

Using gradient descent, the threshold and weight of BP NN is adjusted. When the initial threshold and weight values are appropriate, gradient descent can ultimately achieve good training of the BP NN. However, when the initial value and weight are not appropriate, the BP is easy to stuck in local optima during the training, which reduces its fitting ability.

3.2 Grey Wolf Algorithm

There is a specific hierarchy in the GW. The optimization process of the GW algorithm is to surround, hunt, track and attack the prey. In this algorithm's search process, the best three gray wolves at each iteration are defined as α, β, and σ, which guide the remaining gray wolves ω to catch for prey in the search space.

The model of grey wolves surrounding prey is obtained as

$$\begin{aligned} D &= |C \times X_p(t) - X(t)| \\ X(t+1) &= X_p(t) - A \times D \end{aligned} \qquad (5)$$

where D is the distance between the individual and X_p is the position of the GW, t is the number of iterations. A, C are vector coefficients and can be calculated as

$$\begin{aligned} A &= 2 \times a \times r_1 - a \\ C &= 2 \times r_2 \end{aligned} \qquad (6)$$

where $0 < r_1 < 1$ and $0 < r_2 < 1$ are design parameters, a is convergence factor.

According to the predation strategy, the position of α, β, and σ gray wolves are updated to

$$\begin{cases} D_\alpha = |C_1 \times X_\alpha(t) - X(t)| \\ D_\beta = |C_2 \times X_\beta(t) - X(t)| \\ D_\sigma = |C_3 \times X_\sigma(t) - X(t)| \end{cases} \qquad (7)$$

where $D_i (i = \alpha, \beta, \sigma)$ are the distance between wolves α, β, σ and other wolfs. $X_i (i = \alpha, \beta, \sigma)$ are present position α, β, and σ.

The final update location of the grey wolf individual $X(t + 1)$ is obtained as

$$\begin{cases} X_1(t) = X_\alpha(t) - A_1 \times D_\alpha \\ X_2(t) = X_\beta(t) - A_2 \times D_\beta \\ X_3(t) = X_\sigma(t) - A_3 \times D_\sigma \end{cases} \quad (8)$$

$$X(t+1) = \frac{X_1(t) + X_2(t) + X_3(t)}{3}$$

where $X_i(t), (i = 1, 2, 3)$ are wolf ω individual to update to α, β and σ

3.3 Deck Motion Prediction Based on Improved BP NN by GW

Considering the defects of BP NN, the GW is adapted to optimize the initial thresholds and weights. The DMP strategy based on the improved BP NN is established and the steps are as follows (Table 1):

Table 1. Improved BP based on Grey Wolf algorithm

Step1: Input deck motion sample data, divide it into training set and test set. Then normalize it;
Step2: Determining the BP structure, initialize the threshold and weight;
Step3: Set the parameters of GW, including dimension, parameter upper and lower bounds, population size and maximum number of iterations.
Step4: Initialize the population of gray wolves, putting N wolves to the search space randomly and calculating the fitness $I_{GW} = \frac{1}{N}\sum_{i=1}^{N}(y_i - y_i^*)^2$. y_i and y_i^* represent the actual value and fitted value of deck motion, respectively. The initial parameter of the BP are used for prey.
Step5: The three gray wolves are selected with the highest fitness, which are iteratively calculated the next candidate position and updated the parameters.
Step6: Whether the target accuracy or the maximum of iterations being reached is determined. If it has been reached, output the optimal solution to obtain the threshold and weights of the BP NN. Otherwise, return to step 3.
Step7: Utilize the improved BP NN to predict deck motions.

4 Simulation Analysis

In this section, the deck motion (1) is described the simulation environment and the improved BP NN is tested to illustrate the effectiveness of proposed methods. The noise value, sample time and the seed are selected as 1.0, 0.01s and 213538. Then the other passing function parameter values are described as follow:

$$G_\theta(s) = \frac{0.3341s^2}{s^4 + 0.604s^3 + 0.7966s^2 + 0.2063s + 0.1239}$$
$$G_\phi(s) = \frac{0.2384s^2}{s^4 + 0.2088s^3 + 0.3976s^2 + 0.0386s + 0.0342} \quad (9)$$
$$G_z(s) = \frac{1.16s^2 + 0.0464s}{s^4 + 0.38s^3 + 0.4977s^2 + 0.0836s + 0.0484}$$

The deck motion data of 100s is used as training sample and the input, middle, and output layers of neural network are set to 101, 5 and 21, respectively. The population of Grey Wolf is 300, the maximum of iterations is 100, the convergence factor is $a = 2 - 2t/t_{max}$ and the parameters are $r_1 = 0.6$ and $r_2 = 0.2$, respectively. Deck motion in formula (9) is estimated 2 s ahead. Figure 3 shows that the reference signal of deck motion, BP and the improved BP method(BPGW) are in blue, red and black lines, respectively.

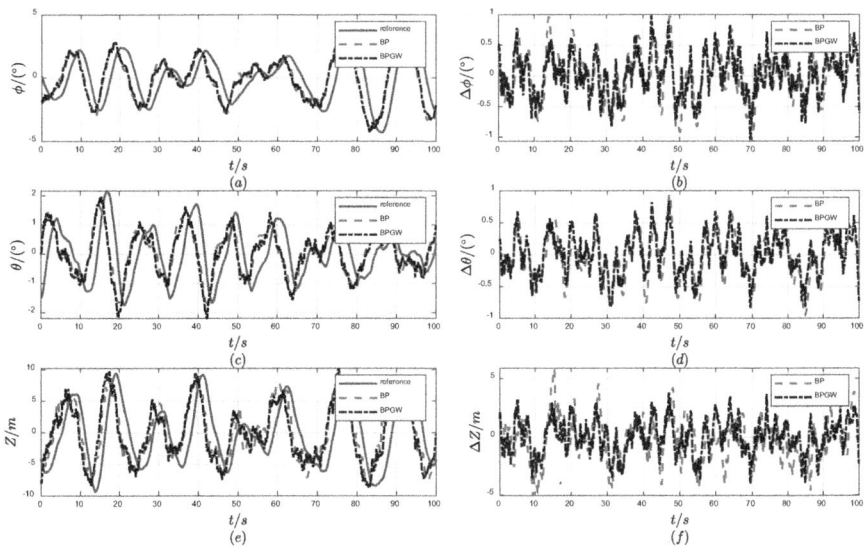

Fig. 3. DMP based on classical BP and the improved BP method

Table 2 shows that the variance of BP and BPGW. It can be obtained that BPGW has smaller variance and more accurate prediction performance.

To verify the effectiveness of the proposed algorithm, the control strategy in reference [12] is used to implement carrier-based aircraft landing simulation under the condition of introducing deck motion interference. Height and sideways deviation by using BP and BPGW to reduce are shown in Fig. 4. Simulation results show that using the BPGW to predict deck motion has a better performance in quickly and accurately tracking desired trajectory and the terminal deviation is smaller. The deck motion information can be obtained in advance and kept consistent as much as possible.

Table 2. The variance of BP and BPGW.

	BP	BPGW
G_θ	0.3715	0.3267
G_ϕ	0.3943	0.3433
G_z	1.9958	1.4729

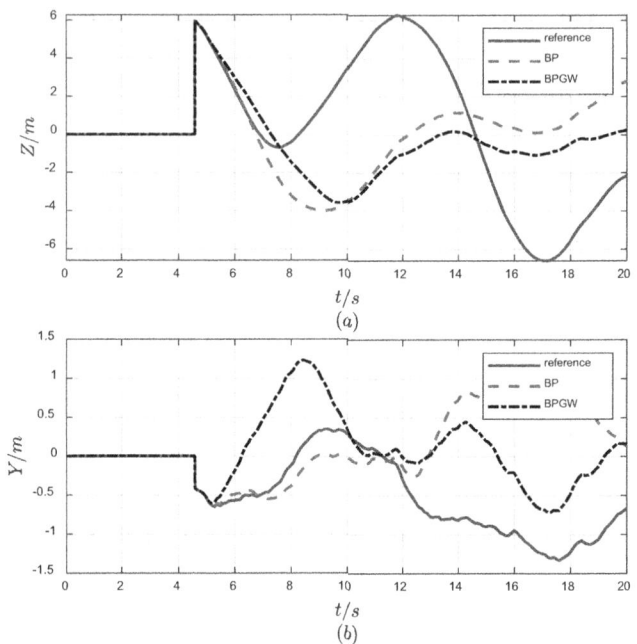

Fig. 4. Height and sideways deviation

5 Conclusion

This paper designs a DMP algorithm based on the improved BP NN with the GW algorithm and tested by a landing control strategy based on the back-stepping architecture. According to the equations, a deck motion model is established. Based on the BP NN, predictions are made and improved with the help of the GW algorithm and it can predict the information after 2s. On this basis, back-stepping control is used to carry out landing verification. Simulation results demonstrate that the proposed algorithm is able to estimate deck motion information more accurately and effectively, enhancing the control accuracy and landing successful rate.

References

1. Wu, Q., Zhu, Q.: Fault-tolerant control for carrier-based aircraft automatic landing subject to multiple disturbances and actuator faults. Int. J. Aerosp. Eng. **2024**(1), 2054883 (2024)
2. Yin, J.C., et al.: Online ship roll motion prediction based on grey sequential extreme learning machine. Neurocomputing **129**, 168–174 (2014)
3. Ghaderi, A.: Physics-Informed Data-Driven Models for Inelastic, Aging. Michigan State University, Failure Behavior of Crosslinked Polymers (2023)
4. Liu, X., et al.: A prediction method for deck motion of aircraft carrier based on particle swarm optimization and kernel extreme learning machine. Sens. Mater. **29** (2017)
5. Zhu, Q., et al.: Carrier-based aircraft landing based on deck motion prediction. In: International Conference on Guidance, Navigation and Control. pp. 265–274. Springer, Singapore (2022)
6. Cai, F., Wan, L., Shi, A.: Extreme short term prediction of ship motions based on phase-space reconstruction. J. Hydrodyn. **20**(6), 780–784 (2005)
7. Duan, H., Yuan, Y., Zeng, Z.: Automatic carrier landing system with fixed time control. IEEE Trans. Aerosp. Electron. Syst. **58**(4), 3586–3600 (2022)
8. Zhen, Z., Jiang, S., Ma, K.: Automatic carrier landing control for unmanned aerial vehicles based on preview control and particle filtering. Aerosp. Sci. Technol. **81**, 99–107 (2018)
9. Guan, Z., Ma, Y., Zheng, Z.: Moving path following with prescribed performance and its application on automatic carrier landing. IEEE Trans. Aerosp. Electron. Syst. **56**(4), 2576–2590 (2019)
10. Durand, T.S., Teper, G.L.: An analysis of terminal flight path control in carrier landing. Systems Technology, Incorporated (1964)
11. Lungu, M., et al.: Inverse optimal control for autonomous carrier landing with disturbances. Aerosp. Sci. Technol. **139**, 108382 (2023)
12. Zhen, Z., Jiang, S., Jiang, J.: Preview control and particle filtering for automatic carrier landing. IEEE Trans. Aerosp. Electron. Syst. **54**(6), 2662–2674 (2018)

Joint Sequencing and Merging Optimization for Airplanes and Helicopters

Dawei Wang, Yi Lyu[✉], Ken Chen, Yiman Zhang, and Chengcheng Wu

Beijing Aircraft Technology Research Institute, COMAC, Beijing, China
etiennely@foxmail.com

Abstract. This paper tackles the challenge of joint sequencing and merging optimization for airplanes and helicopters by proposing an optimized control algorithm aimed at minimizing flight delays. The algorithm comprehensively takes account of various factors such as trajectory conflicts, aircraft priority, speed limitations, and minimum separation constraints to construct the cost function. Simulated annealing is employed to determine the global optimal solution. To evaluate the algorithm, we conducted experiments using radar data from a randomly selected day at Nice Cote d'Azur Airport in France. The results showed that the algorithm could reduce total flight delay time by 50% and save an average of 1.8 L of fuel consumption per helicopter landing, thereby demonstrating its practicality and effectiveness in real-world scenarios.

Keywords: Aircraft guidance · joint sequencing and navigation · global optimization · simulated annealing

1 Introduction

With the development of the aviation industry, the application of helicopters has achieved a steady growth during the past few years, which brings new challenge to the Air Traffic Control (ATC) system.

Upon an airplane entering within the radar range of ATC at an airport, the ATC system is responsible for authorizing and arranging the landing procedure. With the assistance of systems like Arrival Manager (AMAN) and Point Merge (PM), safe and efficient sequencing of airplanes is now possible. However, the new problem occurs due to the introduction of helicopters. Considering that the Final Approach and Take-off Area (FATO) of helicopters is close to the runway, conflicts between airplanes and helicopters can occur when they arrive at a close time. In that case, additional ATC work of sequencing both airplanes and helicopters is inevitable. Currently, this kind of work can only be manually performed by the Air Traffic Controllers (ATCOs), as the existing aiding systems only take the airplanes into account. This manual intervention may increase the risk of safety and efficiency issues with the increasing number of the helicopters.

According to our knowledge, there is no published work about this kind of joint arrival sequencing problem, existing studies such as ALP (Airplane Landing Problem)

[1-4], PM system [5, 6] and Optimization of Terminal Maneuvering Area (TMA) [7] focus mainly on the airplanes. To alleviate workload of ATCOs and enhance capacity of the ATC system, this paper proposes a novel optimization-based sequencing algorithm, which tackles the challenge of joint arrival sequencing. The proposed algorithm takes into account constraints including speed limitation, separation strategies and wake vortex for both airplanes and helicopters, and applies the principle of PM system to optimize the sequencing of helicopters. The effectiveness of the algorithm was evaluated through simulation using real-world data from Nice airport.

2 Problem Statement

In this paper, we selected Nice Cote d'Azur airport as the study target, because both helicopters and airplanes are operating at this airport, with the FATO and the runway situated close to each other.

2.1 Description of the Problem

The arriving problem to be solved could be summarized in the Fig. 1 below.

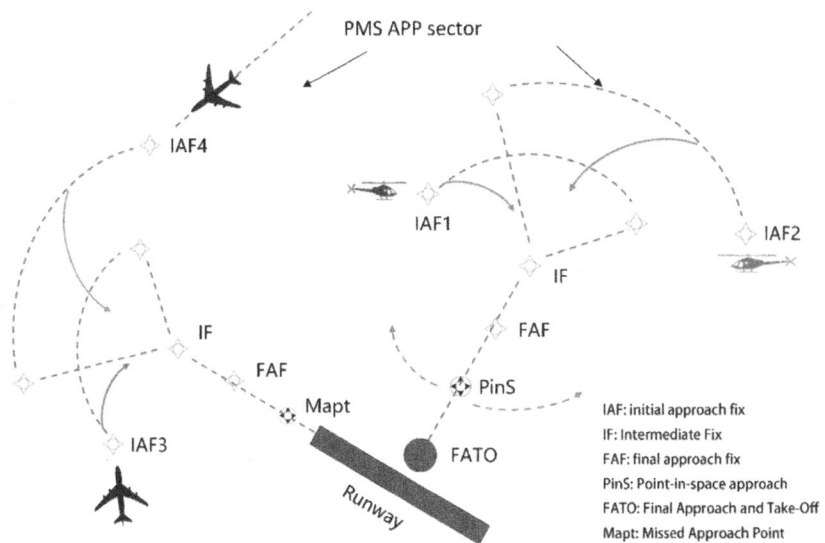

Fig. 1. Problem Description.

As shown in the Fig. 1, we consider a simple scenario with a single runway and a single FATO which locate adjacently and operate simultaneously. During landing, airplanes and helicopters should follow their own designated airways, which are marked by the fixes like IAF, IF and FAF. Considering the separation strategies and the weak turbulence between airplanes and helicopters, they cannot execute their landing procedure at the same time, which means sequencing of airplanes and helicopters during their approach is indispensable. Besides, factors like the priority of airplanes, the separation between airplanes and helicopters, and speed limitations should also be considered during the sequencing.

For this problem, to ensure safety and efficiency for both airplanes and helicopters, we propose the following assumptions as the foundation for our optimization-based sequencing method.

- The FATO and the runway are adjacent and can be treated as one point.
- The influence of wind can be ignored.
- Turning does not add any additional distance in the trajectory.
- The final approach trajectories (IF to Runway and IF to FATO) for both airplanes and helicopters are fixed and perpendicular to each other.

Furthermore, in order to increase the capacity of the algorithm, a PM system for helicopters is designed in this paper.

2.2 Design of the Airway

To create an effective sequencing algorithm, it is crucial to first represent the problem mathematically, taking account of airways for both types of aircraft in the radar range. For airplanes, the design of the airway is based on the Aeronautical Information Publication (AIP) of Nice Cote d'Azur airport which records the landing paths and the related fixes. Furthermore, a PM system is implemented to meet the high traffic volume. For helicopters, however, there is no such document like AIP which can accurately define the airway and fix positions. As such, the related airway should be designed by ourselves with reference to the AIP of the airplanes. In this work, the helicopter airways are of the same type but shorter in distance compared to those of airplanes, taking into account the slower speed of helicopters. The final design is illustrated in Fig. 2. Since the airway has been well designed, the trajectories of both types of aircraft have been fixed and defined by several segments.

To formulate the problem, we make a reasonable assumption that aircraft speed remains constant on each segment of its airway. Under this assumption, given that the airway has been well designed, the position of any aircraft at any time during its approach can be determined through the function below:

$$P_{A_i/H_i}(t) = f\left(P_s, V_{EP} \ldots V_{runway}, T_{PM}\right) \quad (1)$$

This function is an airway depended function including variables such as P_s, the show-up position of aircraft, $V_{EP} \ldots V_{runway}$, the speeds on each segment of airway and T_{PM}, the time spent on the PM system. These variables uniquely determine and predict the entire flying state. In this paper, we define this group of variables except for the P_s, which is given and non-manipulable, as the flight plan. Consequently, joint sequencing can now be considered as a problem of calculating the the flight plan variables for each aircraft. In the following sections, the control algorithm will be illustrated in details.

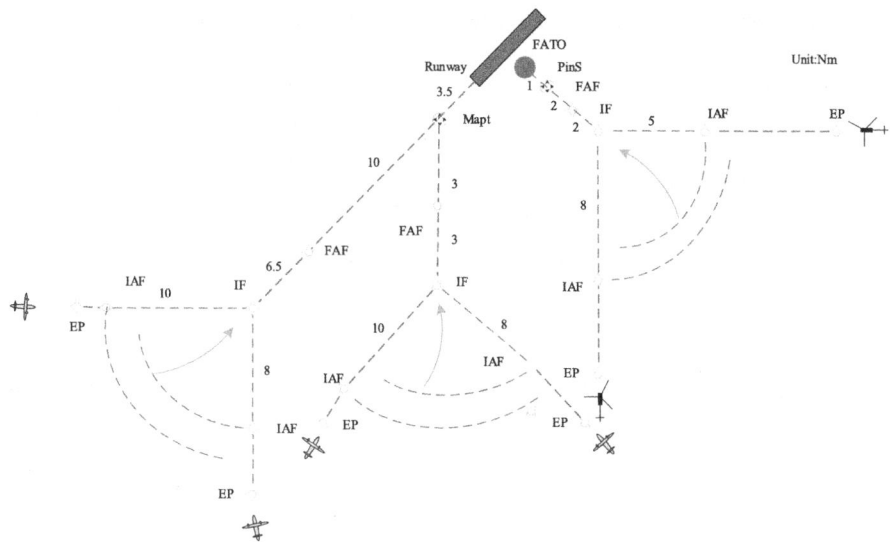

Fig. 2. The design of the airway.

3 Mathematical Modelling for the Joint Sequencing Problem

Based on the analysis of the problem presented above, the sequencing problem can now be formulated as an optimization problem described as follows:

$$\sum_{i}^{N_a} \lambda Obj(Ax_i) + Obj(Hx_i) \tag{2}$$

$$subject\ to\ C(Ax_i), C(Hx_i) \tag{3}$$

Equation (2) is the objective function to be minimized, which can be the delay, conflict, or fuel consumption of the aircraft depending on the application. Variables Ax_i and Hx_i are the flight plans of airplanes and helicopters. λ is the weight factor used to ensure the priority of the airplanes, normally greater than 1. The value of the objective function represents the cost, which indicates the valuation of the flight plan. Equation (3) refers to the constraints to be met during the calculation which are further elaborated in the following sections.

3.1 Objective Function

In this context, the sequencing problem is considered as a Multi Objective Problem (MOP), which involves more than one objective function, increasing the complexity and flexibility of the algorithm. In this paper, two principle objective functions are applied.

Minimization of Delay. Minimization of delay is the crucial objective function in the scheduling problem to ensure punctuality of the flight. The formula is shown as follows:

$$\lambda(Ax_i - AT_i)^2 + (Hx_i - HT_i)^2 \tag{4}$$

In the function above, Ax_i is the actual landing time of the i_{th} airplane, AT_i is the planned landing time of the i_{th} airplane, Hx_i and HT_i refer to the same parameters for the i_{th} helicopter. This function is built to ensure that the aircraft should arrive at their Requested Time of Arrival (RTA). In particular, the function also considers the early arrivals as a peculiar "delay", as both early and late arrivals relative to the planned flight schedules can pose issues to the ATCOs and carriers, such as increased fuel consumption and disruptions to the schedule of other aircraft.

Avoidance of the Conflicts Between the Helicopters and Airplanes. The avoidance of the conflicts is the fundamental objective function which ensures 1) The separation between every two aircraft meets standard requirements and 2) The difference of landing time between airplanes and the helicopters complies with the standards established in our work. This function can be simply expressed as $\lambda_c N_c$, where N_c is the number of conflicts during approach, and λ_c is a weight factor which is sufficient to amplify the influence of conflict.

In this paper, four types of conflicts are identified according to the airway design and previous assumptions, including conflicts on the same segment, conflicts within the PM system, conflicts between segments, and conflicts during landing. According to the previous illustration, we are capable of forecasting the position of aircraft at any time t. Therefore, with the fixed airway, we can predict the conflicts using the provided flight plan for each aircraft, and calculate the value of the objective function accordingly.

3.2 Constraints

The constraints used in our algorithm contain:

The Priority of the Airplanes. Airplanes should be prioritized over helicopters. In our algorithm, the priority of airplanes can be ensured through λ and λ_c.

The Limitation of the Speed. The speed of airplanes and helicopters should be limited in a specific range determined by their types. This constraint is typically used to ensure the possibility of the position shift and to estimate the range of Estimate Time Arrival (ETA) [8].

According to the ICAO Doc 8168, the speed range of airplanes are shown in Table 1. This table specifies the handling speeds range (Indicated Airspeed in kts) for each category of airplane during landing. Class A to E refers to small single-engine, small multi-engine, airline jet, large jet/military jet and special military jet, respectively. The Class H represents the helicopters. The symbol '*' refers to the maximum speed for reversal and racetrack procedures.

The Minimum Separation. The minimum separation is the most important constraint, which is a key factor to guarantee the safety of aircraft. This constraint is influenced by two main factors, separation strategies and wake turbulence.

A. *Separation strategy.* The separation strategy ensures a minimum distance of 3NM between airways during their design. Ideally, this separation should be maintained at every point along each airway. However, when the FATO and runway are near

Table 1. The limitation of the aircraft speed.

Airplane category	Range of speeds for initial approach	Range of final approach speeds	Maximum speeds for intermediate missed approach	Maximum speeds for final missed approach
A	90–150 (110*)	70–100	100	110
B	120–180 (140*)	85–130	130	150
C	160–240	115–160	160	240
D	185–250	130–185	185	265
E	185–250	155–230	230	275
H	70–120	60–90	70–90	70–90

to each other, a crossover between the airplane and helicopter airways may occur during the final approach. In that case, we guarantee this separation only before the Mapt and the PinS, as illustrated in the Fig. 3.

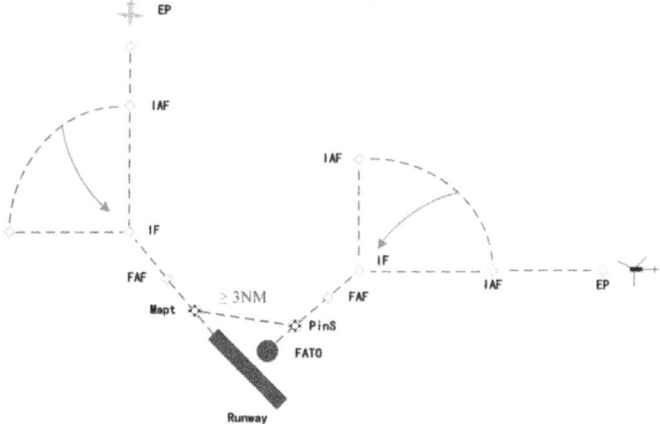

Fig. 3. Separation strategy.

B. *Separation of wake turbulence.* The separation of wake turbulence for airplanes was firstly standardized by ICAO by introducing wake turbulence category (WTC) after several serious accidents. According to the WTC, the airplanes can be classified into three types, L(light), M(medium) and H(Heavy) based on the maximum certificated take-off mass. Furthermore, the minimum separations are defined according to the classification.

Despite being in use for years and widely applied in many studies, the WTC standard has significant drawbacks. It relies solely on airplane weight and categorizes airplanes into only three types, which in some cases are outdated and result in excessive separation, reducing airport capacity and causing unnecessary delays. To address this issue, the RECAT (Re-Categorisation) standard was proposed in a collaborative research between EUROCONTROL and FAA [10]. The study demonstrated that factors beyond weight, such as speed and wingspan, also affected wake turbulence strength. Therefore, airplanes are now reassigned to one of six new categories (A through F), with A representing "Super Heavy", B representing "Upper Heavy", C representing "Lower Heavy", D representing "Upper Medium", E representing "Lower Medium", and F representing "Light", respectively. As such, the new separation is presented below in Table 2.

Table 2. The separation of wake turbulence RECAT.

Follower /Leader	CATA	CATB	CATC	CATD	CATE	CATF
CATA	3NM	4NM	5NM	5NM	6NM	8NM
CATB		3NM	4NM	4NM	5NM	7NM
CATC			3NM	3NM	4NM	6NM
CATD						5NM
CATE						4NM
CATF						3NM

In the table above, a blank cell indicates the minimum radar separation, which is 2.5NM in the normal case prescribed by ICAO Doc 4444.

Helicopters are also vulnerable to the risks associated with wake turbulence. When flying forward, the downwash from their main rotor transforms into trailing vortices, similar to the wing tip vortices produced by airplanes [10]. However, there are no definitive standards like WTC/RECAT for helicopters. In practical operations, helicopters can generate more intense trailing vortices than airplanes of similar weight. Typically, the trailing vortices produced by a helicopter are equivalent to the wing tip vortices generated by an airplane with eight times the maximum take-off weight (MTOW) of the helicopter.

When it comes to the situation involving airplanes and helicopters during landing, we establish a minimum time separation of 1 min between the landing times of the airplane and the helicopter to define the separation between them.

3.3 Simulated Annealing

In our work, we utilize the simulated annealing (SA) method, a widely employed probabilistic technique for finding an approximation of the global optimum, to determine the optimized flight plan for each aircraft in real-time [11]. The procedure of calculation process is illustrated below in Fig. 4.

To begin with, the program will randomly choose a flight plan according to the constraint and calculate the related cost. Then the iteration of searching will begin. For each iteration, the program will calculate the deviation of each variable in the flight plan and generate a new one by adding it to the current one. Each updated variable must adhere to the constraints, otherwise it will be abandoned. The deviation d of each variable satisfies the Eq. (5), where t is the current temperature, *ulimit* and *llimit* are the bounds of the variable.

$$d = (rand(0, 1) * 2 - 1) * t * (ulimit - llimit) \qquad (5)$$

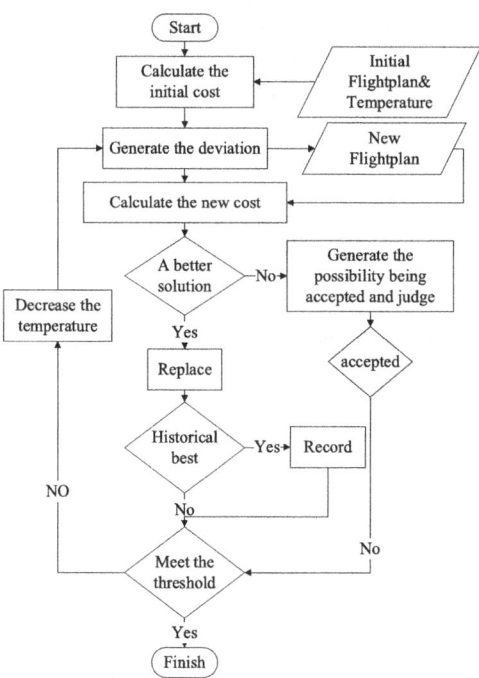

Fig. 4. The calculation flow chart of SA.

Subsequently, the new flight plan will be evaluated by computing its cost. If the new flight plan has lower cost than the old one, the current flight plan will be updated accordingly; otherwise, the current flight plan will be updated with the possibility of,

$$d = p = \frac{1}{1+e^{-\frac{\Delta C}{T}}} \qquad (6)$$

In the function above, T is the initial temperature, ΔC is the difference of cost between new flight plan and the current flight plan. The temperature is reduced by setting $t_{new} = t * \delta$, where δ is the cooling rate. The process move to the next iteration and continues until either the temperature or the cost reaches a specified threshold. Furthermore, to improve the capacity of the algorithm, the historical optimal flight plan is record and updated during calculation.

4 Experiment and Results

4.1 Data Source

The data used in this study was sourced from the Nice Cote d'Azur airport via a radar system, and the format of original data is depicted below in Fig. 5.

```
90397.773 1836 291.078 -204.906 61.500 309.375 0 B734
90405.812 1836 291.719 -204.688 59.750 301.685 0 B734
90413.836 1836 292.359 -204.500 58.500 300.806 0 B734
90421.758 1836 292.984 -204.328 57.250 302.124 0 B734
```

Fig. 5. Sample of the radar data.

The data records the information of arriving aircraft including airplanes and helicopters. Each line represents a detected aircraft. The first column refers to the detecting time in seconds; The second column contains an automatically assigned identifier which in unrelated to the aircraft itself; The third and fourth columns provide the x and y coordinates of the aircraft's position, respectively, with positive x representing east and positive y representing north. These coordinates are given in Nm and can be converted to the runway coordinate system, despite that the original reference point is unknown; The fifth column refers to the flight level; The sixth column refers to aircraft speed in knots; The seventh column refers to the classification of the aircraft, with 0 indicating an airplane and 1 indicating a helicopter; The final column refers to the type of aircraft.

The data presented in a 2-D graph is depicted below in Fig. 6. In the graph, the red line refers to the runway, the violet '+' refers to the helicopters, and the green cross refers to the airplanes.

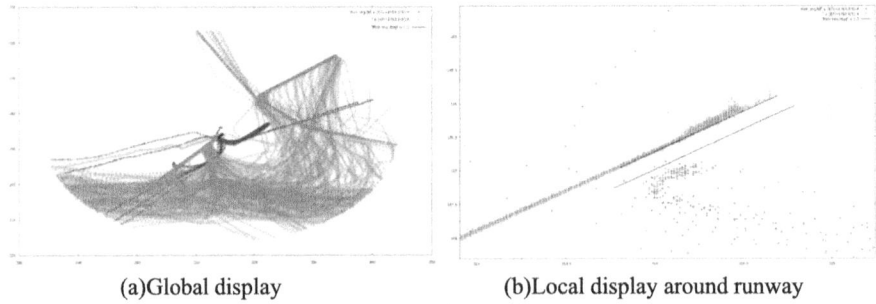

(a)Global display (b)Local display around runway

Fig. 6. Display of the radar data.

As shown in Fig. 6(a), there are 2 runways in the airport, but a closer look at the runway in Fig. 6(b) suggests that only the northern one is in use. Furthermore, although the FATO is not drawn directly in the figure, the distribution of the helicopters suggests that the FATO is very close to the runway. Given that the negligible distance between the runway and the FATO, we choose to ignore the distance and consider that the runway and the FATO as a single point in the map.

4.2 Experiment

The experiment was conducted using the data recorded from the busiest time period of the airport, during which 9 airplanes arrive within about half an hour. Since the real arrival flow of the helicopters is sparse during this period, some instances of helicopter are added to the flow manually to test the performance of our algorithm. Figure 7 shows the snapshot of the algorithm implementation. It is important to note that the left airway was not used in this experiment, but it has been included in the graph to provide a comprehensive view.

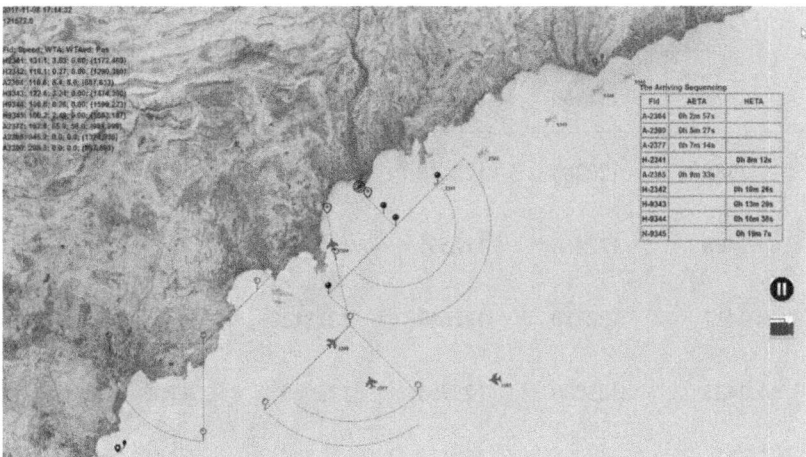

Fig. 7. The snapshot of the algorithm implementation.

The experiment results indicated that no conflicts occurred during landing. Besides, the deviation time for each aircraft is listed, which is defined as the difference between the RTA and the actual arrival time. Furthermore, both the original data and the results using our algorithm are listed in Table 3.

In the table above, all times are measured in seconds, where **FlightID** is the unique identification of each airplane; **RTA** is the Requested Time of Arrival of the airplane; **TA$_{real}$** is the real time of arrival recorded in the radar data; **TA$_{algo}$** is the time of arrival according to the result given by our algorithm; **Dr** is the deviation time of the real data; **Da** is the deviation time of the result given by our algorithm. The table reveals a significant reduction in arrival time deviations, with an average decrease of approximately 50%. As for the helicopter, the average time spent on the PM system is 367s. Without our control strategy, this time would be wasted on hovering, which could increase fuel consumption by approximately 30% compared to a normal flight. Take Bell 47G-2/3 as an example, the fuel consumption is about 60 L/h, which means with our algorithm, each helicopter can save about 1.8 L of fuel on average during landing. The results demonstrate the effectiveness of our algorithm.

While this study proposed an innovative algorithm that demonstrated practical benefits in reducing flight delays, it has several limitations. Firstly, the problem was simplified by neglecting factors such as wind conditions; secondly, the analysis was limited to

Table 3. The comparison of the deviation time.

FlightID	RTA/s	TA$_{real}$/s	TA$_{algo}$/s	Dr/s	Da/s
2366	121360	121296	121388	64	28
2364	121740	121464	121740	276	0
2390	121800	121977	121892	177	92
2377	121860	121600	121988	260	128
2385	122280	122120	122140	160	140
2408	122460	122545	122604	85	144
2417	122780	122649	122812	131	32
2425	123000	122953	123048	47	48
2418	123180	123145	123244	35	64
Average				137	75

data from a busy period at Nice airport, without considering other scenarios; lastly, the algorithm was not compared to other sequencing methods using various metrics, like conflict resolution rates. To overcome these limitations, future research should collect more comprehensive data to strengthen the analysis. With a larger dataset, potential machine learning or AI solutions could be integrated for a comparative evaluation with the proposed algorithm.

5 Conclusion

This study aimed at addressing the problem of joint sequencing and merging optimization for airplanes and helicopters. We proposed an innovative optimization-based control algorithm taking account of factors including trajectory conflicts, aircraft priority, speed limitations, and minimum separation constraints. Additionally, we employed simulated annealing to obtain the optimal solution and applied our algorithm to the data collected from a French airport. The findings demonstrated promising results, achieving approximately 50% less delay compared to the actual schedule, indicating the potential of our approach to significantly improve air traffic efficiency for joint sequencing.

Acknowledgments. This work was supported by the National Key Research and Development Program of China under Grant No. 2023YFB3002800.

Disclosure of Interests. The authors have no competing interests to declare that are relevant to the content of this article.

References

1. Zhang, J., Zhao, P., Zhang, Y., et al.: Criteria selection and multi-objective optimization of aircraft landing problem. J. Air Transp. Manag. **82**, 101734 (2020)
2. Messaoud, M.B.: A thorough review of aircraft landing operation from practical and theoretical standpoints at an airport which may include a single or multiple runways. Appl. Soft Comput. **98**, 106853 (2021)
3. Vié, M.S., Zufferey, N., Leus, R.: Aircraft landing planning under uncertain conditions. J. Sched. **25**(2), 203–228 (2022)
4. Shirini, K., Aghdasi, H.S., Saeedvand, S.: A comprehensive survey on multiple-runway aircraft landing optimization problem. Int. J. Aeronaut. Space Sci. 1–29 (2024)
5. Liang, M., Delahaye, D., Sbihi, M., et al.: Multi-layer point merge system for dynamically controlling arrivals on parallel runways. In: Digital Avionics Syst. Conference (DASC), 2016 IEEE/AIAA 35th. IEEE, pp. 1–9 (2016)
6. Tian, Y., Xu, C., Sun, M., et al.: Study on arrival aircraft sequencing based on optimization of point merge procedure. Discret. Dyn. Nat. Soc. **2021**(1), 6663161 (2021)
7. Gui, D., Le, M., Huang, Z., et al.: Optimal aircraft arrival scheduling with continuous descent operations in busy terminal maneuvering areas. J. Air Transp. Manag. **107**, 102344 (2023)
8. Hu, X.B., Chen, W.H.: Receding horizon control for aircraft arrival sequencing and scheduling. IEEE Trans. Intell. Transp. Syst. **6**(2), 189–197 (2005)
9. EUROCONTROL. RECAT-EU: European Wake Turbulence Categorisation and Separation Minima on Approach and Departure. https://www.eurocontrol.int/publications/recat-eu-european-wake-turbulance-categorisation-and-separation-minima-approach-and Accessed 15 Jul 2015
10. SKYbrary. Mitigation of Wake Turbulence Hazard. https://www.skybrary.aero/index.php/Mitigation_of_Wake_Turbulence_Hazard#Helicopters 4 Aug 2018
11. Kirkpatrick, S., Gelatt, Jr C.D., Vecchi, M.P.: Optimization by simulated annealing. Spin Glass Theory and Beyond: An Introduction to the Replica Method and Its Applications. 339–348 (1987)

Research on UAV SINS/GNSS Integrated Navigation Error Model for Transpolar Flight

Guoqiang Zhang[1], Qi Zhou[2(✉)], and Jinjiang Wang[2]

[1] Northwestern Polytechnical University, Xi'an 710129, China
[2] Flight Automatic Control Research Institute, Xi'an 710076, China
zhouqis@139.com

Abstract. The performance of Strapdown Inertial Navigation System (SINS) and Global Navigation Satellite System (GNSS) integrated navigation largely depends on the formulation of the error equations in polar regions. A key challenge lies in deriving a unified error model that remains effective across the globe. Addressing the limitations of traditional integrated navigation error models for UAVs during transpolar flights, this paper proposes an error equation and correction strategy based on the wander azimuth mechanization of inertial navigation systems. By introducing velocity error as an intermediate variable, the modified Psi-angle error formula is analyzed, reducing update cycles for integrated navigation filters. Furthermore, by examining the projection characteristics of the Psi-angle error state in the computer frame (c), a hypothesis is proposed: aligning the azimuth axis of the platform frame (p) with that of the true frame (t) allows the conversion from the platform frame to the computer frame. This approach resolves the issue of directly utilizing the Psi-angle error state to correct inertial navigation errors. The capabilities of SINS/GNSS integrated navigation in transpolar regions is evaluated through simulation, with results validating the effectiveness of the modified error model based on wander azimuth mechanization. This method avoids the mutual conversion of error parameters between different error models and mitigates filtering instability caused by such conversions, offering a robust solution for polar navigation challenges.

Keywords: Transpolar navigation · integration navigation · error model · INS correction strategy

1 Introduction

With the ongoing effects of global warming, there has been a surge of interest in Arctic scientific exploration. Developing advanced intelligent high-speed unmanned aerial vehicle (UAV) platforms offers the potential to rapidly collect polar atmospheric data, significantly enhancing the efficiency of investigations. The challenges of intelligent UAV control in the complex environments of polar regions have been explored in the literature [1–3], but the navigation issues faced by UAVs in such regions remain a formidable challenge.

Due to its autonomy, stealth, and comprehensive information output, inertial navigation is the preferred method for intelligent UAV transpolar flights. However, in polar regions, the convergence of Earth's meridians causes conventional inertial navigation algorithms based on the true north reference to fail. While the wander azimuth mechanization works effectively on a global scale, it cannot provide heading references or positioning information at high latitudes. To address this, researchers have developed alternative representations of Earth's parameters to circumvent the reliance on true north and longitude lines in polar regions [4–6]. Among the proposed solutions, polar plane navigation [7], polar transverse navigation [8–10], and polar grid navigation [11, 12] are prominent approaches, each with distinct advantages and limitations. Polar plane navigation simplifies computations but suffers from projection distortions at higher latitudes. Polar transverse navigation mitigates heading errors but complicates integration with systems used outside polar regions. Polar grid navigation, on the other hand, redefines Earth's parameters into grid coordinates, overcoming these limitations. Notably, the indirect grid navigation method based on wander azimuth mechanization ensures seamless operation across both middle and low latitudes and polar regions without requiring mechanical modifications, relying only on additional grid navigation parameters [12].

To achieve a globally unified integrated navigation model, the error model must align with the principles of wander azimuth mechanization. The error equation for wander azimuth mechanization traditionally follows the Phi-angle error formulation. However, this formulation introduces wander angle errors related to longitude errors, which become problematic at higher latitudes. These errors can overflow and disrupt the integrated filter, restricting the application of conventional Phi-angle-based error models to low and middle latitudes [5].

In contrast, the Psi-angle error formula [13], while distinct from the Phi-angle equation, offers a simpler and more stable framework. Unlike the Phi-angle error formula, the Psi-angle formulation avoids the complex instruction angular velocity error terms associated with Earth's ellipsoid shape and is independent of velocity and position errors, providing a more concise representation [14–16]. Furthermore, the traditional error equation includes specific force-related terms, which necessitate a higher refresh rate for the integrated filter in high-dynamic environments, increasing system complexity. The Psi-angle error formula does not include this error, and the Psi-angle error is not related to speed and position error, and the form is more concise. Regardless of the Phi-angle or the Psi-angle error formula, the conventional form of the error equation contains a specific force related term. In high dynamic environment, the integrated filter has a faster time required refresh period, thus increasing the difficulty and complexity of system implementation, and therefore need to improve the filter without changing the hardware configuration of the model based on the conventional.

This study proposes a modified Psi-angle error formula that excludes specific force information, enabling stable implementation on a global scale. However, as Psi-angle error formulas are defined in the computational platform frame, the filter parameters must be transformed into an ideal platform frame for effective correction. The lack of a clear physical interpretation for certain parameters in the wander coordinate system poses additional challenges, particularly for the third term of the Psi-angle error, which corresponds to the platform misalignment angle and complicates azimuth correction.

Based on the improved Psi-angle error formula, this paper analyzes a correction strategy for the integrated filter. Simulations of transpolar trajectories using SINS and GPS integration validate the proposed global error model and correction strategy, demonstrating its effectiveness and practicality in addressing the unique challenges of polar navigation.

2 Navigation Reference Coordinate Frame and Relations

To accurately establish an inertial navigation system error model, it is essential to define the position, velocity, and attitude of an aircraft within specific coordinate frames. Several critical frames are introduced for this purpose:

Earth-Centered, Earth-Fixed (ECEF) Frame(e Frame): This frame represents positions relative to the Earth's center. One axis aligns with the Earth's rotational axis, while the other two lie within the equatorial plane, rotating synchronously with the Earth.

Body Frame(b Frame/w Frame): This frame corresponds to the axes of the strapdown inertial sensors, forming a right-handed orthogonal coordinate system based on the sensor orientation.

True Navigation (or Wander) Frame(t Frame): This local frame is level and oriented according to the true navigation position. Its orientation with respect to true north is determined by the wander azimuth angle. The transformation between the wander frame and the body frame is described by a direction cosine matrix (DCM) C_b^t.

ENU Frame(g Frame): The East-North-Up frame is a local geodetic system with "Up" pointing along the ellipsoidal normal, "North" tangent to the meridian, and "East" perpendicular to both.

Grid Frame(G Frame): Designed for grid-based navigation, this frame features a Cartesian structure where "Up" aligns with the ellipsoidal normal, and "North" is parallel to the grid north, defined as the intersection of the local level plane with a Greenwich meridian-aligned grid plane.

Platform Frame(p Frame): This frame is fixed to the vehicle's physical structure, and its orientation relative to the body frame is described by a DCM C_b^p.

Computed Frame(c Frame): This frame represents the coordinate axes derived at the navigation computer's estimated position.

To describe the angular relationships between these frames, the following error angles are defined:

$\delta\theta$: The angle between the true frame and the computed frame.

ϕ: The angle between the true frame and the platform frame.

ψ: The angle between the computed frame and the platform frame.

While the true and platform frames share the same geographic position, their orientations differ. In contrast, the computed frame deviates from the true frame in both position and orientation. These angular relationships form the foundation of the error model, as illustrated in Fig. 1.

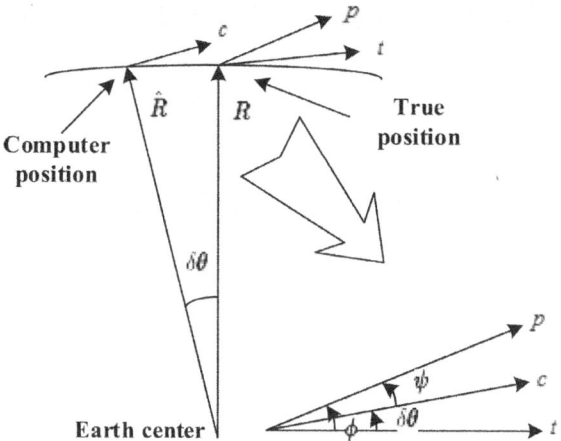

Fig. 1. ϕ, ψ and $\delta\theta$ definitions.

For terrestrial navigation, the $\delta\theta$ error is a direct consequence of longitude and latitude position errors, and the ϕ error is a direct consequence of platform tilt and heading errors. Under the premise of not causing ambiguity, the three sets of coordinate frames are simply referred to as the true frame, platform frame p, and computed frame c. Since these angles are small, they are treated as vectors and can be resolved into any coordinate frame. The DCM can be written:

$$C_t^c = \begin{bmatrix} 1 & \delta\theta_z & -\delta\theta_y \\ -\delta\theta_z & 1 & \delta\theta_x \\ \delta\theta_y & -\delta\theta_x & 1 \end{bmatrix} \quad (1)$$

$$C_t^p = \begin{bmatrix} 1 & \phi_z & -\phi_y \\ -\phi_z & 1 & \phi_x \\ \phi_y & -\phi_x & 1 \end{bmatrix} \quad (2)$$

$$C_c^p = \begin{bmatrix} 1 & \psi_z & -\psi_y \\ -\psi_z & 1 & \psi_x \\ \psi_y & -\psi_x & 1 \end{bmatrix} \quad (3)$$

Noting from definition of the error angles that

$$C_t^p = C_c^p C_t^c \quad (4)$$

We have

$$\phi = \psi + \delta\theta \quad (5)$$

3 Traditional Inertial Navigation System Error Models

3.1 Traditional Phi-Angle Error Formula

Most inertial navigation error equations are formulated using the ϕ and $\delta\theta$ error variables. These equations are derived by binding the inertial navigation equations to a linear perturbation analysis with respect to the true navigation frame, commonly referred to as the traditional Psi-angle error models [17]. The corresponding equations are represented in (6) to (8):

$$\dot{\phi} = -\left(\omega_{ie}^t + \omega_{et}^t\right) \times \phi + \left(\delta\omega_{ie}^t + \delta\omega_{et}^t\right) - \varepsilon^t \tag{6}$$

$$\delta\dot{v}_1^t = f_{sf}^t \times \phi - \left(2\omega_{ie}^t + \omega_{et}^t\right) \times \delta v_1^t + v_t^t \times \left(2\delta\omega_{ie}^t + \delta\omega_{et}^t\right) + \nabla^t + \Delta g_1^t \tag{7}$$

$$\delta\dot{R}^t = -\omega_{et}^t \times \delta R^t + \delta\theta \times v^t + \delta v_1^t \tag{8}$$

Variables and Definitions:
ω_{ie}^t: Earth rate vector in the true navigation frame;
ω_{et}^t: Transport rate vector (angular rate relative to Earth) in the true navigation frame.
$\delta\omega_{ie}^t$: Error in the Earth rate vector in the true navigation frame.
$\delta\omega_{et}^t$: Error in the transport rate vector in the true navigation frame.
f_{sf}^t: Specific force vector in the true navigation frame.
v_t^t: True velocity vector in the true navigation frame.
ε^t: Gyro error vector in the true navigation frame.
∇^t: Accelerometer error vector in the true navigation frame.
δv_1^t: Velocity error vector from perturbation analysis.
Δg_1^t: Gravity anomaly vector from perturbation analysis.
δR^t: Position error vector in the true navigation frame.

Equations (6) to (8) highlight the coupling between position, velocity, and attitude errors due to the inclusion of $\delta\omega_{ie}^t$ and $\delta\omega_{et}^t$ terms. This coupling increases complexity when solving the errors, especially when considering the Earth as an ellipsoid. Under the wander azimuth mechanization, the error solutions are not intuitive and require additional computational effort. To streamline the model, the Earth is often approximated as a sphere, which improves convenience at the cost of reduced accuracy.

3.2 Traditional Psi-Angle Error Formula

Unlike the traditional Phi-angle error formulas, the Psi-angle error analysis derives error equations directly from perturbations of the computed navigation solution. The corresponding Psi-angle error models are expressed in Eqs. (9) to (11).

$$\dot{\psi} = -\left(\omega_{ie}^c + \omega_{ec}^c\right) \times \psi - \varepsilon^c \tag{9}$$

$$\delta \dot{v}_2^c = f_{sf}^c \times \psi - (2\omega_{ie}^c + \omega_{ec}^c) \times \delta v_2^c + \nabla^c + \Delta g_2^c \tag{10}$$

$$\delta \dot{R}^t = -\omega_{ec}^c \times \delta R^t + \delta v_2^c \tag{11}$$

Variables and Definitions:
ω_{ie}^c: Earth rate vector in the computed navigation frame.
ω_{ec}^c: Transport rate vector in the computed navigation frame.
f_{sf}^c: Specific force vector in the computed navigation frame.
δv_2^c: Velocity error vector from Psi-angle error analysis.
Δg_2^c: Gravity anomaly vector from Psi-angle error analysis.

The Psi-angle error formulas differ from the traditional Phi-angle approach in that they decouple the angular errors from velocity and position errors. Consequently, ω_{ie}^c and ω_{ec}^c can be determined directly without assumptions of simplified errors, enhancing the clarity and intuitiveness of the model.

3.3 Relationship Between Phi-Angle Error Formula and Psi-Angle Error Formula

In Eqs. (6) to (8), the variables are defined within the context of the inertial navigation system (INS). Due to inherent errors in the INS, the variables introduced in these equations are as follows:

$$\hat{v} = v_t^t + \delta v_1^t \tag{12}$$

$$\hat{g} = g_t^t + \Delta g_1^t \tag{13}$$

$$\hat{\omega}_{ie}^t = \omega_{ie}^t + \delta \omega_{ie}^t \tag{14}$$

$$\hat{\omega}_{et}^t = \omega_{et}^t + \delta \omega_{et}^t \tag{15}$$

These variables are defined in the true navigation frame, resulting from the perturbation of the true navigation solution. The use of different methods for error analysis can lead to different interpretations of the error variables. In this case, the variables are defined in the computational navigation framework, as shown in Eqs. (9) to (11):

$$\hat{v} = v_t^t + \delta v_2^t \tag{16}$$

$$\hat{g} = g_t^t + \Delta g_2^t \tag{17}$$

$$\hat{\omega}_{ie}^t = \omega_{ie}^c \tag{18}$$

$$\hat{\omega}_{et}^t = \omega_{et}^c \tag{19}$$

Identifying the equivalent variables by comparing equation formula (12), (13), (14), (15) through Eq. (16), (17), (18) and (19) gives

$$\hat{v} = v_t^t + \delta v_1^t = v_t^c + \delta v_2^c \tag{20}$$

$$\hat{g} = g_t^t + \Delta g_1^t = g_t^c + \Delta g_2^c \tag{21}$$

$$\hat{\omega}_{ie}^t = \omega_{ie}^t + \delta \omega_{ie}^t = \omega_{ie}^c \tag{22}$$

$$\hat{\omega}_{et}^t = \omega_{et}^t + \delta \omega_{et}^t = \omega_{ec}^c \tag{23}$$

The relationship between variables in the true navigation frame and the computed navigation frame is determined through the DCM. For small angles, the DCM is defined as shown in Eq. (24):

$$C_t^c = \begin{bmatrix} 1 & \delta\theta_z & -\delta\theta_y \\ -\delta\theta_z & 1 & \delta\theta_x \\ \delta\theta_y & -\delta\theta_x & 1 \end{bmatrix} = I - \delta\theta \times \tag{24}$$

Substituting Eq. (24) into (16) gives

$$\begin{aligned} \delta v_1^t &= v_t^c - v_t^t + \delta v_2^c \\ &= v_t^c - (C_t^c)^T v_t^c + \delta v_2^c \\ &= v_t^c - (1 + \delta\theta \times) v_t^c + \delta v_2^c \\ &= -\delta\theta \times v_t^c + \delta v_2^c \end{aligned} \tag{25}$$

In another form:

$$\begin{aligned} \delta v_1^t &= v_t^c - v_t^t + \delta v_2^c \\ &= (C_t^c) v_t^t - v_t^t + \delta v_2^c \\ &= (1 - \delta\theta \times) v_t^t - v_t^t + \delta v_2^c \\ &= -\delta\theta \times v_t^t + \delta v_2^c \end{aligned} \tag{26}$$

which with Eq. (25) reduces to a approximately relationship:

$$\delta\theta \times v_t^c = \delta\theta \times v_t^t \tag{27}$$

Substituting Eq. (24) into (21) through (22) gives

$$\begin{aligned}
\delta v_1^t &= -\delta\theta \times v_t^c + \delta v_2^c \\
\Delta g_1^t &= -\delta\theta \times g_t^c + \Delta g_2^c \\
\delta\omega_{ie}^t &= -\delta\theta \times \omega_{ie}^c \\
\delta\omega_{et}^t &= -\delta\theta \times \omega_{et}^c
\end{aligned} \quad (28)$$

The analysis reveals the equivalence between the Phi-angle and Psi-angle error formulas. The differences in velocity error and misalignment angle definitions arise naturally from whether the navigation system is viewed as solving the force equation in true or computed frame.

4 Modified Inertial Navigation System Error Models

4.1 Modified Psi-Angle Error Formula

Although the Phi-angle equations are mathematically equivalent to the Psi-angle equations, the Psi-angle formulation offers a notable advantage in simplicity. Unlike the Phi-angle equations, the Psi-angle equations decouple position and velocity errors from the misalignment angle equations, making them less complex to solve. Both angle error equations include specific force terms f_{sf}, which pose challenges in high-dynamic environments. To track rapid carrier dynamics, higher update frequencies are required, significantly increasing implementation complexity. Therefore, the Psi-angle error equations must be improved to address these challenges effectively.

Due to

$$\hat{v} = v_t^t + \delta v_1^t = v_t^c + \delta v_2^c \quad (29)$$

Substituting Eq. (1) into (12) gives:

$$\delta v_1^t = -\delta\theta \times v_t^c + \delta v_2^c \quad (30)$$

To simplify the model and cancel the specific force term, a modified velocity error definition is introduced. This new velocity error differs from the velocity errors in Eqs. (7) and (10). Although it can eliminate the specific force term, it is not a true velocity error in the strict sense of the term, because it cannot be characterized as the calculated velocity minus the true velocity in the common reference frame.

$$\delta v \triangleq \hat{v}_t^p - v_t^t = \delta v_2^c - \hat{v} \times \psi \quad (31)$$

where

δv: The difference between the computational speed resolved on the platform frame (p) and the real speed resolved on the true navigation frame (t).

$\hat{v}$: Solved from the specific force equation in the platform frame.

$$\dot{\hat{v}} = f^p + \hat{g} - (2\omega_{ie}^c + \omega_{ec}^c) \times \hat{v} \quad (32)$$

The formula (32) is differential:

$$\delta \dot{v}_2^c = \delta \dot{v} \times \psi + \hat{v} \times \dot{\psi} \tag{33}$$

Substituting ψ differential Eq. (9) and specific force Eq. (32) into new velocity error differential Eq. (33) yields:

$$\begin{aligned}\delta \dot{v}_2^c &= \delta \dot{v} + \left(f^p + \hat{g} - \left(2\omega_{ie}^c + \omega_{ec}^c\right) \times \hat{v}\right) \times \psi \\ &\quad + \hat{v} \times \left(-\left(\omega_{ie}^c + \omega_{ec}^c\right) \times \psi - \varepsilon^c\right) \\ &= \delta \dot{v} + f^p \times \psi + \hat{g} \times \psi - \hat{v} \times C_b^p \varepsilon^b \\ &\quad - \left(\left(2\omega_{ie}^c + \omega_{ec}^c\right) \times \hat{v}\right) \times \psi \\ &\quad - \hat{v} \times \left(\left(\omega_{ie}^c + \omega_{ec}^c\right) \times \psi\right) - \hat{v} \times C_b^p \varepsilon^b\end{aligned} \tag{34}$$

Substituting Eq. (9) and Eq. (32) into Eq. (33) results in the following modified ψ angle error model:

$$\dot{\psi} = \left(\omega_{ie}^c + \omega_{ec}^c\right) \times \psi - C_b^p \varepsilon^b \tag{35}$$

$$\begin{aligned}\delta \dot{v} &= -\hat{g} \times \psi - \left(2\omega_{ie}^c + \omega_{ec}^c\right) \times \delta v + \Delta g_2^c \\ &\quad - \hat{v} \times \left(\omega_{ie}^c \times \psi\right) + \hat{v} \times \left(C_b^p \varepsilon^b\right) + C_b^p \nabla^b\end{aligned} \tag{36}$$

$$\delta \dot{R} = -\omega_{ec}^c \times \delta R^c + \delta v + \hat{v} \times \psi \tag{37}$$

$$\dot{\varepsilon}^b = -\frac{1}{\tau_\varepsilon} \varepsilon + \omega_\varepsilon \tag{38}$$

$$\dot{\nabla}^b = -\frac{1}{\tau_\nabla} + \omega_\nabla \tag{39}$$

$$\Delta g_2^c = diag[\omega_\varepsilon^2, \omega_\varepsilon^2, 2\omega_\varepsilon^2]\delta R^c \tag{40}$$

The improved ψ angle error formula independent of the specific force term and are only related to the velocity. At the same time, Eq. (35) shows that when position information is used as a measure and the speed of vehicle exists, we can effectively observe the azimuth error, and directly indicate the correction function of position measurement on heading.

4.2 Modified Transpolar Integrated Filter Model

Because of the weak coupling relationship between the high channel and the horizontal channel, the correction of the height channel is often carried out separately. Based on this, only the 12 dimensional error state of the horizontal axis is taken to construct the transpolar integrated filter model. Among them, the definition of the error state is shown in Table 1.

Table 1. Integrated navigation error states

Symbol	MEANING	Unit
ψ_x	x axis misalignment angle in wander frame	Rad
ψ_y	y axis misalignment angle in wander frame	Rad
ψ_z	z axis misalignment angle in wander frame	Rad
δv_x	x axis velocity error	m/s
δv_y	y axis velocity error	m/s
δR_x	x axis position error in wander frame	m
δR_y	y axis position error in wander frame	m
ε_x	x axis gyro drift	Rad/s
ε_y	y axis gyro drift	Rad/s
ε_z	z axis gyro drift	Rad/s
∇_x	x axis accelerometer bias	m/s^2
∇_y	y axis accelerometer bias	m/s^2

5 Error Correction Strategy

5.1 Measurement Equation

1) Position measurement

When the GNSS position is considered as the measurement, the ECEF position $\hat{R}_{GNSS}^e$ of the satellite navigation are used as the measurement so as to satisfy the transpolar integrated navigation.

$$z_P = \hat{R}_{INS}^e - \hat{R}_{GNSS}^e \tag{41}$$

where
$\hat{R}_{INS}^e$: is the strapdown inertial position resolved in the ECEF frame.
For the horizontal channel, the measurement model is as follows:

$$z_{RH} = H_R x + v_R \tag{42}$$

where

$$H_R = [D_t^e \ 0_{2 \times 10}] \tag{43}$$

D_t^e : is the 2×2 matrix of the upper left corner of DCM C_t^e;
v_R: is the GNSS position error.

2) Velocity measurement

When the GNSS velocity is considered as the measurement, the ECEF velocity $\hat{v}^e_{GNSS}$ of the satellite navigation are used as the measurement so as to satisfy the transpolar integrated navigation.

$$z_V = \hat{v} - \hat{C}^w_e \hat{v}^e_{GNSS} \tag{44}$$

$\hat{v}$: Three velocity components indicated by the wander azimuth mechanical navigation system;
$\hat{v}^e_{GNSS}$: Three velocity components displayed by the GNSS receiver system;
$\hat{C}^w_e$: The DCM from the e frame to the w frame.

Due to

$$\hat{C}^w_e = C^c_e \tag{45}$$

$$\hat{v} = v^c_t + \delta v^c_2 \tag{46}$$

$$\hat{v}^e_{GNSS} = v^e_t + \delta v^e_{GNSS} \tag{47}$$

Substituting formula (31), (45), (46) and (47) into the formula (44):

$$\begin{aligned} z_V &= v^c_t + \delta v^c_2 - C^c_e(v^e_t + \delta v^e_{GNSS}) \\ &= \delta v^c_2 - \delta v^e_{GNSS} \\ &= \delta v + \hat{v} \times \psi - \delta v^c_{GNSS} \end{aligned} \tag{48}$$

Therefore, the horizontal channel velocity measurement model is as follows:

$$z_{VH} = H_V x + v_V \tag{49}$$

where

$$H_V = \begin{bmatrix} 0 & -v^w_z & v^w_y & 1\ 0\ 0_{1\times 7} \\ v^w_z & 0 & -v^w_x & 0\ 1\ 0_{1\times 7} \end{bmatrix} \tag{50}$$

5.2 Error Correction

The error of ϕ angle error estimation is relatively intuitive, which can be directly used as a correction amount of navigation error, such as velocity error δv_1, indicating the real velocity error of navigation system. The error of the ψ angle error model is in the calculation frame, and it is necessary to convert it into the true navigation frame.

First, the horizontal position error δR_x and δR_y estimated by the integrated navigation filter can directly construct the horizontal component of $\delta\theta$ error vector:

$$\delta\theta_x = -\delta R_y/(R_E + h) \tag{51}$$

$$\delta\theta_y = -\delta R_x/(R_E + h) \tag{52}$$

For the wander mechanization INS, it is assumed that the azimuth axis of the platform system at any time is the azimuth axis of the ideal platform system, that is, the assumption of $\phi_z = 0$, then $\delta\theta_z = -\psi_z$. Combining the ψ_x, ψ_y and ψ_z angles of the filter, the ϕ angle error vector can be obtained.

$$\begin{bmatrix} \phi_x \\ \phi_y \\ \phi_z \end{bmatrix} = \begin{bmatrix} \psi_x + \delta\theta_x \\ \psi_y + \delta\theta_y \\ 0 \end{bmatrix} \quad (53)$$

1) Position error correction

The position of wander azimuth mechanization INS is determined indirectly by position DCM $\hat{C}_t^e$, so the position error is eliminated by indirect correction of position DCM $\hat{C}_t^e$.

The error of the position DCM $\hat{C}_t^e$ under the t frame is δC_t^e, which satisfies:

$$\delta C_t^e = C_c^e (C_t^c - I) \quad (54)$$

The correction process of the position DCM $\hat{C}_t^e$ is:

$$C_t^e = C_c^e + \delta C_t^e = \delta \hat{C}_t^e + \delta C_t^e \quad (55)$$

where, $\delta\theta_x, \delta\theta_y$ and $\delta\theta_z$ are obtained by formula (51), (52) and (53), and the error correction matrix δC_t^e of the position DCM $\hat{C}_t^e$ can be obtained by combining the formula (55).

2) Velocity error correction

In the transpolar integrated navigation error model, the velocity error δv is an intermediate variable, which needs to be converted to the true frame (t).

Substituting (14) into (13), then the velocity error correction under the true frame (t) can be obtained by formula (55).

$$\delta v_1^t = -\delta v + \phi \times \hat{v} \quad (56)$$

If only the velocity error in the horizontal channel is corrected, the formula (56) can be simplified to:

$$\delta v_x^t = \delta v_x + \phi_y v_z \quad (57)$$

$$\delta v_y^t = \delta v_y - \phi_x v_z \quad (58)$$

3) Attitude Error Correction

The misalignment angle ϕ is determined by the formula (53).

6 Simulation Analysis

The initial position of the trajectory is $[83°N \ 108°E \ 5000m]$ and is flying at the $108°E$ meridian for two hours at $300m/s$ speed. The performance parameters of inertial devices used in simulation are: gyro random drift 0.01 o/h, random walk coefficient 0.002 o/$\sqrt{h}$, accelerometer random constant bias 40 μg, random walk coefficient 8 $\mu g/\sqrt{Hz}$. The initial error of simulation is set: the initial attitude error [0.003,0.003, $-0.06]°$, the initial velocity error [0.01,0.01,0.01] m/s.

Figure 2 and 3 is the transpolar integrated navigation error using the conventional ϕ angle error formula. Figure 4 and 5 is the transpolar integrated navigation error using the modified ψ angle error formula.

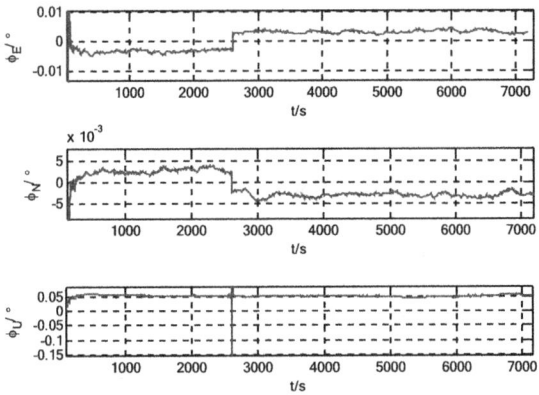

Fig. 2. Misalignment angle based on conventional Phi-angle error models.

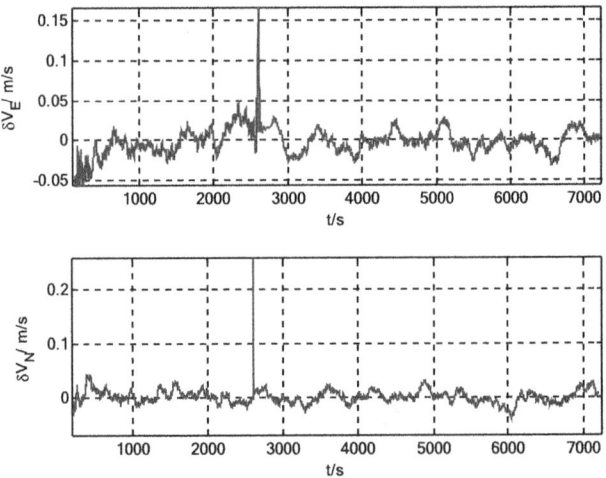

Fig. 3. Velocity error based on conventional Phi-angle error models.

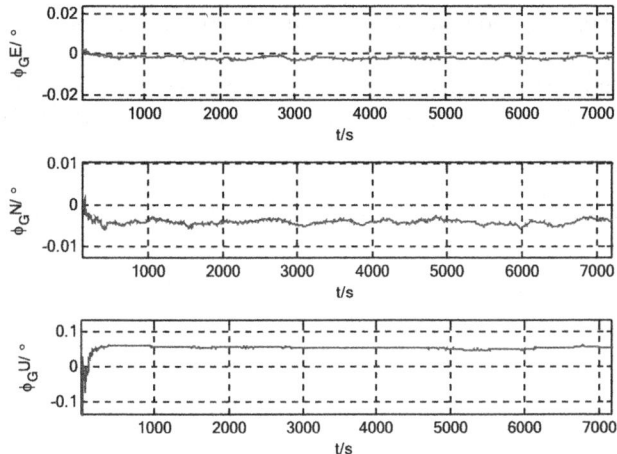

Fig. 4. Misalignment angle based on modified Psi-angle error model.

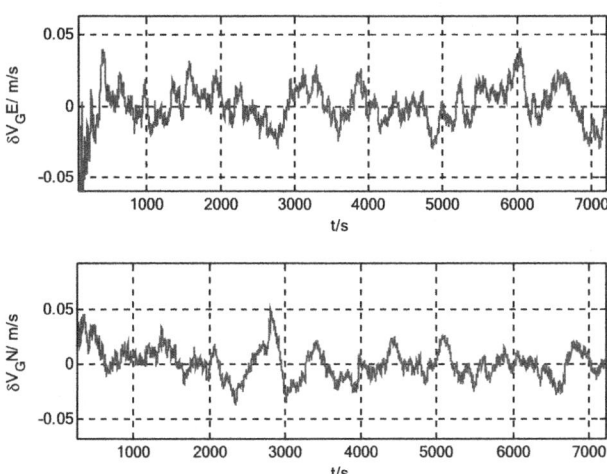

Fig. 5. Velocity error based on modified Psi-angle error model.

Simulation results show that the traditional Phi-angle integrated error model has a singular value near the pole, which affects the navigation performance of the transpolar flight. The modified Psi-angle error model near the poles there is no singular value, to meet the navigation requirements across the region.

7 Conclusion

This study tackles the challenges of integrated navigation in transpolar regions by conducting a comparative analysis of the traditional Phi-angle and Psi-angle error models. Building on the identified strengths and weaknesses of these models, an improved Psi-angle error model is developed to specifically address the unique demands of navigation in polar environments. Additionally, a tailored correction strategy for error estimation is designed based on the improved model's characteristics. Simulation results validate the enhanced Psi-angle error model, showcasing its superior performance and reliability in transpolar navigation scenarios, effectively overcoming the limitations of conventional models.

Acknowledgments. This work was supported by the Aeronautical Science Fund of China under Grant 20165853041.

Disclosure of Interests. The authors have no competing interests to declare that are relevant to the content of this article.

References

1. Yang, C., Teng, T., Xu, B., Li, Z., Na, J., Su, C.-Y.: Global adaptive tracking control of robot manipulators using neural networks with finite-time learning convergence. Int. J. Contr. Autom. Syst. (IJCAS), vol. 15, no. 4, pp. 1916–1924. https://doi.org/10.1007/s12555-016-0515-7
2. Yang, C., Zeng, C., Liang, P., Li, Z., Li, R., Su, C.-Y.: Interface design of a physical human robot interaction system for human impedance adaptive skill transfer. IEEE Trans. Autom. Sci. Eng. https://doi.org/10.1109/TASE.2017.2743000
3. Yang, C., Wang, X., Cheng, L., Ma, H.: Neural-learning based telerobot control with guaranteed performance. IEEE Transa. Cybern. vol. 47, no. 10, pp.3148–3159. https://doi.org/10.1109/TCYB.2016.2573837
4. Greenaway, K.R., Gates, M.D.: Polar Air Navigation-A Record. Canada: Art Bookbindery (2009)
5. Savage, P.G.: Strapdown analyticsII. Maple Plain, Minnesota: Strapdown Associates (2009)
6. Department of the Air Force. Air Navigation (US. Air Force Pamphlet 11–216). CreateSpace Independent Publishing Platform (2013)
7. 周琪. 大飞机全球惯性导航算法研究[D].西安:西北工业大学 (2013)
8. Li, Q., Sun, F., Ben, Y.Y., et al.: Polar navigation of strapdown inertial navigation system based on transversal frame in polar region. J. Chinese Inertial Technol. **22**(3), 288–295 (2014)
9. Li, Q., Sun, F., Ben, Y.Y., Yu, F.: Transversal strapdown INS and damping design in polar region. Syst. Eng. Electron. **36**(12), 2496–2503 (2014)
10. Yao Yiing, X., Xiaou, T.: Indirect transverse inertial navigation algorithm in polar region. J. Chinese Inertial Technol. **23**(1), 29–34 (2015)
11. Zhou, Q., Qin, Y.Y., Fu, Q.W., et al.: Grid mechanization in inertial navigation systems for transpolar aircraft. J. Northwestern Polytechn. Univ. **31**(2), 210–217 (2013)
12. Zhou, Q., Yue, Y.Z., Zhang, X.D., Tian, Y.: Indirect grid inertial navigation mechanization for transpolar aircraft. J. Chinese Inertial Technol. **22**(1), 18–22 (2014)
13. Chen, K., Chang, G., Chen, C., et al.: An improved TDCP-GNSS/INS integration scheme considering small cycle slip for low-cost land vehicular applications. Meas. Sci. Technol. **32**(5), 055006 (2021)
14. Benson, D.O.: A comparison of two approaches to pure-inertial and doppler-inertial error analysis. Trans. Aerosp. Electron. Syst. **AES-11**(4) (1975)
15. Goshen-Meskin, D., Bar-Itzhack, I.Y.: Unified approach to inertial navigation system error modeling to inertial navigation system error modeling. J. Guidance Control Dyn. **15**(3) (1992)

16. West-Vukovich, G., Zywiel, J., Scherzinger, B., Russell, H., Burke, S.: The honeywell/DND helicopter integrated navigation system(HINS). IEEE AES Magazine (1989)
17. Scherzinger, B.M., Reid, D.B.: Modified strapdown inertial navigator error models. In: Proceedings of 1994 IEEE Position, Location and Navigation Symposium-PLANS'94. IEEE, pp. 426–430 (1994)
18. Benson, D.O.: A comparison of two approaches to pure-inertial and doppler-inertial error analysis. IEEE Trans. Aerosp. Electron. Syst.Aerosp. Electron. Syst. **4**, 447–455 (1975)

Technical Research on Helicopter Blind Landing Under Degraded Visual Environments (DVE)

Yanwei Du[1,2(✉)], Mingdong Qi[2], and Heng Zhang[2]

[1] School of Automation, Northwestern Polytechnical University, Xi'an 710072, China
dyw2369@163.com
[2] AVIC Xi'an Flight Automatic Control Research Institute, Xi'an 710065, China

Abstract. Based on the analysis of the mission requirements of helicopter field combat scenario, starting with the field blind landing mission in complex environment, this paper first analyzes the domestic and foreign solutions for helicopter landing site, then designs the search and sequencing method of helicopter landing landing sites based on safety constraints, and designs the route plan of blind landing Finally, the solution simulation of the guidance instruction is carried out, and the integrated verification of the comprehensive visual scene is carried out. The experiment shows that this technology is reasonable and effective, which can reduce the operating pressure of the aircrew, improve the operational efficiency and maneuverability of military helicopter in field battlefield, and lay a foundation for the further development of military helicopter.

Keywords: Helicopter Landing · Visual Guided · Route Planning

1 Introduction

In modern three-dimensional battlefield, helicopters are widely used in dangerous tasks such as information monitoring, reconnaissance, battle damage assessment, communication relay and fire attack [1]. With the continuous upgrading of modern warfare methods, the requirements for helicopters in military battlefields have increased gradually, including consideration of their own safety performance, operational efficiency, and pilot-assisted piloting capabilities [2]. Therefore, how to reduce the pilot's operating load and improve its own autonomous flight capability and flight safety is one of the key technical problems encountered in the process of helicopter performance upgrading [3].

Combined with the application scenario of military helicopter, aiming at improving its defects, the technology of blind landing is particularly important [4]. The helicopter with the capability of blind landing has higher maneuverability and reliability than the ordinary helicopter, can adapt to the combat environment better, and can effectively improve the survival rate of the aircraft itself and mission success rates [5]. Therefore, helicopters with capabilities of blind landing will play a more and more important role in the future military battlefield [6].

In order to improve the success rate of helicopter's autonomous landing in unknown environment under low visibility condition and improve the battlefield survivability, the research on helicopter blind landing assistance project and technology under low visibility has been carried out to varying degrees abroad.

1.1 TALOS Project

In April 2014, the U.S. Naval Research Agency selected TALOS (Tactical Autonomous Aviation Logistics System), the Aurora Flight Science Company, to develop a generic autonomous air cargo system that frees pilots from heavy and dangerous driving tasks and reduces the reliance on personnel on the ground for landing processes [7]. Forward ground forces provide faster and more accurate logistics and battlefield assistance. TALOS is a generic autonomous control system that integrates with unmanned/manned helicopters and interacts with the aircraft's platform management system and flight control system with information and instructions to accuse the aircraft of autonomous flight, transport required materials to designated locations, safely and autonomously return. The airborne system consists of sensors such as laser radar and photoelectric/infrared camera, as well as calculation units such as path planner and trajectory planner, which solves two key technical problems:

The TALOS system adopts low-altitude real-time path planning, differing from other waypoint-based autonomous flight control systems. It generates flight instructions in real time and submits them to the flight control system. At the beginning of the mission, the path planner generates a preset waypoint; during flight, a laser radar continuously images and identifies obstacles in front of the helicopter. Based on the waypoint, mission status, perception information, and aircraft status, the trajectory planner uses "greedy algorithms" and "safety algorithms" to generate real-time flight instructions within 1 s, guiding the aircraft to avoid obstacles or no-fly zones. Additionally, TALOS achieves autonomous landing decision-making by scanning the terrain, evaluating landing areas, selecting suitable landing points, and analyzing approach and evacuation routes, while using infrared cameras for navigation assistance.

TALOS is highly autonomous, versatile, and easy to operate, allowing it to autonomously handle tasks like obstacle avoidance and landing-point evaluation with minimal ground crew intervention, integrate with various helicopters that meet NATO standards, and be operated by ordinary soldiers after just 15 min of training.

1.2 HALS Project

In mid-2013, the Aeronautical Applications Technology Division of the Research, Development, and Engineering Command of the American Armies contracted Sierra Nevada Corporation to integrate and test the company's Helicopter Autonomous Landing Systems (HALS) on board the Army-6 UH-60A/ L helicopter [8].

Sierra Nevada's HALS helicopter avionics use 3D imagery with 94 GHz pulsed radar, GPS, and inertial sensors to help pilots navigate through dust, fog, and snow by providing graphical representations of terrain and obstacles on cockpit displays. This system enhances situational awareness, enabling safe takeoff, landing, and flight in poor visibility by allowing pilots to avoid wires and terrain, follow landmarks, and transition smoothly from visual to instrument flight conditions, while also featuring dust symbol software for precise landing guidance. HALS Project Effect Picture is shown in Fig. 1.

Fig. 1. HALS Project Effect Picture

1.3 DEV-M Project

DEV-M project is verified on Yuma helicopter, aiming at improving the success rate of helicopter landing site under adverse visual conditions such as dust and night. By means of integrating multi-system radar, updating flight control system and navigation equipment, the helicopter can realize the perception of the surrounding environment of the carrier aircraft under bad visual conditions, identify the potential landing areas independently, plan the landing paths, speed and altitude limitation independently, and improve the vertical ascent. On-the-fly landing safety. The test results show that the helicopter successfully completed 64 landing times and 6 return flights under bad visual conditions, with a success rate of 91%, which is greatly improved compared with that without installation of equipment [9]. The flight test diagram of DEV-M project is shown in Fig. 2.

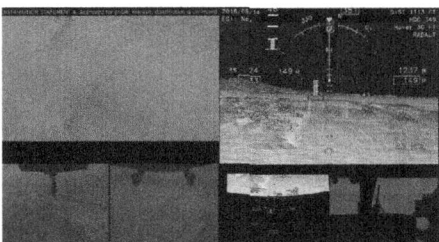

Fig. 2. Flight Test Diagram of DEV-M Project

This paper summarizes the typical helicopter blind landing assistance projects abroad, which improve the success rate of helicopter landing by means of radar, high precision map and landing assistance guidance, etc.

The Summary of Demonstration and Verification Items for Blind Landing of Typical Helicopter Abroad is shown in Table 1. The other studies focus on advancements in mitigating challenges posed by degraded visual environments (DVE) for rotor-craft operations, mainly related to Real-Time Imaging [10], Pilot Cueing Systems [11], Paradigm Shift [12], Flight Trials [13, 14] and so on. Overall, these studies collectively contribute to the ongoing efforts to enhance hovercraft operational capabilities in adverse visual conditions through innovative technologies and methodologies.

Table 1. Summary of Demonstration and Verification Items for Blind Landing of Typical Helicopter Abroad

PROJECTS	Project Scheme	Verification Results
TALOS Project	1) Low altitude real-time path planning 2) Scan terrain and generate geo-morphologic map 3) Landing-zone identification 4) Assessment of Zhizhi-6 Area 5) Approach and withdrawal route planning	It improves the ability of landing-court safety, reduces the reliance on ground personnel, and can command the plane to land site or return in case of emergency
HALS Project	1) LIDAR ASSIST 2) 3D environment rendering 3) Obstacle avoidance assistance 4) Landing assistance	Complete test verification on UH-60A/ L helicopter to improve landing abilities of situational awareness and autonomous handling under adverse visual conditions
DEV-M Project	1) Laser/photoelectric radar assistant 2) Autonomous identification of the landing zones 3) Independent planning 4) Visual auxiliary guidance	The verification was carried out on Yuma helicopter, and the success rate reached 91% under complex weather conditions

2 Programme Structure

According to the identified landing sites, the assistant decision-making of blind landing completes the optimization of landing sites, the planning of landing trajectory, the generation of safe landing pipeline, the calculation of landing guidance command, and realizes the visualization of the field blind landing task based on the integrated visual software.

The landing sites are preferably aimed at the potential landing sites identified by the detection sensor. The distance from the helicopter's current position to the landing fields, the terrain flatness of the landing sites, the size of the landing sites, the surrounding obstacles and terrain threat constraints, etc. shall be taken into consideration to comprehensively calculate each of the landing sites. Then the optimal landing site will be obtained.

Landing trajectory planning divides the landing into glide process and approach process; The gliding process plans the landing trajectory from the initial descent point to the final approach point. The helicopter adjusts the altitude and heading according to the reference landing trajectory to intercept the reference trajectory of the approach section at the final approach point in a fixed state; The approach process plans the landing trajectory from the final approach point to the decision height. In order to ensure the landing safety, the approach section is provided with a protection zone and a standard path is constructed.

The guidance module is the common module of obstacle avoidance assistant decision and blind landing assistant decision, and it is used to solve the landing guidance instruction during blind landing assistant decision.

The integrated visual software realizes the visualization of the field blind landing mission and realizes the display of the safe flight pipeline.

According to the above architecture design, the specific flow of intelligent obstacle avoidance auxiliary decision-making is shown in Fig. 3.

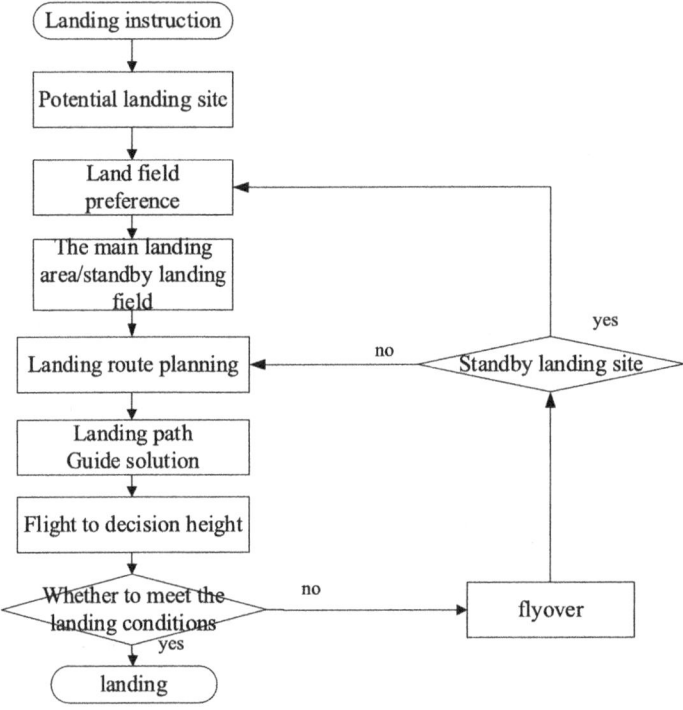

Fig. 3. Flow Chart of Decision-making Aid for Blind Fall

In order to accomplish the task of obstacle avoidance/blind landing in complex environment, it is necessary to solve the authority assignment of pilot and assistant decision algorithm, so as to achieve the best assistance to pilot under the premise of ensuring safety.

3 Detailed Design

3.1 A Landing-Field Search and Sorting Method Based on Security Constraints

Landing-Field Sliding Window Search Method. A sliding window search method based on safe landing constraints is proposed. Set the search width as the area required for the helicopter landing, sliding one pixel at a time from top to bottom and left to right.

The terrain index is designed for the purpose of analyzing the flatness of terrain as shown in Fig. 4. For the digital elevation map of the area to be measured, the local slope, local variance and local roughness are calculated respectively. Local slope, local variance, and local roughness together determine the terrain flatness. Based on expert experience and experimental validation, a flatness threshold is provided. Then, regions with flatness values within the threshold range are identified in the terrain elevation map. The areas that meet the threshold requirements are considered as candidate landing points. The specific method is as follows.

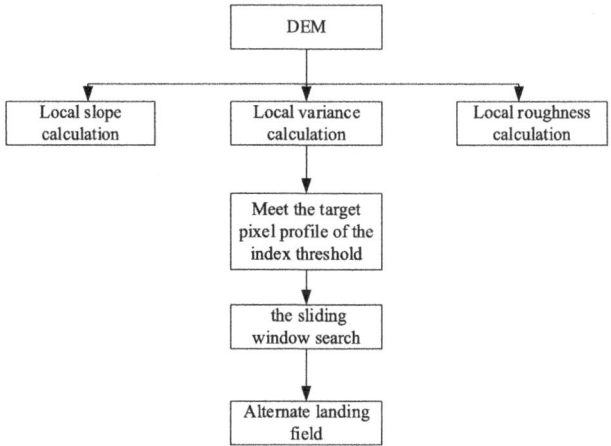

Fig. 4. Flow chart of method

The calculation formula of Landing-field is as follows.

$$Mask(area_{m_s \times m_s}) = \begin{cases} \text{reach the standard} & \#\{(i,j) \in area_{m_s \times m_s} | f_{ys}(i,j) = 1\} > \alpha \cdot m_s^2 \\ \text{not up to the standard} & \#\{(i,j) \in area_{m_s \times m_s} | f_{ys}(i,j) = 1\} \leq \alpha \cdot m_s^2 \end{cases} \quad (1)$$

$$m_s = \left\lfloor \frac{\lceil 200/d \rceil}{2} \right\rfloor \quad (2)$$

m_s is half length of sliding window, $f_{ys}(i,j)$ is described as follows:

$$f_{ys}(i,j) = \begin{cases} 1 & \{S(i,j) < UT_S\} \wedge \{\sigma^2(i,j) < UT_{\sigma^2}\} \wedge \{R(i,j) < UT_R\} \\ 0 & others \end{cases} \quad (3)$$

Mask is the output result, $area_{m \times n}$ it's the sliding window area, α is pixel occupy ratio parameter reach the standard, $f_{ys}(i,j)$ is a constraint condition, UT_S, UT_{σ^2}, UT_R these symbols represent the threshold values for slope, variance, and roughness indicators, respectively.

Landing Alternative Sequences. The alternative landing area obtained by terrain analysis based on digital elevation map is selected by the crew to obtain landing fields of several optional helicopters, as shown in Fig. 5. Firstly, judge the limiting factors of each landing fields, and then determine whether the landing fields meet the limiting conditions, and then according to the landing site object and size. Such as shape, geographical location, terrain index, etc. The evaluation value of terrain flatness can be obtained by judging the secondary factors (slope, variance and roughness) combined with the weight of secondary factors, and then the evaluation value of the first level factors (terrain flatness, area, terrain influence, distance and ground object category) shall be evaluated. The evaluation results in a single factor evaluation matrix. Combined with the evaluation model for the optimum selection of landing matches of helicopter, the comprehensive evaluation and evaluation score ordering, output optimization and sub-optimal evaluation are carried out.

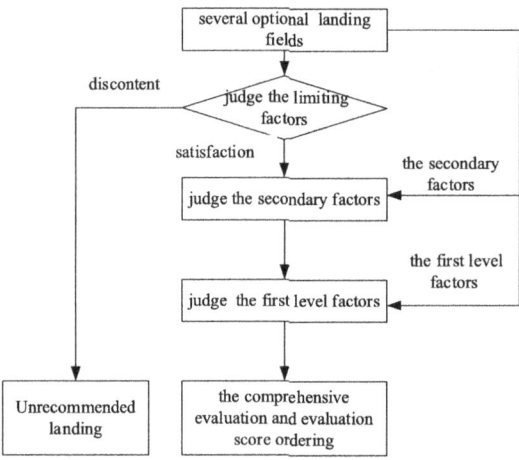

Fig. 5. Framework of helicopter landing field optimization method based on comprehensive evaluation method

The process of comprehensive evaluation method is as follows:

(1) Establish the optimal evaluation factor set of landing field $U = \{u_1, u_2, ..., u_n\}$ and decision set $V = \{v_1, v_2, ..., v_m\}$;
(2) Establish a comprehensive evaluation matrix, i.e. for each factor u_i establish single factor judgment first $(r_{i1}, r_{i2}, ..., r_{im})$ and then $r_{ij}(0 \leq r_{ij} \leq 1)$ representational factor u_i and v_j the single factor evaluation matrix can be obtained from the judgment made $R = (r_{ij})_{n \times m}$;
(3) Comprehensive evaluation according to factor weight $A = (a_1, a_2, ..., a_n)$, $\sum a_i = 1$ The relation matrix can be obtained by comprehensive evaluation $B = A \oplus R = (b_1, b_2, ..., b_m)$.

3.2 Route Planning for Blind Landing Based on Dichotomy Method

When receiving the information of landing points with planning, the software will automatically start blind landing route planning, and the end point of planned route is landing games. Flight guidance will follow the latest planned route. According to the relative position between the landing points to be planned and the aircraft, the landing profile is segmented. Route planning shall be carried out according to different zones, and the waypoint information and route pipeline information shall be output. Figure 6 shows the standard division of blind landing route.

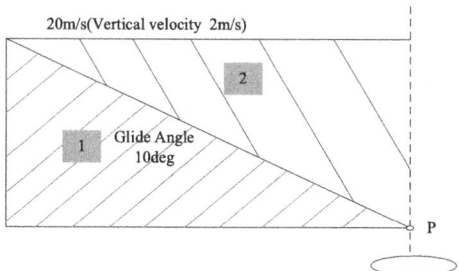

Fig. 6. Airway Division of Blind Landing

According to the zoning standard, the system will partition according to the relative position of the helicopter relative to the landing sites. When the helicopter is between 1/2 zones and the flying height is lower than the standard glide profile, enter into the planning criterion of zone 1, as shown in Fig. 7. Then the helicopter will fly directly to point P in the figure after heading adjustment, then the helicopter will be located at Hovering over landing games high (default 20 m). The course adjustment process is the course of horizontal flight and turning. The turning adopts the method of horizontal flight by sections. The course of each section changes monotonically (fixed increment) until the helicopter (planned track) heading and the heading angle from the helicopter to the landing sites are less than the fixed increment, then the heading adjustment is considered to be over.

When the helicopter is between 1/2 zone and flying at an altitude lower than the standard gliding profile, it enters the planning standard of zone 2, descends to zone 1 first, and then adjusts its heading, as shown in Fig. 8.

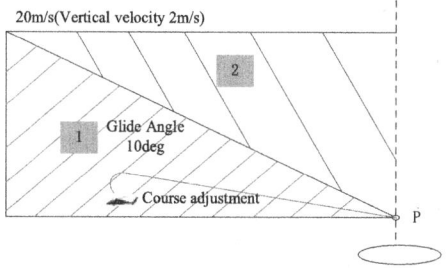

Fig. 7. Blind landing of zone 1

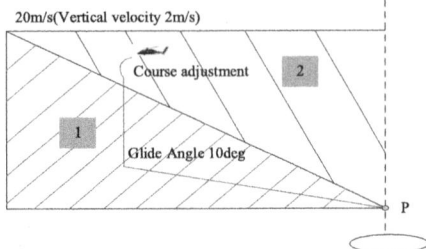

Fig. 8. Blind landing in zone 2

With reference to the direct flight procedure in civil aviation, the heading can be manually and gradually adjusted during manual flight, and finally the helicopter's heading is aimed at landing flights. The heading adjustment as shown in Fig. 9.

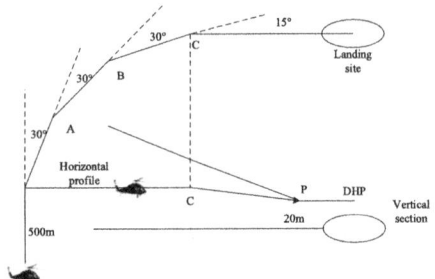

Fig. 9. Heading adjustment strategy

3.3 Visual Guided Solution

Boot Instruction Solving. High-precision flight guidance module is needed for helicopter to avoid obstacle/blind landing. As a bridge between trajectory planning task layer and flight control execution layer, the design of high-precision flight guidance module is very important. According to the classification of helicopter flight profile, flight guidance can be divided into two sub-functions: horizontal guidance (LNAV) and vertical guidance (VNAV).

The horizontal/vertical guidance module will guide the helicopter along the complete flight path from take-off to landing site. The flight path is landing consecutive from take-off to destination. The whole flight guidance process shall meet all altitude limits, speed limits and gradient limits specified in the flight plan, taking into account such factors as aircraft performance, selected speed and speed transition, altitude correction of barometric pressure, environmental factors, planned control mode, transition mode of flight segment, etc.

Horizontal guidance is a kind of helicopter flight guidance module. In helicopter obstacle avoidance/blind landing assistant decision-making system, parameters such as reference trajectory output by trajectory prediction module are input information for horizontal guidance, and at the same time, according to current flight phase, position and speed output by integrated navigation Degree, heading and other information, calculate the horizontal guidance instruction, and output to the integrated display system, and the pilot controls the helicopter to fly along the predetermined horizontal profile according to the guidance instruction.

Vertical guidance is another function of helicopter flight guidance module, which controls trajectory deviation and vertical velocity in vertical profile according to helicopter performance constraint, terrain constraint, pilot's altitude interference in FCU and other factors, so as to realize helicopter climbing and leveling according to predetermined trajectory Fly or descend. Its main tasks include guiding the helicopter to climb or descend to a given target point or target area (transitional hovering), cruise at fixed altitude, vertical speed control and glide control of helicopter's return/approach.

Visual Guidance. After receiving the route planning information and the current state information of the helicopter, the flight director compares the actual flight path of the helicopter with the target flight path, calculates the control amount required for entering the target flight path, and displays it on the bottom display and the head display in the form of visual indicator. The guidance signal directly displays whether the command to be operated currently is up, down, left or right. After the pilot sees it, the pilot can directly track the pilot rod to operate the helicopter to ensure that the helicopter is correctly cut into or kept on the predetermined route, as shown in Fig. 10.

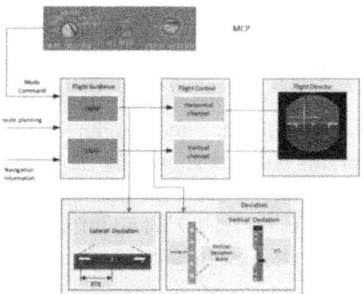

Fig. 10. Operating Principle of Flight Guidance

4 Experimental Verification

In this paper, landing programming experiments were carried out to verify the simulation in a certain area. Firstly, the sliding window search and ranking method was used to generate landing candidate verification, and then based on the sequencing results, the landing candidate tests were sorted.

The flight route planning and guidance test were carried out by selecting the descending area randomly, planning landing sites, simulating the helicopter's real-time position, and solving the guidance instructions to avoid obstacles.

4.1 Sliding Window Search and Sorting Method

The longitude and latitude of the helicopter are (91.65, 31.50). The longitude described as east direction is positive, the signal range is –180°~180°, the latitude description is north direction is positive, the signal range is –90~90, the aircraft heading angle is 90°, the radio height is 100, the integrated absolute pressure altitude is 5000, the wind speed is 5 m/ s, and the wind direction is 45° (signal). Range: 0–360° to define future wind, based on true north), select 400 * 400 mesh with pixel resolution of 4.45 m. Set the local slope threshold parameter as 5, use 0.01 for local variance and 0.1 for local roughness, as shown in Fig. 11.

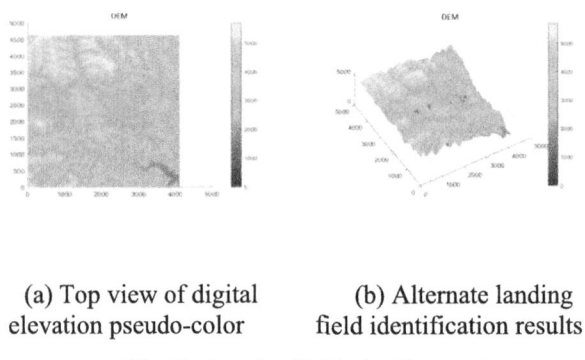

(a) Top view of digital elevation pseudo-color

(b) Alternate landing field identification results

Fig. 11. Landing Field Identification

Given the engineering constraints of helicopter landing on terrain, and transform the constraints into three indexes of slope, variance and roughness, then calculate and search the three indexes of slope, variance and roughness, we can get landing alternative flight fields of helicopter in this area, as shown in Table 2. Ten groups of landing data packets obtained by sliding window search, the output azimuth is described as positive clockwise, and the signal range is –90° –90°. The information of landing sites is sorted according to the indicators, and the results are shown in Table 3.

4.2 Route Planning Simulation of Blind Landing

Simulation condition 1 for blind landing route planning about the priority landing Site: The latitude of helicopter is 31.50°, longitude is 91.65°, altitude is 200 m, initial speed 30 m/ s, initial heading 11°, decision height 20 m. The landing site rounds latitude 27.98°, longitude 91.71°, altitude 0 m, simulation blind landing route as Fig. 12.

4.3 Integration Verification Based on Visual Guided

Based on the high-precision digital map, the comprehensive visual scene realizes the precise reconstruction of the situation environment around the aircraft. The information such as digital map, safe landing zone and safe flight pipeline oriented to the blind landing task will be superimposed and displayed in the integrated visual software to

Technical Research on Helicopter Blind Landing Under DVE 225

Table 2. Sliding Window Search Result Data

Landing Field ID	Longitude/ deg	Latitude/ deg	Height/ m	Length/ m	Width/ m	Average grade/ deg	Ground flatness
1	91.93	27.03	4383	99	99	1.58	0.17
2	92.13	27.00	5003	99	99	1.65	0.18
3	91.91	27.97	4459	99	99	1.91	0.23
4	91.96	27.99	4325	164	164	1.99	0.24
5	91.71	27.98	4470	114	114	1.59	0.17
6	91.89	27.97	4518	104	104	1.55	0.16
7	91.96	27.94	4266	110	110	1.41	0.17
8	91.90	27.93	4331	99	99	1.70	0.18
9	91.90	27.92	4341	114	114	2.67	0.32
10	91.84	27.68	4103	104	104	1.88	0.20

Table 3. Landing Sorted Output Data

Landing Field Sorted Output Sequence	Landing Field ID
The Priority Landing Site	5
The Alternative Landing Site1	10
The Alternative Landing Site2	9
The Alternative Landing Site3	7
The Alternative Landing Site4	6
The Alternative Landing Site5	8

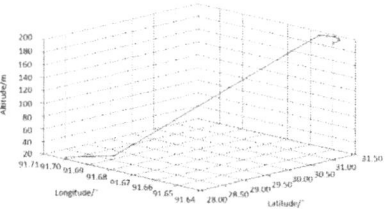

Fig. 12. Horizontal section and vertical section of blind landing route in zone 1

improve the pilot's ability to perceive the surrounding situation and environment, as shown in Fig. 13 and Fig. 14.

It is confirmed that the pilot can complete such tasks as temporary landing area selection, landing route planning and safe flight guidance under unfavorable visual environment.

Fig. 13. Display effect of comprehensive visual software

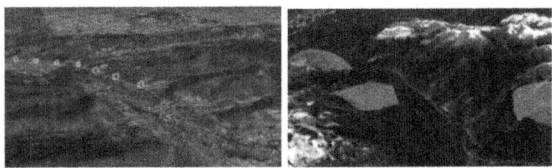

Fig. 14. Global 3D Situation Display Effect

5 Summary

The field blind landing mission is to improve the helicopter's ability to land autonomously under non-visual field conditions. This technology is based on the research of foreign technical schemes, designs the blind landing architecture, and then simulates the landing-field search, blind landing route planning and visual guidance solution, and verifies the scheme. The results show that this study can reduce the burden of pilots and improve flight safety, which is of great significance to improve helicopter combat capability.

References

1. Lüken, T., Döhler, H.U.: Increasing situation awareness for a helicopter pilot by data acquisition and fusion of multiple sensors. Proceedings of ODAS (2009)
2. Dreslin, B., Mersinger, M.C., Patel, S., Chaparro, A.: Flying blind: exploring the visual cues used by helicopter pilots in degraded visual environments. Hum. Fact. Transp. **95**, 656 (2023)
3. Moralez, E., Braddom, S., Grunwald, A., Hovev-Rottem, M.: In-flight evaluation of forward-looking integrated symbology for 4-D reroutable helicopter approach to blind landing. In: AIAA Guidance, Navigation, and Control Conference, p. 6685 (2011)
4. Iiyama, K., Tomita, K., Jagatia, B.A., Nakagawa, T., Ho, K.: Deep reinforcement learning for safe landing site selection with concurrent consideration of divert maneuvers. arXiv preprint arXiv:2102.12432 (2021)
5. Schroeder, J.A., Merrick, V.K.: Control and display combinations for blind vertical landings. J. Guid. Control. Dyn. **15**(3), 751–760 (1992)
6. Wang, Q., Cheng, Y.M., Qi, M.D., et al: An adaptive method for landing zone Search with terrain index constraint, 2022 China Automation Congress (CAC), pp. 2170–2175 (2022)
7. Paduano, J.,et al.: TALOS: an unmanned cargo delivery system for rotorcraft landing to unprepared sites. Am. Helicopter Soc. 1–17 (2015)

8. Cross, J., Schneider, J., Cariani, P.: MMW radar enhanced vision systems: the helicopter autonomous landing system (HALS) and radar-enhanced vision system (REVS) are rotary and fixed wing enhanced flight vision systems that enable safe flight operations in degraded visual environments. Degraded Visual Environments: Enhanced, Synthetic, and External Vision Solutions 2013, SPIE, vol. 8737, pp. 122–134 (2013)
9. Miller, J.D., Godfroy-Cooper, M., Szoboszlay, Z.P.: Degraded visual environment mitigation (DVE-M) program, bumper radar obstacle cueing flight trials 2020 (2021)
10. Dillon, T.E., et al.: Passive, real-time millimeter wave imaging for degraded visual environment mitigation. Degraded Visual Environments: Enhanced, Synthetic, and External Vision Solutions 2015, SPIE, vol. 9471, pp. 10–18 (2015)
11. Szoboszlay, Z., et al.: The design of pilot cueing for the degraded visual environment mitigation (DVE-M) system for rotorcraft. Vertical Flight Society 77th Annual Forum (2021)
12. Bratt, R.M., Walker, A.D.: Degraded visual environment paradigm shift from mission deterrent to combat enabler. Situation Awareness in Degraded Environments 2019, SPIE, vol. 11019, p. 2 (2019)
13. Fujizawa, B.T., Szoboszlay, Z.P., Flanigen, P.R., Minor, J.S., Morford, Z.G., Davis, B.M.: Degraded visual environment mitigation program NATO flight trials: US army flight test and results. Proc. 43 rd Eur. Rotorcraft Forum, pp. 1–13 (2017)
14. Stambler, A., Spiker, S., Bergerman, M., Singh, S.: Toward autonomous rotorcraft flight in degraded visual environments: experiments and lessons learned. Degraded Visual Environments: Enhanced, Synthetic, and External Vision Solutions 2016, SPIE, vol. 9839, pp. 19–30 (2016)

Learning and Systems

A Contactless Demonstration Learning Method for Robotic Systems Based on Dynamic Parametric Regression Modeling

Meng Li[1], Jinzhu Peng[1], Jixian Gao[1(✉)], Nan Zhao[2], Yaonan Wang[3], and Mingkuo Wu[1]

[1] School of Electrical and Information Engineering, Zhengzhou University, Zhengzhou 450001, China
gaojixian@zzu.edu.cn
[2] Department of Artificial Intelligence Business Zhengzhou JD Cloud Computing Co., Ltd., Zhengzhou, China
[3] National Engineering Research Center for Robot Visual Perception and Control Technology, Hunan University, Changsha 410082, China

Abstract. Currently, contactless demonstration learning faces significant challenges in data acquisition, and its trajectory generalization capability is limited when adapting to new task parameters. To address these issues, this paper proposes a method for learning action primitive trajectories based on a dynamic parametric regression model. The proposed approach first integrates a Gaussian mixture regression model to reduce redundant features arising from individual demonstrations by synthesizing multiple demonstration trajectories. Subsequently, building on dynamic motion primitives (DMPs), the dynamic parametric regression model is used to capture the dynamic relationship between feature parameters and task parameters, thereby improving the model's generalization performance. Additionally, this study introduces the Fréchet distance as a similarity measure to evaluate generalization ability. Finally, the effectiveness of the proposed method is demonstrated through pick-and-place experiments involving objects, using the Kinova collaborative robotic arm.

Keywords: Dynamic Motion Primitives · Dynamic Parameter Regression Model · Trajectory Generalization

1 Introduction

In recent years, robotics technology [1] has made significant strides across industries, healthcare, services, and other sectors, accompanied by increasing com-

---

This work was supported in part by the National Natural Science Foundation of China (62273311, 61773351), in part by Henan Provincial Science Foundation for Distinguished Young Scholars (242300421051), in part by Zhengzhou JD Cloud Computing Co., Ltd.

plexity in operational environments. To address these challenges, learning from demonstration (LfD) methods [2] have gained widespread adoption, enabling robots to acquire flexible manipulation capabilities. LfD can be divided into contact and non-contact demonstrations. Contact demonstrations, often facilitated by kinesthetic teaching, offer high precision but are complex to implement and present safety risks. Consequently, non-contact demonstration methods [3,4] are receiving growing attention. Simultaneously, high-quality data collection techniques have become a critical aspect of demonstrative learning. Chen et al. [5] employed inertial measurement units (IMUs) and motion capture markers to monitor foot movements. With advances in computer vision, cameras are progressively taking the place of traditional sensors in recording human demonstration activities. One key advantage of using cameras is the elimination of sensors, which simplifies the data collection process. Cai et al. [6] utilized a single camera to efficiently monitor the positions of objects manipulated by humans, enabling robotic systems to accurately replicate these movements. In recent years, camera-based skeletal detection technologies have made significant advancements. In 2017, Cao et al. [7] introduced OpenPose, an innovative tool designed for the real-time extraction of human skeletal structures. Although it is computationally and hardware demanding, it can precisely handle motion data of humans performing tasks. In 2023, Fang et al. [8] improved upon OpenPose, enabling high-precision tracking of individuals. However, for tasks requiring high accuracy, tracking only the 3D coordinates of the human body is insufficient; understanding the orientation and posture of body parts, such as the wrists and arms, is equally critical for achieving optimal performance. As an optical motion capture system, Vicon's camera series offers high-speed and high-resolution capabilities [9], enabling precise tracking of each reflective marker's movement to capture the required motion data.

Calinon et al. [10] transformed the problem of estimating dynamic system parameters into Gaussian Mixture Regression (GMR) in different coordinate systems and used Gaussian Mixture Models (GMMs) to model motion trajectories. Although this method can effectively handle large-scale data, the generated trajectories may exhibit discontinuities. In 2002, Schaal et al. [11] proposed a learning from demonstration method based on dynamic systems, known as the Dynamical Movement Primitives (DMPs) algorithm. This algorithm models simple linear dynamic systems through a set of linear differential equations and converts them into weakly nonlinear systems with designated attractors by incorporating autonomously learned nonlinear terms. In 2013, Ijspeert et al. [12] improved this algorithm by reconstructing the equations for the nonlinear terms, modulating the nonlinear functions with start positions, goal positions, and time, and introducing a phase variable to replace time, thereby avoiding explicit time dependence. In recent years, the integration of Deep Reinforcement Learning (Deep RL) with Dynamic Movement Primitives (DMPs) has attracted considerable attention. [13] Noohian et al. [14] proposed a method that combines the TD3 algorithm with DMP. While this approach improves the generalization accuracy of the trajectories, it reduces the system's real-time performance and computational efficiency.

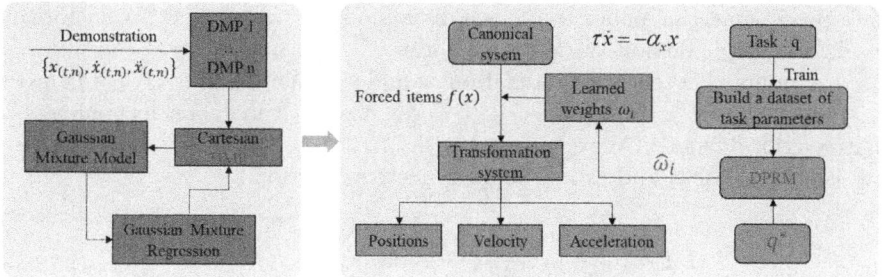

Fig. 1. Overall framework diagram of the algorithm

Despite the significant progress of the current data acquisition methods and DMP trajectory learning methods in the field of robotics, there are still problems such as redundancy of data features and low accuracy of trajectory generalization for the non-contact demonstrative learning aspect.The main contributions of this paper are:

- This paper addresses the issue of redundant features that may arise from a single demonstration trajectory input in the DMPs model by combining the GMM/GMR model. This approach enhances the reliability of the model and achieves contactless learning of human arm endpoint movements.
- A generalization method for Dynamical Movement Primitives (DMPs) based on a dynamic regression model is proposed. This method introduces DPRM into DMPs, significantly improving trajectory learning capability and reproduction accuracy.

The detailed algorithmic framework of this paper is shown in Fig. 1. First, an optimal trajectory of the demonstration is input into the DMP model for learning, based on pre-processing methods applied to the demonstration dataset, such as DTW, GMM, and GMR. Next, the weight parameters generated during the learning process are used to establish a dynamic parametric regression model (DPRM) with the relevant task parameters, followed by training. Finally, the system can utilize the DPRM to efficiently solve for the weight parameters when different task parameters are altered. This approach significantly improves the system's generalization accuracy and computational efficiency in the vicinity of the task parameters.

2 Methodology

2.1 Preprocessing Methods for Demonstrating Teaching Data Sets

Dynamic Time Warping Algorithm. When humans demonstrate an action primitive multiple times, it is difficult to ensure that each demonstration is exactly the same length. Due to the misalignment on the time step, similar spatial points cannot be aligned on the length sequence. Therefore, for subsequent data processing, this paper employs a dynamic time warping (DTW) algorithm to align multiple demonstration trajectories.

Assuming that the demonstration sample sequences are $Q : \{q_1, q_2, \cdots q_i, \cdots, q_n\}$ and $C : \{c_1, c_2, \cdots, c_j, \cdots, c_m\}$, respectively, construct an $n \times m$ matrix grid, define DTW regularized paths $H = \{h_1, h_2, \cdots h_\omega, \cdots h_W\}$, where h_W is a path point, and establish the objective function:

$$DTW(Q,C) = \frac{1}{\sum_{w=1}^{W} \Psi_w} \min\{\sum_{w=1}^{W} \Psi_w d(i_w, j_w)\} \tag{1}$$

$$H^* = \arg_P \min\{\sum_{w=1}^{W} \Psi_w d(i_w, j_w)\} \tag{2}$$

where ψ_w is the weighting factor, H^* is the regularized path, and $d(i_w, j_w)$ is the Euclidean distance between two sequences q_i and c_j (representing the similarity between the two points, the smaller the value the more similar):

$$d(i_w, j_w) = \sqrt{(\xi_{s,i} - \xi_{s,j})^2} \tag{3}$$

The dynamic time warping algorithm optimally matches the reference template and the test template in time through dynamic planning, the process is shown in Fig. 2-(a), based on the lattice points that the path passes through to derive the two sequences aligned in such a way that the two bar trajectories correspond to similar points are aligned as shown in Fig. 2-(b).

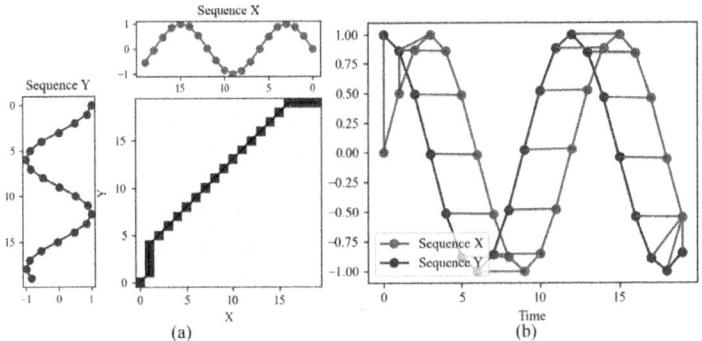

Fig. 2. DTW algorithm dynamic planning process

Gaussian Mixed and Gaussian Mixed Regression Models. In this paper, we combine DMP with Gaussian Mixture Modeling (GMM) to obtain better trajectory estimates from multiple demonstration data via GMR, which is then used as the demonstration trajectory input to the DMP model. This avoids demonstration errors and errors in sensor acquisition data, while better preserving the generalization ability of the model.

Each sample point in the didactic trajectory is denoted as $x_i = \{x_{v,t}, x_{t,i}\}$, where $x_{v,t}$ is the spatial vector of the didactic sample point and $x_{t,i}$ is the corresponding time step. Suppose we have N didactic sample points $i = \{1, 2, \cdots, N\}$, where N is the length of the didactic trajectory. To compute the probability distribution $p(x_i)$ of the sample points x_i, we first need to compute the prior probability $p(k)$ of each Gaussian kernel, and then use the conditional probability to compute the probability distribution of x_i as shown in the following equation.

$$P(x_i) = \sum_{k=1}^{k} p(k)p(x_i|k) \qquad (4)$$

The calculation method of $p(k)$, $p(x_i|k)$ is shown in the following formula.

$$p(k) = \pi_k \qquad (5)$$

$$p(x_i|k) = \frac{1}{\sqrt{(2\pi)^D |\Sigma_k|}} \exp(-\frac{1}{2}(x_i - \mu_k)^T \Sigma_k^{-1}(x_i - \mu_k)) \qquad (6)$$

where Σ_k represents the covariance matrix, μ_k is the mean vector of the probability density function for the kth Gaussian kernel, and π_k denotes the prior probability of a sample point belonging to the kth Gaussian kernel. The Expectation-Maximization (EM) algorithm is used for parameter estimation, yielding the optimal parameters of the GMM after multiple iterations.

The GMR model is then employed to obtain a smooth trajectory that incorporates the characteristics of the demonstrated motion, using weighted Gaussian kernels.

2.2 Action Primitive Trajectory Learning Based on Dynamic Parametric Regression Modeling

Principles of Dynamic Motion Primitive Trajectory Learning Methods. The core idea of DMP is to transform a simple and easy-to-understand attractor system into the desired attractor system using a learnable force-field function, thus enabling the guidance of point attractors and extremal ring attractors of almost any level of complexity. The method of training the demonstration trajectories to obtain weight values by means of dynamic motion primitives, and reproducing and generalizing them as needed, provides excellent coordination and stability. The basic idea is to achieve the desired attractor behavior by modulating an easily understood dynamical system using nonlinear terms to modulate it. For this purpose, a spring-damped system is usually chosen to be represented, using first-order differential equations to characterize its dynamics:

$$\tau \dot{z} = \alpha_z(\beta_z(g-y) - z) + f(x) \qquad (7)$$

$$\tau \dot{y} = z \tag{8}$$

where τ is the time constant, $\dot{y}$ and $\dot{z}$ are the first and second derivatives respectively of y, g is the desired target pose, and x is a quantity independent of time. By selecting the appropriate α_z and β_z, the system can reach the critical damped state, which is generally set to $\beta_z = \alpha_z/4$. Such a system could implement a stable but insignificant pattern generator with g as a single point attractor. The nonlinear forcing term $f(x)$, which is used to model the trajectory's shape, is defined as:

$$f(x) = \frac{\sum_{i=1}^{N} \psi_i(x)\omega_i}{\sum_{i=1}^{N} \psi_i(x)} x(g - y_0) \tag{9}$$

where y_0 represents the starting point of the coding curve, N denotes the total number of basis functions, $\psi_i(x)$ refers to the basis function, and ω_i indicates the corresponding weight of each basis function. In the whole process, the basis function adopts Gaussian basis function, namely:

$$\psi_i(x) = \exp(-\frac{1}{2\sigma_i^2}(x - c_i)^2) \tag{10}$$

where c_i and σ_i are the center and width of the i Gaussian basis function.

Action Primitive Trajectory Learning Based on Dynamic Parametric Regression Modeling. Dynamic Parametric Regression Model (DPRM) is used to estimate the feature parameters (weight parameters) in a way that can capture the dynamic changes of the task parameters through the time series data. The DPRM model assumes that the current feature parameters are not only dependent on the current task parameters, but also on the task parameters of several previous moments. The specific regression model form is:

$$\omega_i(t) = \beta_{i0} + \beta_{i1}q(t-1) + \beta_{i2}q(t-2) + \cdots + \beta_{ip}q(t-p) + \varepsilon_i(t) \tag{11}$$

where $\omega_i(t)$ is the ith feature parameter at time t; $q(t-1), q(t-2), \cdots, q(t-p)$ is the task parameter at the previous p moments; $\beta_{i0}, \beta_{i1}, \cdots, \beta_{ip}$ is the regression coefficient; and $\varepsilon_i(t)$ is the noise term.

The DPRM is trained by constructing a time series dataset of task parameters and the corresponding feature parameter dataset as in eqns. (12)(13) below:

$$Q = \begin{pmatrix} q_1(t) & q_1(t-1) & \cdots & q_1(t-p) \\ q_2(t) & q_2(t-1) & \cdots & q_2(t-p) \\ \vdots & \vdots & \ddots & \vdots \\ q_M(t) & q_M(t-1) & \cdots & q_M(t-p) \end{pmatrix} \tag{12}$$

$$\Omega = \begin{pmatrix} \omega_1(t) \\ \omega_2(t) \\ \vdots \\ \omega_M(t) \end{pmatrix} \tag{13}$$

A linear regression model was applied to establish the relationship between task parameters and characteristic parameters, resulting in the regression coefficient matrix B as shown in eqn (14):

$$B = (Q^T Q)^{-1} Q^T \Omega \tag{14}$$

When given a new task parameter q^*, use the trained DPRM model to predict a new feature parameter $\hat{\omega}_i$:

$$\hat{\omega}_i(t) = \beta_{i0} + \beta_{i1} q^*(t-1) + \beta_{i2} q^*(t-2) + \cdots + \beta_{ip} q^*(t-p) \tag{15}$$

The new motion trajectory is then reconstructed using the new feature parameters $\hat{\omega}_i$ and the DMP model.

$$\hat{f}(s) = \sum_{i=1}^{N} \Psi_i(s) \hat{\omega}_i \tag{16}$$

Substituting $\hat{f}(s)$, the target point g and the time scaling τ into the (7)(8) equations, iteratively compute the new motion trajectory $\{y(t)\}$.

3 Experiment

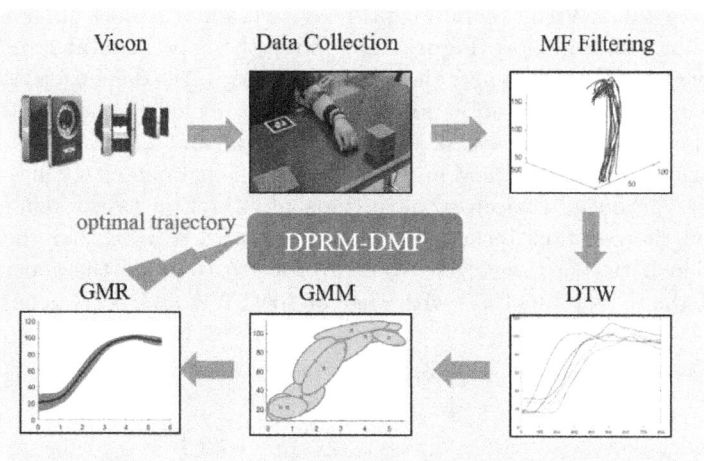

Fig. 3. Flowchart of data set collection and processing

In this paper, a Vicon optical motion capture system is used to collect high-accuracy human hand trajectory data. Although contactless teaching is more labor-saving and convenient than direct teaching, it is still affected by sensor environment conditions and tracking algorithm errors, which result in lower quality raw data. Therefore, it is necessary to process the data, and the specific steps are shown in Fig. 3. After processing, the trajectory obtained from the end-of-arm motion data is input into the DPRM-DMP model, from which trajectory weights are obtained through training. The new task parameters are then set to perform trajectory generalization, and the output trajectory is finally input into the Kinova robotic arm for execution via ROS communication.

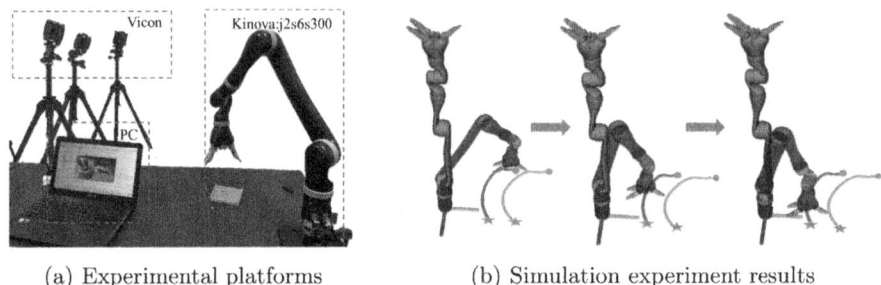

(a) Experimental platforms   (b) Simulation experiment results

Fig. 4. The composition of the experimental platform and the results of the simulation experiments

Figure 4 shows the composition of the experimental platform and the results of the simulation experiments. In particular, the experimental platform mainly consists of a PC, a Vicon motion capture system and a Kinova six-axis robotic arm, as shown in Fig. 4-(a). Figure 4-(b) shows the experimental results of one trajectory generalization, where the green trajectory is the demonstration trajectory at the end of the human hand, and the red trajectory is the new trajectory generated by setting the new task parameters (initial and termination point positions) learned from the method in this paper. In the parameter setting, let $\tau = 1$ make the reproduced trajectory have the same duration as the demonstrated trajectory; the constant terms $\alpha_y = 25, \beta_y = \alpha_y/4, \alpha_x = 1$, and the number of Gaussian basis functions $N = 100$. In order to compare the generalization ability of this paper's method with that of the LWR and GPR generalization

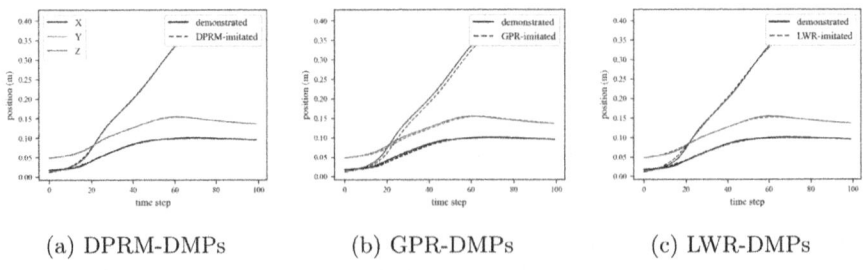

(a) DPRM-DMPs   (b) GPR-DMPs   (c) LWR-DMPs

Fig. 5. Comparison of schematic trajectories and generalization trajectories for each method

methods in the vicinity of the demonstrated trajectory, the comparison of the results of the demonstration and generalization of the trajectory is shown in Fig. 5, given the fixed task parameters q^*.

In order to accurately describe the generalization performance of the proposed method, in this paper, the Fréchet distance is used as a similarity metric to compare the schematic trajectory with the generalized trajectory, with the following formula:

$$d_F(P,Q) = \min_{\alpha,\beta} \max_{t\in[0,1]} ||P(\alpha(t)) - Q(\beta(t))|| \qquad (17)$$

where α and β are strictly increasing continuous functions, called reparameterizations, from the interval $[0, 1]$. They denote the paths and speeds of travel on curves P and Q.

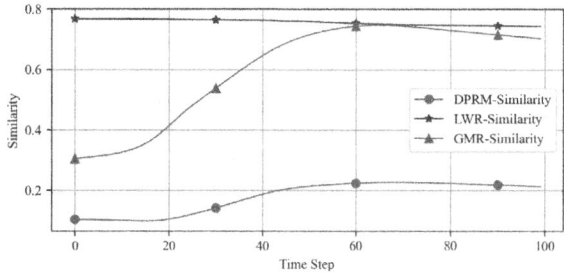

Fig. 6. Comparison of the similarity of the three methods under the same task parameters

Figure 6 shows the similarity comparison of the three methods from Fig. 5, evaluated under the same task parameters. According to the criterion that lower values indicate higher similarity, the DRPM-DMP method demonstrates greater similarity than both LWR and GPR, suggesting a significant improvement in the generalization accuracy of the trajectory.

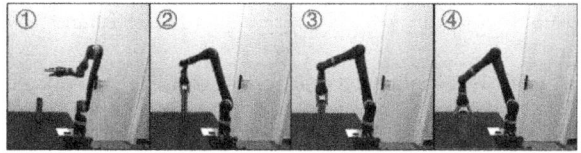

Fig. 7. Working steps of the robotic arm pick-and-place

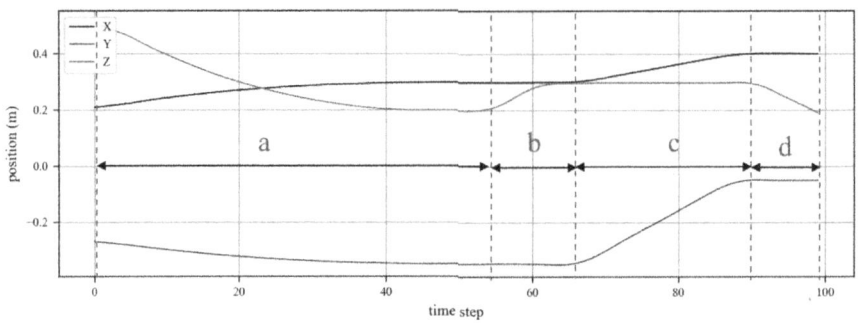

Fig. 8. Cartesian trajectory of the end of the robot arm

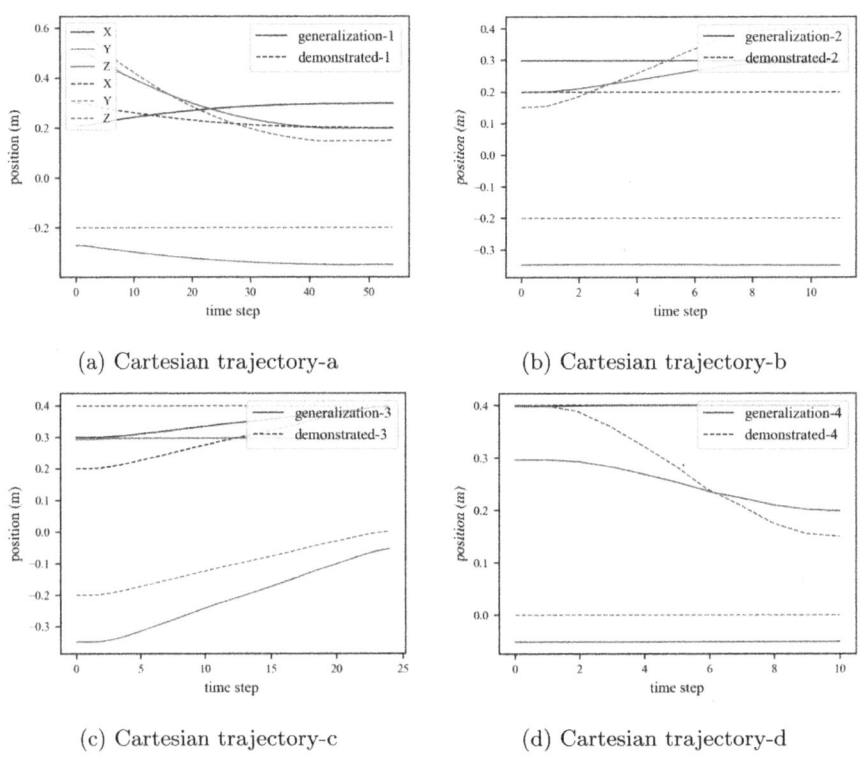

(a) Cartesian trajectory-a

(b) Cartesian trajectory-b

(c) Cartesian trajectory-c

(d) Cartesian trajectory-d

Fig. 9. Comparison of the schematic trajectories with the generalized trajectories for each segment

In addition, this paper sets up a robotic arm pick-and-place task to verify the generalization performance of the algorithm in specific task scenarios. As Fig. 7 shows the working steps of robotic arm pick-and-place to target object (fixed initial position).

Figure 8 shows the real motion trajectory of the end effector of the robotic arm in the Cartesian coordinate system during a pick-and-place task using the method proposed in this paper. Sections a, b, c, and d correspond to the teaching trajectories shown in Fig. 9- (a), (b), (c), and (d), respectively, which are obtained by generalizing the task parameters of the robotic arm during task execution. Figure 9 shows the comparison results between the teaching trajectories of each segment and the generalized trajectories based on actual task parameters. It can be seen that this method can effectively improve the generalization ability of the robotic arm in real working environments.

4 Conclusion

This paper addresses the current problem of low generalization ability and computational efficiency of trajectories under new task parameters in contactless demonstration learning. We propose an action primitive trajectory learning method based on a dynamic parametric regression model, which combines dataset preprocessing methods, such as the dynamic time regularization algorithm and the dynamic hybrid regression model, to eliminate the problem of redundant features in contactless demonstration trajectories. The method is compared with existing methods using similarity metrics, and the results demonstrate a significant improvement in generalization performance. Finally, experiments are conducted with a Kinova robot on a pick-and-place task to validate the effectiveness of the proposed method.

References

1. Auddy, S., Hollenstein, J., Saveriano, M., et al.: Continual learning from demonstration of robotics skills. Robot. Auton. Syst. **165**, 104427 (2023)
2. Luo, J., Zhou, X., Zeng, C., et al.: Robotics perception and control: key technologies and applications. Micromachines **15**(4), 531 (2024)
3. Pan, Y., Chen, C., Zhao, Z., et al.: Robot teaching system based on hand-robot contact state detection and motion intention recognition. Robot. Comput. Integrated Manufact. **81**, 102492 (2023)
4. Chen, X., Zhang, K., Liu, H., et al.: A probability distribution model-based approach for foot placement prediction in the early swing phase with a wearable imu sensor. IEEE Trans. Neural Syst. Rehabil. Eng. **29**, 2595–2604 (2021)
5. Kundrat, D., et al.: Preclinical performance evaluation of a robotic endoscope for non-contact laser surgery. Annals Biomed. Eng. **49**(2), 585–600 (2021)
6. Cai, C., Somani, N., Knoll, A.: Orthogonal image features for visual servoing of a 6-DOF manipulator with uncalibrated stereo cameras. IEEE Trans. Rob. **32**(2), 452–461 (2016)
7. Cao, Z., Simon, T., Wei, S. E., et al.: Realtime multi-person 2d pose estimation using part affinity fields. In: Proceedings of the IEEE Conference on Computer Vision and Pattern Recognition, pp. 7291–7299(2017)
8. Fang, H.S., Li, J., Tang, H., et al.: Alphapose: whole-body regional multi-person pose estimation and tracking in real-time. IEEE Trans. Pattern Anal. Mach. Intell. **45**(6), 7157–7173 (2022)

9. Lee, Y., Lama, B., Joo, S., et al.: Enhancing human key point identification: a comparative study of the high-resolution VICON dataset and COCO dataset using BPNET. Appl. Sci. **14**(11), 4351 (2024)
10. Calinon, S., Li, Z., Alizadeh, T., et al.: Statistical dynamical systems for skills acquisition in humanoids. In: 2012 12th IEEE-RAS International Conference on Humanoid Robots (Humanoids 2012). IEEE, pp. 323–329 (2012)
11. Schaal, S.: Dynamical movement primitives-a framework for motor control in humans and humanoid robotics. Adaptive Motion of Animals and Machines, pp. 261–280 (2006)
12. Ijspeert, A.J., Nakanishi, J., Homann, H., et al.: Dynamical movement primitives: learning attractor models for motor behaviors. Neural Comput. **25**(2), 328–373 (2013)
13. Yuan, Y., Yu, Z.L., Hua, L., et al.: Hierarchical dynamic movement primitive for the smooth movement of robots based on deep reinforcement learning. Appl. Intell. **53**(2), 1417–1434 (2023)
14. Noohian, A., Raisi, M., Khodaygan, S.: A framework for learning dynamic movement primitives with deep reinforcement learning. In: 2022 10th RSI International Conference on Robotics and Mechatronics (ICRoM). IEEE, pp. 329–334 (2022)

A Multi-scale Fusion and Dynamic Upsampling Model for Road Surface Snow Detection

Lipeng Du[1], Xinhao Zhou[1], Qili Chen[2], Tong Wang[1], Lin Zhao[1(✉)], and Guangyuan Pan[1]

[1] Linyi University, Linyi 276000, Shandong, China
zhaolin@lyu.edu.cn
[2] Beijing Information Science and Technology University, Beijing 100096, China

Abstract. The issue of road surface snow accumulation during winter is one of the leading causes of traffic accidents. Therefore, real-time road surface snow detection is a crucial research topic in the field of traffic management. Currently, manual sensor-based and traditional machine learning detection methods face challenges such as high labor consuming and resource costs, low accuracy, substantial computational load and poor robustness. To address these issues, this study proposes a real-time road surface snow detection model, YOLOv8-FCD, which incorporates multi-scale fusion and dynamic upsampling, aiming for efficient winter road snow detection. The model integrates the C2f_Faster module into the backbone network, achieving a lightweight design while ensuring detection accuracy. Additionally, the SPPF module incorporates a C3STR module based on Swin Transformer to enhance backdrop differentiation and snow feature extraction. The neck network then uses the DySample upsampling operator to increase the accuracy and resilience of the model while successfully lowering the number of parameters. Finally, utilizing a real-world road surface snow dataset gathered by in-vehicle cameras operated in Canada throughout the winter seasons, testing results show that the proposed YOLOv8-FCD achieves a significant gain in accuracy, mAP@50%, and mAP@50–95% when compared to other benchmark models.

Keywords: Road Surface Snow · Object Detection · Deep Learning · Machine Vision

1 Introduction

Road surface snow detection is one of the critical issues in traffic safety management, directly impacting driving safety and the incidence of accidents. Over the past several years, weather-related vehicle accidents in the U.S. have consistently accounted for a significant proportion of reported incidents, leading to numerous fatalities and injuries [1]. When Basyouny and Kwon examined 11 years' worth of daily collision and meteorological data from Edmonton, Alberta, Canada, they found that for every centimeter of snowfall, there was a 3.3% increase in serious crashes and a 5.8% increase in minor collisions [2]. These findings highlight the critical need to advance road snow detection technologies to ensure the safe functioning of transportation systems.

Methods for achieving road surface snow detection can generally be divided into two main categories: conventional object detection techniques and detection algorithms based on deep learning. In traditional methods, manual inspections and sensor-based detection are primarily employed [3]; however, these approaches exhibit limitations in timeliness and accuracy [4]. The road snow problem is usually addressed by classification techniques in deep learning-based detection algorithms. Even while these techniques provide some degree of accuracy, they sometimes include too much redundant data, including information from nearby roads, which raises the computational requirements [5].

This study proposed an object detection model to tackle the road surface snow detection problem in order to address these problems. By integrating cutting-edge technologies and optimizing structural design, the model seeks to improve the precision and resilience of snow detection on road surfaces while maintaining computational efficiency. In order to achieve effective and precise road snow condition monitoring, it is made to handle complicated snow circumstances and enhance detecting capabilities in a variety of snow scenarios. The following are the primary contributions:

(1) To improve detection accuracy, the C3STR module is developed to enhance the network's capacity for modeling multi-scale features and capturing global context, thereby improving the object detection network's performance in identifying road surface snow in complex scenarios.
(2) In order to facilitate lightweight computation and enhance real-time snow detection while enhancing the stability of the model, this study presents the C2f_Faster module in backbone and the DySample module in neck.
(3) To validate the effectiveness and advantages of the proposed framework, adequate experimental validation is planned, utilizing both comparative and ablation experiments. In terms of precision, mAP, and floating-point computations, the comparative trials show that the model performs noticeably better than other relevant models. It also shows improved detection capabilities without a major increase in computational cost. The integration of the C2f_Faster, C3STR, and DySample modules is crucial for enhancing model performance, especially with regard to accuracy and robustness in complicated snow conditions, as the ablation tests further demonstrate.

The structure of this document is as follows: Sect. 2 provides a comprehensive review of previous research relevant to this subject. Section 3 describes the architecture and structure of the proposed model. Section 4 provides a description of the experimental results and their evaluation. Section 5 represents the study's final conclusion.

2 Related Works

2.1 Traditional Detection Methods

Traditional detection methods for road surface snow largely rely on manual inspections and sensor-based technologies. These approaches, while historically significant, exhibit notable limitations in terms of timeliness and accuracy. Manual detection is labor-intensive and can suffer from human error, leading to inconsistent results [6]. Sensor-based methods, on the other hand, often struggle with environmental variability, including changes in lighting and weather conditions, which can compromise detection performance [7]. As such, these traditional methods may not provide the requisite reliability needed for effective traffic management during winter conditions.

2.2 Deep Learning Approaches

In recent years, deep learning-based methods for road surface snow detection have garnered increasing attention, primarily categorized into classification methods and target detection methods. Classification methods typically focus on a holistic analysis of images to determine the presence of snow. However, these methods have limitations, as they often fail to adequately consider the spatial information and specific locations of objects within the images [8]. Consequently, in complex road environments, classification methods may be hindered by surrounding redundant information, leading to suboptimal detection results [9]. In contrast, target detection methods exhibit greater potential for road snow detection [10]. These approaches not only identify the presence of snow within images but also accurately localize the snowy areas, thereby providing more detailed information. This precision is crucial for winter traffic management, as it enables decision-makers to take timely actions to ensure driving safety [11]. Furthermore, target detection algorithms demonstrate significant advantages in handling multi-target scenarios, effectively identifying multiple snowy areas in crowded and complex environments, thereby enhancing overall detection accuracy. Therefore, the application of target detection methods for snow detection is of paramount importance for improving efficiency and reliability, aiding in overcoming the numerous challenges currently faced in snow detection.

Target detection models can be broadly classified into two groups: single-stage detectors, like the YOLO series, which can directly predict class labels and bounding box locations, allowing for faster detection [13]; and two-stage detectors, like the R-CNN series [12], which offer high accuracy but comparatively slower processing speeds and larger parameter counts. One of the most sophisticated object detection models on the market right now is YOLOv8 [14], which strikes a better balance between speed and accuracy because of advancements in architecture and loss functions. YOLOv8 was chosen for this investigation in large part due to its extensive use in real-time detection circumstances.

3 Methodology

Figure 1 shows the architecture of the proposed model, while Fig. 2 shows the detailed structures of PConv and DySample.

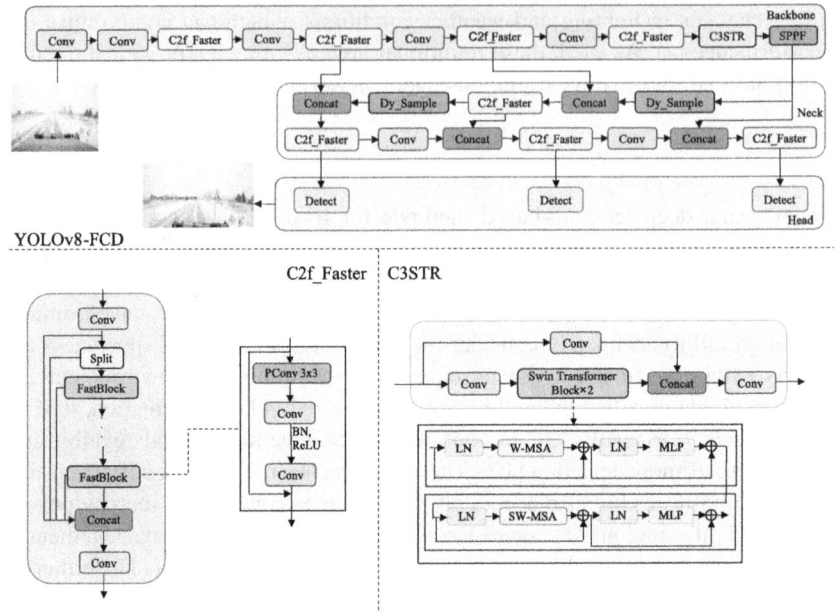

Fig. 1. Proposed framework of YOLOv8-FCD.

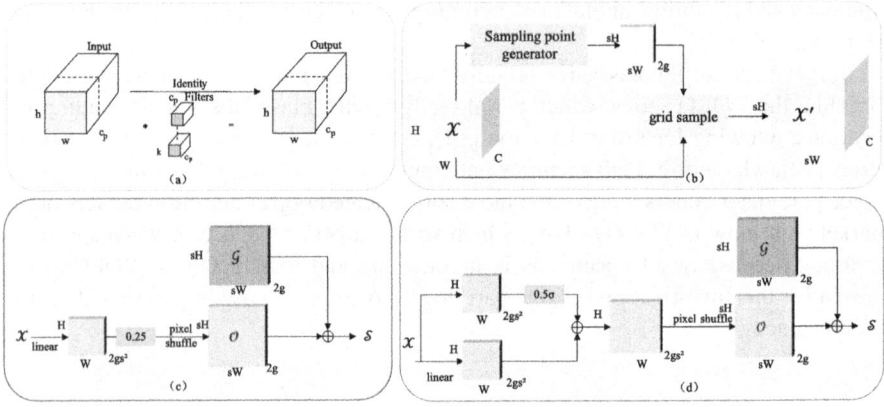

Fig. 2. (a) PConv structure. (b) Sampling based dynamic upsampling. (c) Static Scope Factor. (d) Dynamic Scope Factor.

3.1 Backbone Network Improvement

3.1.1 C2f Module Improvement

FasterNet is a highly efficient neural network architecture aimed at improving computational speed while maintaining accuracy, particularly for visual tasks. It incorporates an innovative technique referred to as PConv, as shown in Fig. 2(a). For continuous or standard memory access, the first or last contiguous c_p channels represent the whole feature map during calculation [15]. This method guarantees that the input and output feature maps retain the same number of channels, ensuring flexibility without compromising generality. Consequently, the computational cost of PConv, measured in FLOPs, is only.

$$h \times w \times k^2 \times c_p^2 \qquad (1)$$

When $r = \frac{c_p}{c} = \frac{1}{4}$, the computational cost of PConv, measured in FLOPs, is reduced to merely one-sixteenth of that associated with a standard convolution.

The leftover $(c - c_p)$ channels are fed to successive convolutions rather than being eliminated, even though only c_p channels are utilized for spatial feature extraction in PConv. By maintaining feature integrity and guaranteeing the flow of information, this method ensures that feature information is sent across all channels.

A PConv layer and two 1x1 convolutional layers arranged in a residual block structure make up the FasterBlock module. The channel dimension is increased in the intermediate layer, and shortcut connections are incorporated to efficiently reuse the input characteristics.

When the FasterBlock module is used in place of the Bottleneck in the YOLOv8 C2f module, the detection speed is significantly increased and the road snow detection accuracy is significantly improved.

3.1.2 Backbone Network Fusion: Swin Transformer Block

This work presents the Swin Transformer Block from the Swin Transformer model [16] to increase the generalization ability of the model and the accuracy of road snow feature recognition. In order to reduce computational complexity, this module first divides the input feature map into separate windows that provide independent computations within each window. In order to improve the integration of snow features from various sections of the road, it also makes use of a sliding window technique to enable cross-window feature information exchange.

The Swin Transformer Block uses window partitioning and multi-head self-attention as its two primary techniques. First, multiple non-overlapping small windows of the feature map are used for independent multi-head self-attention (W-MSA) [17]. This technique reduces computational complexity while increasing the ability to capture local features. Additionally, to improve cross-window feature interaction and ensure that features are properly integrated at both the local and global levels, a shifting window multi-head self-attention (SW-MSA) [18] technique is employed.

$$\text{Attention}(Q, K, V) = \text{SoftMax}\left(QK^T/\sqrt{d} + B\right)V \qquad (2)$$

After the attention calculations, the features within each window undergo a nonlinear transformation through a multi-layer perceptron (MLP) [19] to enhance their representational capability. Additionally, the Swin Transformer Block incorporates a residual connection mechanism, ensuring that the original information is not lost during the deep transmission of features.

In order to improve the model's capacity for generalization, the C3STR module—which is based on the Swin Transformer—was introduced into the backbone network of this study. This allowed snow features to interact across various windows and lessened the adverse effects of variations in snow feature distribution. The C3STR module gradually expanded the receptive field by enhancing the resolution of the input feature maps through the combination of convolutional operations and the Swin Transformer. The model's capacity to detect road snow was also much improved by the multi-head self-attention method, which improved global interactions among features.

3.2 Neck Network Improvement: DySample Upsampling Operator

In YOLOv8, the UpSample layer is employed to extend the spatial dimensions of feature maps, integrating deep semantic features with shallow high-resolution features to enhance multi-scale object detection capabilities. DySample [20], on the other hand, is an efficient and lightweight dynamic upsampling operator. While traditional methods such as CARAFE, FADE, and SAPA have shown significant improvements, they entail substantial computational overhead due to complex dynamic convolutions and the additional subnetwork, which limits their applicability in scenarios requiring high-resolution features.

DySample uses grid-guided point-based sampling to accomplish upsampling. The PyTorch grid sampling function uses the coordinates in $\mathcal{S}$ to perform bilinear interpolation on $\mathcal{X}$ given a feature map $\mathcal{X}$ with dimensions $C \times H_1 \times W_1$ and a sampling point set $\mathcal{S}$ sized $2 \times H_2 \times W_2$, where the two channels represent the x and y coordinates of the sampling points. An upsampled feature map $\mathcal{X}\prime$ with dimensions $C \times H_2 \times W_2$ is produced by this technique.

$$\mathcal{X}' = \text{grid\_sample}(\mathcal{X}, \mathcal{S}) \tag{3}$$

In a simple implementation, a linear layer calculates an offset $\mathcal{O}$ of size $2s^2 \times H \times W$ for an upsampling factor $\mathcal{S}$ and a feature map $\mathcal{X}$ with dimensions $C \times H \times W$. Pixel shuffling is used to transform this offset into an offset grid with dimensions of $2 \times sH \times sW$. The offset $\mathcal{O}$ is then added to the initial sampling grid $\mathcal{G}$ to obtain the sampling set $\mathcal{S}$.

$$\mathcal{O} = \text{linear}(\mathcal{X}) \tag{4}$$

$$\mathcal{S} = \mathcal{G} + \mathcal{O} \tag{5}$$

Subsequently, the upsampled feature map is generated using the sampling set $\mathcal{S}$ and Eq. (3) through grid sampling.

DySample further optimizes the above implementation by introducing a dynamic sampling point generation mechanism. This mechanism addresses the issues of shared initial offset positions for upsampling points, neglecting spatial relationships between points, and the unconstrained offset range leading to unordered distribution of sampling points, which negatively impacts the upsampling quality. By adjusting the range of the offsets, DySample reduces the overlap between sampling points and ensures a reasonable distribution of sampling points on the feature map.

The DySample upsampling process dynamically generates offsets based on local features, using a sigmoid function to control the range of offsets. This enhancement improves the precision and flexibility of the sampling process, effectively elevating the quality of the features after upsampling in the context of road snow detection.

4 Experimental Analysis

4.1 Dataset Preparation

The Ontario Ministry of Transportation contributed the dataset for this study, which was gathered from onboard cameras mounted on snowplows in Ontario, Canada. The 4323 photos in the dataset are separated into 449 validation images, 409 test images, and 3465 training images. Full coverage, partial coverage, and no snow are the three categories of snow coverage.

The study's training environment is set up with an RTX 3060 graphical processor and Windows 10 as the operating system. Using PyTorch 2.2.1 as the deep learning framework and Python 3.8 as the programming language, the development process was carried out utilizing the PyCharm IDE as the development environment.

4.2 Comparative Analysis of Different Models for Road Surface Snow Detection

To achieve optimal model performance, the object detection hyperparameters for YOLOv8-FCD were set as follows: the image size was 640x640, the number of training epochs was 100, and the batch size was set to 16. All other parameters remained at their default settings.

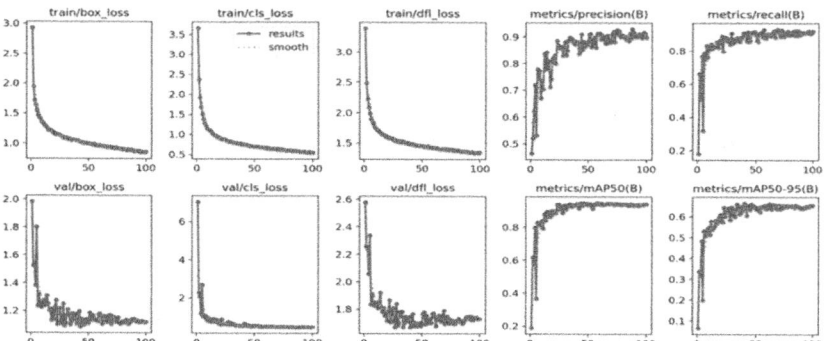

Fig. 3. Training result.

Figure 3 illustrates the results obtained during model training. As the number of epochs increases, the losses for both training and validation datasets steadily decrease, indicating a continuous optimization process. Simultaneously, the precision continuously improve, and the mAP@50% and mAP@50–95% values on the validation set steadily increase, demonstrating that the model's detection performance is consistently improving.

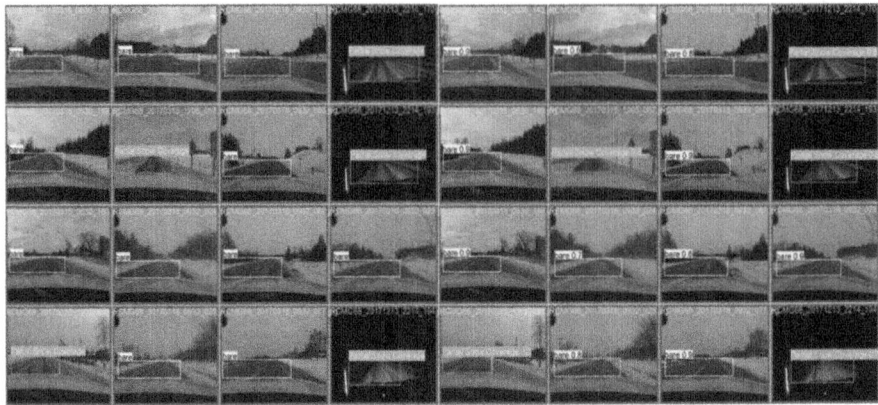

Fig. 4. Snow detection result.

Figure 4 displays the road surface snow detection results of the proposed model. As shown, the left image represents the labeled data from the dataset, while the right image shows the detection results from the model. The results indicate that the model achieves high accuracy even under challenging conditions such as complex road surfaces and poor lighting.

Comparative tests were conducted against several models in order to assess the proposed model's performance in this investigation. These comprised the Mask R-CNN model, different YOLO versions, and YOLOv8 variants combined with ScConv, WTConv, and FASFF-P6 detection heads. The outcomes, which are displayed in Table 1, demonstrate how well the proposed model performs on a variety of measures. Notably, this model reduced the number of parameters to some degree while simultaneously achieving notable increases in precision, mAP, and GFLOPs. The lowest FLOPs were just 7.0, the best precision was 0.898, and the mAP@50% and mAP@50–95% scores were 0.962 and 0.705, respectively. The model's improved abilities in road surface snow detection were highlighted by an average performance increase of 2%.

Table 1. Comparison between models.

Model	P	mAP@0.5	mAP@0.5–0.95	FPS	Params	GFLOPs
Yolov5	0.89	0.944	0.701	26.7	25	7.1
Yolov6	0.871	0.939	0.693	27.6	42.3	11.8
Yolov9	0.878	0.948	0.684	20.4	19.7	7.6
Yolov10	0.868	0.931	0.688	21.5	27	8.2
Yolov8-ScConv	0.817	0.937	0.684	19.7	31.8	8
Yolov8-WTConv	0.876	0.939	0.684	19.6	29.7	7.4

(*continued*)

Table 1. (continued)

Model	P	mAP@0.5	mAP@0.5–0.95	FPS	Params	GFLOPs
Yolov8-FASFF	0.82	0.932	0.687	19.8	64.3	8.8
Improved Mask RCNN*[12]	/	89.8	/	15.4	72.4	158.35
Yolov8	0.835	0.941	0.679	24.7	30	8.1
Yolov8-FCD	**0.898**	**0.962**	**0.705**	**23.3**	**28.3**	**7**

Note: * indicates data are original from literature.

4.3 Ablation Experiments

To evaluate the effect of the proposed sub-modules, ablation experiments were carried out. Precision, mAP@50%, and mAP@50%-95% were greatly increased by combining the C2f_Faster and C3STR modules in the backbone and DySample in the neck, as indicated in Table 2. The best performance was obtained by combining all three modules, especially when subjected to the stringent mAP@50–95% criterion. This demonstrated how well the modules improved feature extraction, upsampling, and interaction for the identification of snow on the road surface.

Table 2. Ablation experiment.

Yolov8	FasterNet	C3STR	DySample	P	mAP@0.5	Improvement	mAP@0.5–0.95
√	×	×	×	0.835	0.941	Baseline	0.679
√	√	×	×	0.9	0.948	0.7%	0.691
√	×	√	×	0.823	0.944	0.3%	0.697
√	×	×	√	0.853	0.935	/	0.674
√	√	√	×	0.877	0.95	0.9%	0.704
√	√	×	√	0.838	0.945	0.4%	0.693
√	×	√	√	0.832	0.948	0.7%	0.683
√	√	√	√	**0.898**	**0.962**	**2.1%**	**0.705**

5 Conclusion

This work proposed YOLOv8-FCD to address the issue of road snow detection in complex scenarios. By combining the C2f_Faster module, the C3STR module, and the DySample upsampling operator, the model effectively reduced the number of parameters while improving detection accuracy and robustness in road snow areas, all while maintaining exceptional real-time performance. Experimental results showed that the model outperformed several benchmark models across multiple assessment metrics. Thanks to the innovative integration of cutting-edge technologies, the model demonstrated robust performance even in complex scenarios and backgrounds, showcasing significant detection advantages. This research provided a reliable and efficient solution for intelligent winter traffic management and road safety monitoring, with promising practical value and potential for wide adoption.

Acknowledgments. This research is supported by National Natural Science Foundation of China under grant no. 62103177.

References

1. Walker, C.L., Hasanzadeh, S., Esmaeili, B., Anderson, M.R., Dao, B.: Developing a winter severity index: a critical review. Cold Reg. Sci. Technol. **160**, 139–149 (2019)
2. El-Basyouny, K., Kwon, D.W.: Assessing time and weather effects on collision frequency by severity in Edmonton using multivariate safety performance functions. In: Transportation Research Board 91st Annual Meeting, no. 12–0494, pp. 1–15. Washington, D.C. (2012)
3. Rasol, M., et al.: Progress and monitoring opportunities of skid resistance in road transport: a critical review and road sensors. Remote Sens. **13**(18), 3729 (2021)
4. Zhou, H., Xu, C., Tang, X., Wang, S., Zhang, Z.: A review of vision-laser-based civil infrastructure inspection and monitoring. Sensors **22**(15), 5882 (2022)
5. Xie, Q., Kwon, T.J.: Development of a highly transferable urban winter road surface classification model: a deep learning approach. Transp. Res. Rec. **2676**(10), 445–459 (2022)
6. Bogaert, J., de Marneffe, M.C., Descampe, A., Standaert, F.X.: Automatic and manual detection of generated news: case study, limitations and challenges. In: Proceedings of the 1st International Workshop on Multimedia AI against Disinformation, pp. 18–26 (2022)
7. Zhao, W., Xu, L., Bai, J., Ji, M., Runge, T.: Sensor-based risk perception ability network design for drivers in snow and ice environmental freeway: a deep learning and rough sets approach. Soft. Comput. **22**, 1457–1466 (2018)
8. Yang, S., Lei, C.: Research on the classification method of complex snow and ice cover on highway pavement based on image-meteorology-temperature fusion. IEEE Sens. J. (2023)
9. Pan, G., Fu, L., Yu, R., Muresan, M.I.: Winter road surface condition recognition using a pre-trained deep convolutional neural network. No. 18–00838 (2018)
10. Bahrampour, S., Ray, A., Sarkar, S., Damarla, T., Nasrabadi, N.M.: Performance comparison of feature extraction algorithms for target detection and classification. Pattern Recogn. Lett. **34**(16), 2126–2134 (2013)
11. Kwon, T.J., Fu, L., Jiang, C.: Effect of winter weather and road surface conditions on macroscopic traffic parameters. Transp. Res. Rec. **2329**(1), 54–62 (2013)
12. Pan, G., Bai, Z., Fu, L., Zhao, L., Xiao, Q.: Road meteorological state recognition in extreme weather based on an improved Mask-RCNN. In: International Conference on Neural Information Processing, pp. 3–15. Springer Nature Singapore, Singapore (2023)
13. Zhou, X., Zhao, L., Liu, Z., Fu, L., & Pan, G.: An adaptive machine learning framework for multi-scenes road surface weather condition monitoring. Canadian J. Civil Eng. (2024)
14. Ultralytics: YOLOv8: real-time object detection. https://github.com/ultralytics/ultralytics Accessed 2023
15. Chen, J., et al.: Run, don't walk: chasing higher FLOPS for faster neural networks. In: Proceedings of the IEEE/CVF Conference on Computer Vision and Pattern Recognition, pp. 12021–12031 (2023)

16. Liu, Z., Lin, Y., Cao, Y., Hu, H., Wei, Y., Zhang, Z., Guo, B.: Swin transformer: hierarchical vision transformer using shifted windows. In: Proceedings of the IEEE/CVF International Conference on Computer Vision, pp. 10012–10022 (2021)
17. Wei, Q., Tian, X., Cui, L., Zheng, F., Liu, L.: WSAFormer-DFFN: a model for rotating machinery fault diagnosis using 1D window-based multi-head self-attention and deep feature fusion network. Eng. Appl. Artif. Intell. **124**, 106633 (2023)
18. Pacal, I., Alaftekin, M., Zengul, F.D.: Enhancing skin cancer diagnosis using swin transformer with hybrid shifted window-based multi-head self-attention and SwiGLU-based MLP. J. Imaging Inform. Med. 1–19 (2024)
19. Tolstikhin, I.O., et al.: MLP-Mixer: An All-MLP architecture for vision. Adv. Neural. Inf. Process. Syst. **34**, 24261–24272 (2021)
20. Liu, W., Lu, H., Fu, H., Cao, Z.: Learning to upsample by learning to sample. In: Pro-ceedings of the IEEE/CVF International Conference on Computer Vision, pp. 6027–6037 (2023)

A Scoring System for Single and Parallel Bars Actions Based on Multi-view 3D Pose Estimation

Yuntong Kang, Mingwei Cao, Haoran Yao, Sen Qiu(✉), and Zhelong Wang

Dalian University of Technology, Dalian 116024, China
qiu@dlut.edu.cn

Abstract. This paper proposes a scoring system to evaluate the correctness of movements performed by trainees during single and parallel bar exercises. First, the system employs a two-step multi-view 3D human pose estimation algorithm to capture the 3D key point coordinates of all individuals in the training scene. The process begins with 2D pose estimation for each individual view, followed by cross-view matching and subsequent 3D reconstruction. Next, a feature selection strategy is introduced to extract the 3D pose information of a specific individual (the trainee) based on his positional features and movement characteristics. The system then assesses and scores the trainee's 3D poses according to standard action scoring criteria. In multi-person scenarios, the proposed scoring system demonstrates high accuracy in pose estimation and in selecting and scoring the trainees' movements.

Keywords: 3D pose estimation · multi-view reconstruction · action scoring

1 Introduction

Human pose estimation [20] has been a persistent challenge in computer vision, with diverse applications spanning fields like video surveillance and sports training [17]. In sports training, single and parallel bar exercises are essential, primarily consisting of movements such as Single Bar I, Single Bar III, and Parallel Bar III. However, due to a lack of digital recording methods and analysis tools, the scoring of single and parallel bar movements is still done manually. Experts can only rely on subjective assessments based on the trainees' posture to evaluate the quality of their movements, which is not only labor-intensive but also prone to bias. Moreover, trainees are often unaware of their mistakes. There are three main approaches to human pose estimation: depth image-based [5], monocular RGB image-based [7], and multi-view RGB image-based [4,14,15,18]. Depth images offer depth information that is absent in RGB images and are less influenced by environmental conditions such as lighting. However, depth cameras have limited range, and their accuracy decreases as distance increases, making them unsuitable for single and parallel bar scenarios. Monocular RGB images are highly affected by occlusion and are less accurate than multi-view pose estimation, which is the most suitable for single and parallel bar exercises.

Multi-view 3D human pose estimation involves using multiple calibrated RGB cameras to capture data from the trainees. Each camera's video feed undergoes object detection to identify human targets in every frame, followed by 2D pose estimation for each individual. Deep learning algorithms have reached a high level of maturity in fields such as emotion recognition [11], action recognition [12,13] and pose estimation. Cross-view matching is then performed, and triangulation methods are used to obtain each person's 3D coordinates, digitizing the key points of the trainees in single and parallel bar training. Since there are often multiple individuals in the training environment, this paper introduces a feature selection strategy to filter out unnecessary data and retain only the 3D data of the relevant trainees. The system then scores the 3D poses based on movement scoring criteria.

In this scoring system for single and parallel bar exercises, incorrect movements are recorded, and the image frames showing the errors are displayed to clearly indicate the cause of the trainee's mistakes. Additionally, the system uses SMPL (Skinned Multi-Person Linear) [8,9] skinning to create 3D models of the trainee's poses, providing a more vivid representation of their movements.

In summary, the main contributions of this work are:

- A two-step process that combines 2D keypoint detection and 3D pose reconstruction to obtain the 3D world coordinates of trainees, tailored to the characteristics of the single and parallel bar environment.
- A feature selection strategy is proposed to filter the trainee's posture data, with movements being scored based on the established scoring criteria.
- The development of a scoring system for single and parallel bar exercises, which automates scoring, highlights incorrect movements, records reasons for deductions, and digitizes the training process.

2 Methods

Figure 1 illustrates the workflow of our method. Given a multi-person scene during single and parallel bar training, captured by multiple RGB cameras, 2D human pose detection is first applied to obtain the bounding boxes and 2D key point locations of each individual in each view. A matching algorithm is then used to establish correspondences between individuals across different views. Triangulation is employed to calculate the 3D key point coordinates for each person. Next, a feature selection strategy filters out the trainee's pose, and finally, the movements are scored based on the established scoring criteria.

2.1 Camera Calibration

Camera calibration aims to determine the internal and external parameters of a camera for accurate mapping from 2D pixel coordinates to 3D world coordinates. External parameters, including the rotation matrix, Euler angles, and

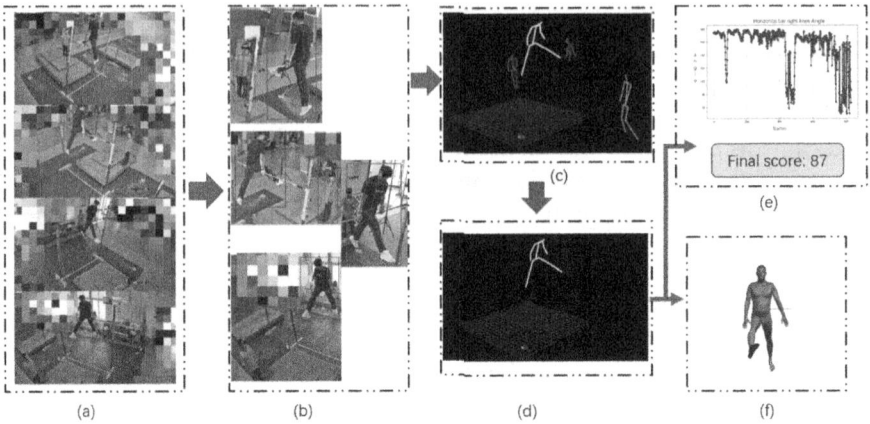

Fig. 1. Overview of our system. (a) Images captured from four different views. (b) 2D human detection and pose estimation. (c) 3D reconstruction. (d) Feature selection strategy results. (e) Joint angle calculation and scoring. (f) SMPL human mesh model.

translation vector, describe how real-world points are transformed into the camera coordinate system. Internal parameters, such as focal length, principal point coordinates, and distortion coefficients, define how points in the camera coordinate system are projected onto the image.

The Zhang calibration method [19], proposed by Zhengyou Zhang, is a classic technique for camera calibration. A commonly used evaluation metric in camera calibration is the reprojection error, which is crucial for assessing the accuracy and stability of the calibration results. The reprojection error measures the difference between the observed pixel coordinates of each calibration point and the pixel coordinates reprojected onto the image plane using the estimated 3D points and the calibration parameters. The formula is shown in Eq. 1, where n is the number of calibration points; (x_i, y_i) are the observed pixel coordinates, and (x'_i, y'_i) are the pixel coordinates calculated based on the calibration parameters. The reprojection results for the single bar scenario are shown in Fig. 2, with an error of less than 1 pixel, which meets the application requirements.

$$E = \frac{\sum_{i=1}^{n} \sqrt{(x_i' - x_i)^2 + (y_i' - y_i)^2}}{n} \tag{1}$$

2.2 2D Pose Detection

The process of detecting 2D poses can be broken down into two stages. First, a target detector is used to perform human detection on each video frame, generating bounding box information for each individual in the scene. Then, 2D pose estimation is conducted for each bounding box to obtain the human body's

Fig. 2. Reprojection results of the four-camera calibration for the single bar scenario.

2D key points. This approach follows a top-down architecture, which provides higher pose estimation accuracy for each individual, though the computational load increases with the number of people. In the single and parallel bar training scenario, where the number of individuals (including trainees and spotters) is relatively small, this increase in computation time is minimal. However, pose estimation accuracy significantly impacts 3D reconstruction, making the top-down architecture an optimal choice.

YOLOv5 [10] is used as the target detector due to its strong feature extraction capabilities and high detection accuracy. The network's feature extraction component fuses both high-level and low-level features, leveraging the high resolution of low-level features and the rich semantic information of high-level features to perform independent predictions across multiple scales. For 2D pose estimation, HRNet [16] is employed to detect single-person poses within each bounding box. By utilizing high-resolution feature maps, the model can more precisely recognize human poses, while multi-scale feature fusion enhances the model's recognition capability.

2.3 Cross-View Matching and Triangulation

Since the single and parallel bar training scenario involves multiple individuals, it is necessary to match the 2D poses of each person across different views before reconstructing the 3D key points. Cross-view matching [2] is achieved by optimizing the objective function M, as shown in Eq. 2, where S_b represents the bounding box similarity, S_g denotes the epipolar constraint similarity, and λ refers to the coefficients for each term.

$$M = \lambda_b S_b + \lambda_g S_g \tag{2}$$

The bounding box similarity is calculated by using the two corner points of the bounding box as feature vectors and computing the Euclidean distance between the feature vectors of the human regions in the two views to measure their similarity. The specific formula is shown in Eq. 3, where B_u^k denotes the bounding box feature vector of the k-th person in view u, u and v represent views u and v respectively, and Sig represents the sigmoid function used to map the Euclidean distance to a value between 0 and 1.

$$S_b = 1 - Sig(\left\|B_u^k - B_v^k\right\|_2) \tag{3}$$

The epipolar constraint similarity utilizes the epipolar constraint between the 2D joint positions, meaning that the joint in the first view should lie on the epipolar line corresponding to the joint in the second view. The specific formula is shown in Eq. 4,where x_n^u represents the 2D coordinate of the n-th joint in view u, u and v represent views u and v respectively, $l_{uv}(x_n^v)$ denotes the epipolar line associated with x_n^v from view v, D_g represents the distance from a point to the epipolar line, and Sig represents the sigmoid function used to map the Euclidean distance to a value between 0 and 1.

$$S^g = 1 - Sig(\frac{1}{2N}\sum_{n=1}^{N}(D_g(x_n^u, l_{uv}(x_n^v)) + D_g(x_n^v, l_{vu}(x_n^u)))) \tag{4}$$

The triangulation algorithm [6] is employed to reconstruct a matched set of 2D key point coordinates into 3D coordinates. The primary concept involves connecting the camera center with the 2D key points in each view to form a straight line. In an ideal scenario, the lines from all views converge at a single point, representing the 3D key point. However, due to various sources of noise in the experiments and potential calibration errors in the camera parameters, the optimal solution is typically determined by using the least squares method to find the point that is closest to all the epipolar lines.

2.4 Filtering and Scoring

After obtaining the 3D key points for all individuals in the scene, it is essential to retain only the information of the single and parallel bar trainees for further analysis and scoring [3]. First, we analyze the training scene, which includes trainees, protectors, and other individuals who may cause interference. Prior knowledge indicates that the height of the trainees on the single bar is significantly greater than that of the individuals on the ground. Additionally, if the world coordinate origin is positioned directly below the single bar during calibration, the Euclidean norm in the x-y plane for the trainees will be quite small. Based on these observations, we propose a feature selection strategy, with the specific formula provided in Eq. 5.

$$Slc = \sum_{n=1}^{N} \lambda_{hn} z_n - \sum_{n=1}^{N} \lambda_{xyn} \|(x_n, y_n)\|_2 \tag{5}$$

where (x_n, y_n, z_n) represents the 3D coordinates of the n-th key point, λ denotes the coefficient, and N indicates the number of selected key points. Slc represents the score from the feature selection strategy, with a higher score indicating a greater likelihood that the data belongs to the trainee. Depending on the different movements, selecting various key points for the calculation of the feature selection strategy can slightly affect accuracy. In the case of the single-bar I (pull-up), the trainee's arms remain hanging on the bar, causing the heights of the six key points related to the arms to be significantly higher than those of the protector. As a result, the six key points of the arms, together with the key points of the head and neck, are utilized to compute the feature score, with greater emphasis placed on the arm key points. In the case of the single-bar III, the trainee rotates around the bar, resulting in considerable body movement and significant changes in the $x-y$ coordinates. Thus, key points with smaller movement amplitudes, such as the torso and wrists, are selected for feature score calculation. For the double-bar III, because the overall height of the double bar is not high, it is essential to select key points that can clearly distinguish the trainee from the protector. During the execution of double-bar movements, the trainee's lower limbs are mostly on the double bar, and their hands remain grasped on the bar. Therefore, the six key points of the lower limbs, along with the wrist and elbow key points, are used to calculate the feature score.

The scoring criteria are established through consultations with experts and the review of educational materials. The primary method of scoring involves assessing whether the angles of the human joints meet the requirements of the movement. In the case of the single-bar III, the specific evaluation metrics are outlined in Table 1, where $\angle(1,8,10)$ represents the angle formed by points 1, 8, and 10 in Pose18 [1], H_p represents the horizontal plane, V_p represents the vertical plane, h_{11} represents the height of the foot joint, and H_{bar} represents the height of the single bar.

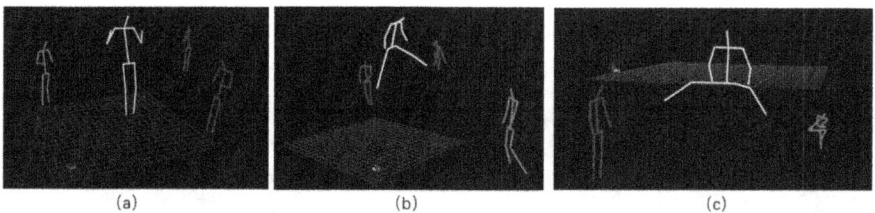

Fig. 3. Unfiltered multi-person scene. (a) Single Bar I. (b) Single Bar III. (c) Parallel Bar III.

3 Experiments

3.1 3D Pose Estimation

The multi-view 3D pose estimation algorithm ultimately identifies 18 human key points, effectively representing the pose information. According to the scoring criteria, the key points that significantly impact the action scores are primarily those related to the torso and limbs. Therefore, in the visualizations, to emphasize these critical areas, we have chosen to omit the finer details of the facial pose (such as the eyes and ears). The unfiltered visualization results from the single and parallel bar scene are shown in Fig. 3. Figure 4 shows the process of the Single Bar III action, where the first row represents the original images and the third row displays the selection results.

Table 1. Evaluation Criteria for Single Bar III Exercises

Requirements	Evaluation Indicators
Trunk Hang	$175° < \angle(1,8,10) < 185°$
Knees Locked	$175° < \angle(9,10,11) < 185°$
	$175° < \angle(12,13,14) < 185°$
Legs Together	$0° < \angle(10,8,13) < 5°$
Arms Locked	$175° < \angle(2,3,4) < 185°$
	$175° < \angle(5,6,7) < 185°$
Horizontal Bar Support	$30° < \angle(1,11,H_p) < 45°$
One-leg Ride Position	$30° < \angle(10,8,13)$
Back Turning	$15° < \angle(11,8,V_p) < 75°$
Back-swing Leg Straight	$175° < \angle(9,10,11) < 185°$
Dismount	$h_{11} > H_{bar}$
Stick the Landing	$120° < \angle(1,8,11)$
	$120° < \angle(9,10,11)$

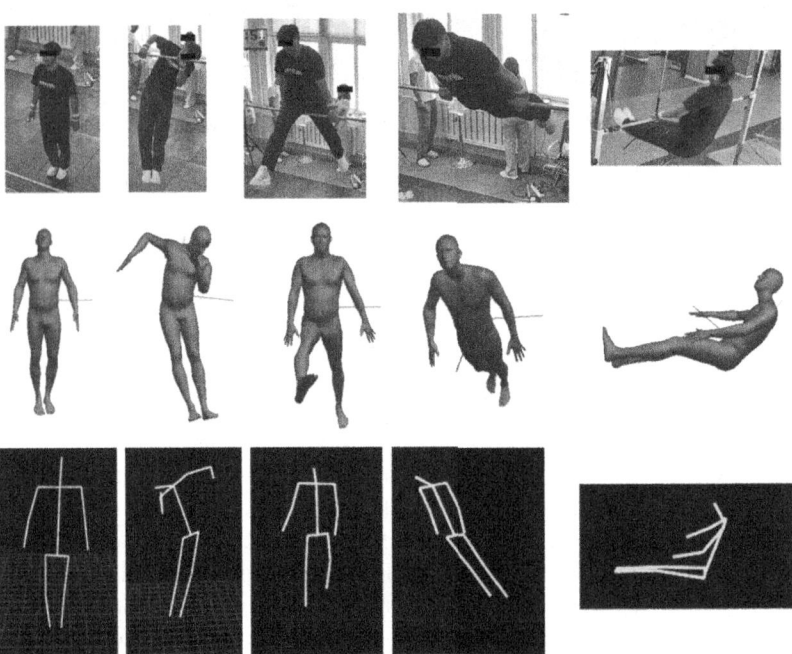

Fig. 4. The process of the single bar III. The first row shows the real images, the second row presents the SMPL human model, and the third row visualizes the selected 3D key point coordinates.

3.2 Filtering and Scoring

Based on the feature selection strategy, trainees can be effectively identified in the majority of frames, with the filtering results visualized in Fig. 4, row 3. Comparative experiments were conducted using different calculation methods for the feature selection strategy, and the results are presented in Table 2. Here, $score-h$ represents the strategy that considers only height; $score-all$ takes into account both height and the Euclidean norm of the horizontal coordinates, selecting all key points; and $score-slct$ considers height and the Euclidean norm of the horizontal coordinates while selecting specific key points. The results indicate that incorporating the horizontal coordinate Euclidean norm and selecting specific key points can enhance the accuracy of the filtering process.

Table 2. Selection accuracy with different strategies.

Action	$score-h$	$score-all$	$score-slct$
single bar I	93.2%	96.1%	97.2%
single bar III	93.5%	95.9%	96.7%
parallel bar III	88.1%	90.2%	92.6%

Based on the estimated 3D pose key points, various joint angles required by the scoring criteria can be calculated. Figure 5 illustrates the changes in joint angles for the left knee and the left elbow during the double bar action. According to the action scoring criteria, the corresponding sub-actions are evaluated to determine whether the joint angles meet the standards. If they do not meet the requirements, the corresponding scores are deducted, and the incorrect actions along with their causes are recorded. Figure 6 shows the interface of the scoring system.

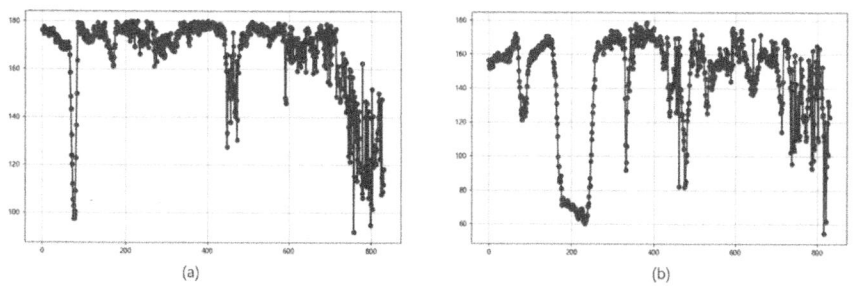

Fig. 5. Joint angle line chart. The horizontal axis represents the frame number, and the vertical axis represents the joint angle. (a) Left knee joint. (b) Right elbow joint.

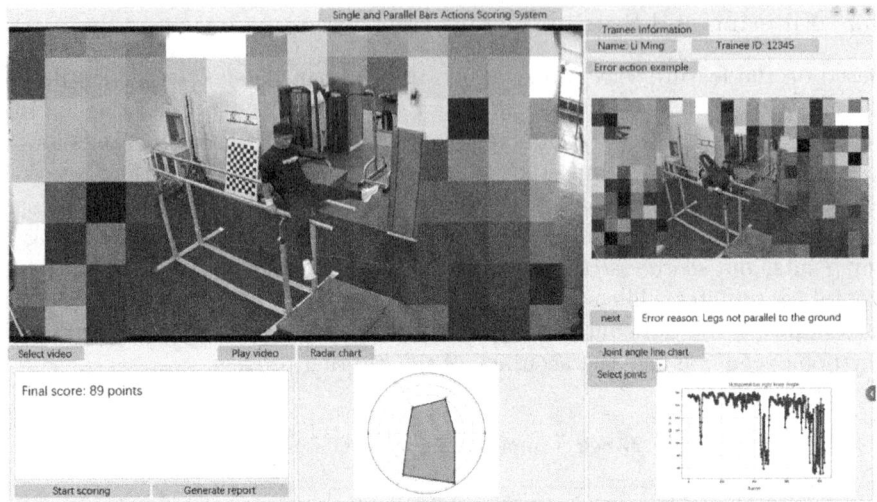

Fig. 6. The scoring interface for the single and parallel bars system. The interface includes features such as selecting and playing videos, scoring, displaying incorrect actions and their causes, showing joint angles, and a radar chart of point deductions.

3.3 Human Skinning Model

The human skinning model is based on the SMPL model, a parameterized representation with 6890 triangular mesh vertices, driven by 72 body pose parameters (θ) and 10 shape parameters (β). Pose parameters are derived from 3D key points via optimization, with shape parameters fixed to a standard form. Figure 4 shows the original images of several movements alongside the SMPL human model. As a visualization method, the advantage of the SMPL model is that it can vividly and clearly illustrate human actions, making it easier to observe specific postures.

4 Conclusion

In this paper, we applied multi-view 3D pose estimation to the single and parallel bar training scenario and addressed several issues in practical applications. We used YOLOv5 as the human object detector in the 2D pose reconstruction to output bounding boxes, which can provide feature vectors for bounding box similarity in cross-view matching. We proposed a feature selection strategy to extract the relevant trainee data after obtaining the 3D pose key points. We established a scoring system based on multi-view 3D pose estimation for single and parallel bar exercises, which digitizes the actions. The system can automatically score the exercises and provide error action notifications, alleviating the burden on evaluators.In future work, we will refine the pose estimation algorithm to reduce computation time and explore further integration of data acquisition, pose estimation, and skinning visualization. By enhancing the overall workflow, we aim to achieve real-time 3D human motion capture.

Acknowledgments. This work was jointly supported by the National Natural Science Foundation of China under Grant No. 62272081 and 61903062, Xingliao Talent Project under Grant XLYC2203033, and in part by the Kweichow Moutai Group under Grant MTGF2023038. The authors would like to express their thanks to these funding bodies.

References

1. Chen, C., Zhuang, Y., Nie, F., Yang, Y., Wu, F., Xiao, J.: Learning a 3d human pose distance metric from geometric pose descriptor. IEEE Trans. Visual Comput. Graphics **17**(11), 1676–1689 (2011)
2. Chen, L., Ai, H., Chen, R., Zhuang, Z., Liu, S.: Cross-view tracking for multi-human 3d pose estimation at over 100 fps. In: 2020 IEEE/CVF Conference on Computer Vision and Pattern Recognition (CVPR). pp. 3276–3285 (2020)
3. Chen, Y., Qiu, S., Wang, Z., Zhao, H., Cao, X.: Multiperceptive region of spatial-temporal graph convolutional shrinkage network for arrhythmia recognition. IEEE Trans. Instrum. Meas. **73**, 1–11 (2024). https://doi.org/10.1109/TIM.2024.3376017
4. Dong, J., et al.: Fast and robust multi-person 3d pose estimation and tracking from multiple views. IEEE Trans. Pattern Anal. Mach. Intell. **44**(10), 6981–6992 (2022). https://doi.org/10.1109/TPAMI.2021.3098052
5. Ge, L., Liang, H., Yuan, J., Thalmann, D.: 3d convolutional neural networks for efficient and robust hand pose estimation from single depth images. In: 2017 IEEE Conference on Computer Vision and Pattern Recognition (CVPR). pp. 5679–5688 (2017). https://doi.org/10.1109/CVPR.2017.602
6. Hua, G., Liu, H., Li, W., Zhang, Q., Ding, R., Xu, X.: Weakly-supervised 3d human pose estimation with cross-view u-shaped graph convolutional network. IEEE Trans. Multimedia **25**, 1832–1843 (2023)
7. Liu, W., Bao, Q., Sun, Y., Mei, T.: Recent advances of monocular 2d and 3d human pose estimation: a deep learning perspective. ACM Comput. Surv. **55**(4) (2022)
8. Loper, M., Mahmood, N., Romero, J., Pons-Moll, G., Black, M.J.: Smpl: a skinned multi-person linear model. ACM Trans. Graph. **34**(6) (2015). https://doi.org/10.1145/2816795.2818013
9. Loper, M., Mahmood, N., Romero, J., Pons-Moll, G., Black, M.J.: SMPL: A Skinned Multi-Person Linear Model, 1st edn. Association for Computing Machinery, New York, NY, USA (2023)
10. Nguyen, H.C., Nguyen, T.H., Scherer, R., Le, V.H.: Unified end-to-end yolov5-hr-tcm framework for automatic 2d/3d human pose estimation for real-time applications. Sensors **22**(14) (2022). https://doi.org/10.3390/s22145419
11. Qiu, S., et al.: A review on semi-supervised learning for eeg-based emotion recognition. Inf. Fusion **104**, 102190 (2024). https://doi.org/10.1016/j.inffus.2023.102190
12. Qiu, S., et al.: A novel two-level interactive action recognition model based on inertial data fusion. Inf. Sci. **633**, 264–279 (2023). https://doi.org/10.1016/j.ins.2023.03.058
13. Qiu, S., et al.: Sensor combination selection strategy for kayak cycle phase segmentation based on body sensor networks. IEEE Internet Things J. **9**(6), 4190–4201 (2022). https://doi.org/10.1109/JIOT.2021.3102856

14. Qiu, S., et al.: Multi-sensor information fusion based on machine learning for real applications in human activity recognition: state-of-the-art and research challenges. Inf. Fusion **80**, 241–265 (2022). https://doi.org/10.1016/j.inffus.2021.11.006
15. Shuai, H., Wu, L., Liu, Q.: Adaptive multi-view and temporal fusing transformer for 3d human pose estimation. vol. 45, pp. 4122–4135 (2023). https://doi.org/10.1109/TPAMI.2022.3188716
16. Sun, K., Xiao, B., Liu, D., Wang, J.: Deep high-resolution representation learning for human pose estimation. In: 2019 IEEE/CVF Conference on Computer Vision and Pattern Recognition (CVPR). pp. 5686–5696 (2019). https://doi.org/10.1109/CVPR.2019.00584
17. Wang, J., Qiu, K., Peng, H., Fu, J., Zhu, J.: AI coach: deep human pose estimation and analysis for personalized athletic training assistance. In: Proceedings of the 27th ACM International Conference on Multimedia. pp. 374–382. MM '19, Association for Computing Machinery, New York, NY, USA (2019)
18. Zhang, L., et al.: Esmformer: error-aware self-supervised transformer for multi-view 3d human pose estimation. vol. 158, p. 110955 (2025). https://doi.org/10.1016/j.patcog.2024.110955,
19. Zhang, Z.: A flexible new technique for camera calibration. IEEE Trans. Pattern Anal. Mach. Intell. **22**(11), 1330–1334 (2000)
20. Zheng, C., et al.: Deep learning-based human pose estimation: a survey. ACM Comput. Surv. **56**(1) (2023)

Adaptive PID Controller for Industrial Process Based on Reinforcement Learning

Yicong Yang and Qili Chen(✉)

Beijing Information Science and Technology University, Beijing 1001192, China
2023020393@bistu.edu.cn

Abstract. This study proposes an adaptive PID control strategy for industrial processes by integrating Deep Reinforcement Learning (DRL) with conventional PID controllers, particularly focusing on wastewater treatment. By leveraging the strengths of DRL, specifically Proximal Policy Optimization (PPO) and Deep Deterministic Policy Gradient (DDPG), the proposed DRL-PI framework adaptively adjusts PID parameters to enhance control performance under nonlinear and time-varying conditions. Experiments using the BSM2 simulation model demonstrated that the DRL-PI framework outperformed traditional PI controllers in terms of stability, response accuracy, and adaptability. The results highlight the potential of DRL-based adaptive control in complex industrial environments.

Keywords: Deep Reinforcement Learning · Process Control · Wastewater Treatment Process

1 Introduction

In industrial control, nonlinear systems often present complex dynamics and multivariable coupling, making traditional control methods, which rely on precise mathematical modeling, challenging for highly nonlinear, time-varying systems. PID control remains widely used due to its simplicity, ease of implementation, and robustness, but it struggles when faced with nonlinear and multivariable scenarios [1].

Reinforcement Learning (RL) has gained attention as a promising approach to address complex control challenges [2]. RL can learn control strategies through interaction with the environment without requiring a precise model, demonstrating advantages in handling complex environments. Deep Reinforcement Learning (DRL), which integrates deep learning's feature extraction with RL's strategy optimization, has become a focus in industrial automation for its adaptability and ability to manage high-dimensional tasks. In particular, DRL has the potential to enhance PID performance by tuning parameters adaptively, thereby achieving better control outcomes even under varying conditions.

Significant progress has been made in recent years in combining DRL with PID control. For example, some researchers proposed an RL-PID framework, where an RL agent serves as a supervisor to automatically adjust PID parameters, improving the system's response and accuracy. This adaptive control approach offers a new direction for industrial control, maintaining stable and efficient performance despite environmental changes [3–5].

Most wastewater treatment control strategies in prior works are based on the BSM1 model, which has limitations in representing real processes. [7, 8] The BSM2 model, on the other hand, provides a more comprehensive simulation of wastewater treatment, including sludge treatment and more complex workflows, offering a better platform for evaluating and optimizing control strategies. This study focuses on applying DRL to the BSM2 environment to more accurately replicate the control of wastewater treatment.

We propose a novel DRL-PI framework by improving the classic PI controller in the BSM2 environment. By integrating DRL's feature extraction capabilities with the conventional PI structure, the DRL-PI controller maintains stability in complex conditions while adapting to environmental changes, significantly improving system performance. Experimental results show that the DRL-PI framework outperforms traditional PI controllers in terms of dynamic response and stability. Furthermore, we evaluated Proximal Policy Optimization (PPO) and Deep Deterministic Policy Gradient (DDPG), demonstrating their varying strengths in control performance and computational complexity.

2 Methods

2.1 Deep Reinforcement Leaning

Deep Reinforcement Learning (DRL) integrates deep learning's feature representation with reinforcement learning's decision-making abilities, making it effective for complex, nonlinear, and time-varying scenarios in automatic control. Model-free DRL algorithms are divided into value gradient and policy gradient methods. Value gradient methods select actions based on state-action values, while policy gradient methods directly model and optimize policies to maximize rewards. Policy gradient methods are particularly suitable for continuous action spaces and high-precision industrial control [8]. This study focuses on policy gradient DRL, including both on-policy and off-policy methods.

PPO

PPO is an improved version of TRPO that simplifies the optimization process through a clipping mechanism that limits policy updates [9]. Instead of the hard constraint of TRPO, PPO introduces a penalty for drastic changes by clipping the policy ratio, making it easier to implement and computationally more efficient.

DDPG

DDPG is an off-policy algorithm combining deterministic policy gradient and actor-critic structure, particularly suitable for continuous action spaces [10]. The actor network determines the optimal action for a given state, while the critic network estimates the value of state-action pairs.

2.2 DRL-PI Framework

The DRL-PI framework combines the advantages of traditional PID control and Deep Reinforcement Learning. For nonlinear, non-affine systems, we propose a DRL-PID framework, as shown (see Fig. 1). During training, the reinforcement learning agent predicts the optimal PID tuning parameters K_P, K_I and K_D based on the condition that maximizes the reward. During the control process, the fixed RL agent updates the tuning parameters of the controller according to the current state of the environment.

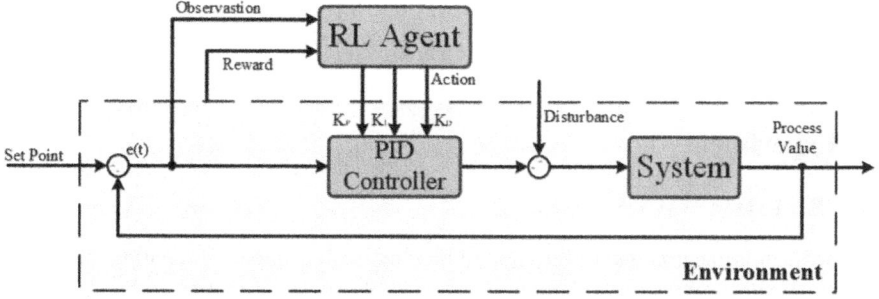

Fig. 1. DRL-PID framework

PID control, as a classical control method, adjusts the system output through a feedback mechanism consisting of Proportional (P), Integral (I), and Derivative (D) components, bringing the system output closer to the desired value. For a nonlinear system, the total output of a typical PID controller is represented as:

$$u(t) = K_P \times e(t) + K_I \times \int e(t)dt + K_D \times \frac{de(t)}{dt} \tag{1}$$

$u(t)$ is the control output at time t, and $e(t)$ is the error between the system output and the setpoint at time t. When combining reinforcement learning with PID control, a triplet $o(t) = [o_1, o_2, o_3]$ can be used as the state value for reinforcement learning at time step t, Where o_1, o_2 and o_3 are the normalized values of the proportional, integral, and derivative components of the error $e(t)$. Using the tanh normalization method, the agent's input and output can be aligned for more effective learning. The action value for reinforcement learning is represented by another triplet $a(t) = [a_1, a_2, a_3]$, After scaling the three action values, we can obtain the PID parameters K_P, K_I and K_D, where k_1, k_2 and k_3 are constants:

$$K_p = k_1 a_1 + k_1 \quad K_I = k_2 a_2 + k_2 \quad K_D = k_3 a_3 + k_3 \tag{2}$$

Reinforcement learning learns an optimal policy network by outputting action values $a(t)$ and adjusting based on the state $s(t)$ and the reward $R(t) = R(a_t|s_t)$ at the current time step, where θ represents the parameters of the policy neural network $\pi_\theta(a|s)$. For nonlinear industrial control, we define the reward value as:

$$r_t = \tanh(r_e + r_{add}) \tag{3}$$

where r_e is a negative value directly related to the error, encouraging the agent to select actions that minimize the error. r_{add} is a stepwise positive value, introduced to guide the agent towards choosing actions that meet the requirements. Finally, the two values are summed and normalized to help the reinforcement learning agent effectively learn the reward features. The two reward values are defined as follows:

$$r_e = -k_1 e^2 - k_2 |e| \qquad (4)$$

$$r_{add} = \begin{cases} 0 & \text{if } 0.1 < |e| \\ 1 & \text{if } 0.01 \leq |e| \leq 0.1 \\ 2 & \text{if } |e| < 0.01 \end{cases} \qquad (5)$$

3 Experiment

3.1 RL-PID in WWTP

Due to the nonlinearity and uncertainty of biochemical reactions, simulating the wastewater treatment process are a challenging industrial nonlinear control problem. To fairly evaluate various control methods used in wastewater treatment, the Benchmark Simulation Model No. 2 (BSM2) was developed as an improved and more comprehensive version of the previous BSM1. BSM2 defines plant layout, influent load, experimental procedures, and evaluation criteria [11]. In this study, we use BSM2 to experiment with the RL-PID controller. As shown (see Fig. 2), BSM2's activated sludge reactor and secondary sedimentation tank are core components of wastewater treatment. The reactor uses microorganisms to degrade organic substances and remove nitrogen and phosphorus, consisting of an anaerobic tank, an anoxic tank, and three aerobic tanks to treat various pollutants. Adjusting the aeration rate (KLa) in the aerobic tanks enhances treatment efficiency and effluent quality, with the stability of dissolved oxygen concentration in the fourth tank $S_{O,4}$ being a key indicator. Treated wastewater enters the secondary sedimentation tank for settlement, with remaining sludge either recycled to the reactor or sent for further processing.

For the experiment's objective, in order to ensure effluent quality, it is necessary to maintain $S_{O,4} = 2(g \cdot m^{-3})$. Thus, the observation value is set as a triplet $[\tanh(e(t)), \tanh(\int e(t)), \tanh(\frac{de(t)}{dt})]$. By normalizing the error and amplifying the tracking error close to zero, we avoid performance degradation due to excessively large neural network inputs.

Since BSM2 uses a PI controller, the action value of the reinforcement learning agent is $[a_1(t), a_2(t), 0]$, with its range being $[-1, 1]$. For the PID controller, the input is $[K_P = 25 \cdot a_1(t) + 25, K_I = 12500 \cdot a_2(t) + 12500, K_D = 0]$, and the reward value is output based on the error range, divided into three cases:

$$r_t = \begin{cases} \tanh(-10e^2 - 10|e|) & \text{if } 0.1 < |e| \\ \tanh(-10e^2 - 10|e| + 1) & \text{if } 0.01 \leq |e| \leq 0.1 \\ \tanh(-10e^2 - 10|e| + 2) & \text{if } |e| < 0.01 \end{cases} \qquad (6)$$

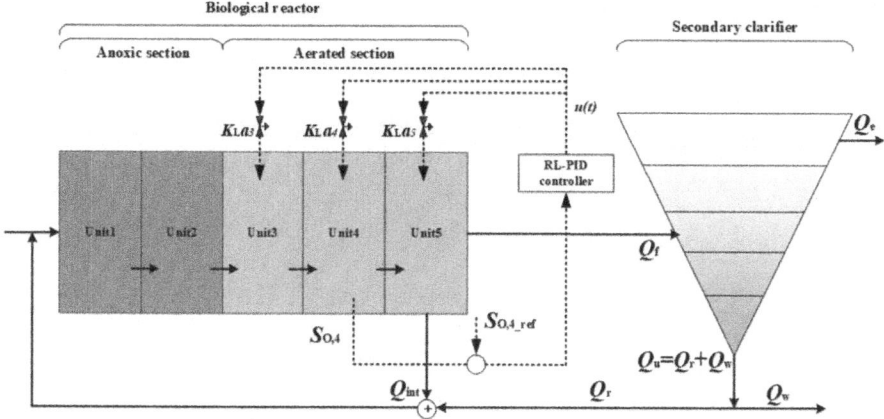

Fig. 2. RL-PID control of Biological reactor in BSM2

3.2 DRL Training

For the DRL-PI structure's reinforcement learning agent, training is performed using the aforementioned state, action, and reward design. After 500 episodes of training, with each episode spanning 7 days, each agent achieved relatively stable performance. The average reward per episode for DDPG was -467.7, while for PPO it was-473.1. The hyperparameters for the reinforcement learning agent are set as Table 1:

Table 1. Hyperparameters for the reinforcement learning agent.

Algorithm	Actor learn Rate	Critic learn Rate	Actor Networks	Critic Networks
PPO	8e-3	8e-3	3-128-1-32-3	3-128-1
DDPG	1e-3	1e-8	3-32-3	(3-64-128,3-128)-1

3.3 Model Testing

After training, the trained PPO and DDPG models were tested in the BSM2 simulation environment for a runtime of 30 days, the input of BSM2 consists of dynamically unstable values. The following are the experimental results comparing both models to the original PID model(see Fig. 3(a)(b)(c)).

As shown in the figure, the original PID controller's ability to adapt to disturbances is clearly inferior to that of the RL-PID controller. In some areas, the original PID controller's error exceeds 1, whereas the RL-PID controller's error mostly fluctuates around 0.5, with the absolute error not exceeding 0.8. It can also be observed that the control performance of DDPG-PID is slightly better than that of PPO-PID, which is consistent with the average reward results per episode. The specific evaluation metrics for the three control methods are shown in the Table 2 below.

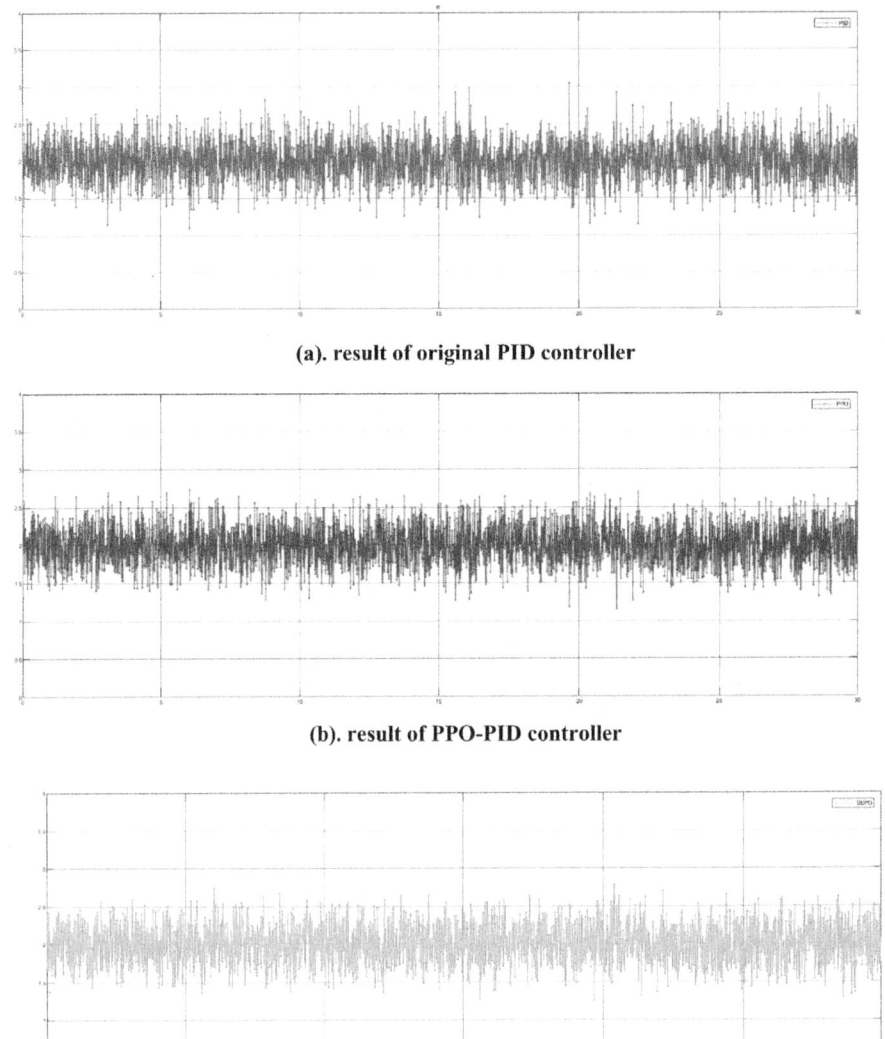

Fig. 3. (a). Result of original PID controller (b). Result of PPO-PID controller (c). Result of DDPG-PID controller

Table 2. Evaluation metrics of three control methods.

methods	MAE	IAE	ISE
Origin PID	0.237	0.236	0.092
PPO-PID	0.199	0.144	0.072
DDPG-PID	0.198	0.131	0.071

4 Conclusion

This research demonstrates the effectiveness of integrating Deep Reinforcement Learning (DRL) with PID control to create an adaptive control framework for complex industrial processes, specifically wastewater treatment. The DRL-PI framework, utilizing PPO and DDPG algorithms, exhibited superior performance compared to traditional PI controllers, particularly in dynamic response and stability. The findings show that the adaptive tuning of PID parameters through reinforcement learning significantly enhances control quality in nonlinear, non-affine systems. Future work may explore other advanced DRL algorithms and their integration with PID to further improve control accuracy and efficiency in various industrial applications.

References

1. Dubey, V., Goud, H., Sharma, P.C.: Role of PID control techniques in process control system: a review. In: Nanda, P., Verma, V.K., Srivastava, S., Gupta, R.K., Mazumdar, A.P. (eds.) Data Engineering for Smart Systems. Lecture Notes in Networks and Systems, vol. 238. Springer, Singapore. https://doi.org/10.1007/978-981-16-2641-8_62 (2022)
2. Kai A., Marc P.D., et al.: Deep reinforcement learning: a brief survey. In: IEEE Signal Processing Magazine, vol. 34, no. 6, pp. 26–38 (2017)
3. Shuprajhaa, T., Shiva, K.S., Srinivasan, K.: Reinforcement learning based adaptive PID controller design for control of linear/nonlinear unstable processes. Appl. Soft Comput. **128** (2022)
4. Sierra-Garcia, J.E., Matilde, S., Ravi, P.: Wind turbine pitch reinforcement learning control improved by PID regulator and learning observer. Eng. Appl. Artif. Intell. vol. 111 (2022)
5. Ignacio, C., Mariano, D.P., Gerardo, G.A.: An adaptive deep reinforcement learning approach for MIMO PID control of mobile robots, ISA Transactions, vol. 102 (2020)
6. Ruyue, Y., Ding, W., Junfei, Q.: Policy gradient adaptive critic design with dynamic prioritized experience replay for wastewater treatment process control. In: IEEE Transactions on Industrial Informatics, vol. 18, no. 5, pp. 3150–3158 (2022)
7. Ding, W., Hongyu, M., Jin, R., Ning, G., Junfei, Q.: Adaptive critic design with weight allocation for intelligent learning control of wastewater treatment plants. Eng. Appl. Artif. Intell. vol. 133, Part C (2024)
8. Rui, N., Jinfeng, L., Biao, H.: A review on reinforcement learning: Introduction and applications in industrial process control. Comput. Chem. Eng. 139 (2020)
9. John, S., Filip, W., et al.: Proximal policy optimization algorithms. arXiv:1707.06347 (2017)
10. Timothy, P.L., Jonathan, J.H., et al.: Continuous control with deep reinforcement learning. arXiv:1509.02971 (2019)
11. Jens, A., et al.: Benchmark Simulation Model no. 2 (BSM2) London, U.K.: IWA Task Group on Benchmarking of Control Strategies for WWTPs (2008)

Computational Analysis of Synaptic Plasticity in Echo State Network

Xinyu Shen, Shaoqi Cheng, Fanjun Li(✉), and Jiayue Feng

University of Jinan, Jinan 250022, China
ss_lifj@ujn.edu.cn

Abstract. Topological structure of the reservoir significantly impacts the network performance of Echo State Network (ESN). Synaptic plasticity learning rules influence the network performance of ESN by altering the topological structure of the reservoir, but this effect is unclear. To investigate this problem, five synaptic plasticity learning rules are applied to the ESN to change the topology of the reservoir. Through the experimental analysis, we explore the relationship between the change of the topology structure of the reservoir and the network performance. The experimental results show that in terms of predictive performance, anti-Oja rules perform well on nonlinear autoregressive moving average (NARMA) system and Oja rule perform well on Mackey-Glass system.

Keywords: Echo state network · Reservoir · Topological structure · Synaptic plasticity · Comparative analysis

1 Introduction

ESN, a subclass of Recurrent Neural Networks (RNNs), operate on the reservoir computing framework and are utilized for time-series analysis within the realm of machine learning [1]. The core structure of ESN is a randomly generated and constant reservoir, which is the main difference between ESN and traditional RNN [2].

In ESN, alterations to the topology of the reservoir can significantly affect the performance of the network. Therefore, studying and designing reservoirs based on brain networks represents an essential research direction. Many researchers began to explore the design and optimization of the topology of the reservoir [3]. However, there still exists a notable disparity between the topology of the ESN reservoir and that of brain networks. Currently, the influence of the topological configuration of reservoir on ESN performance remains an open question in academic research.

Recently, the field of synaptic plasticity has garnered increasing attention in research endeavors. In biology, synaptic plasticity refers to the modifiable strength of synapses between neurons, constituting a form of neural plasticity [4]. It is crucial for the learning of brain and memory functions [5] and hold significant implications in artificial neural networks [6]. Based on the research in neuroscience, some researchers have introduced synaptic plasticity rules into ESN to enhance its predictive performance [7].

Among the synaptic plasticity learning rules, the most fundamental and noteworthy is the unsupervised learning rule proposed by Hebb in 1949, known as the Hebbian rule [8]. This rule postulates that the connection between neurons A and B is strengthened when both are simultaneously activated. However, this rule solely addresses synaptic potentiation, and its negative counterpart, the anti-Hebbian rule, solely diminishes synaptic strength [9]. Building upon the Hebbian rule, Oja proposed the Oja rule. Weight normalization is introduced in Oja rule to improve the stability of the model [10]. Like the Hebbian rule, the Oja rule focuses exclusively on synaptic potentiation. In contrast, the anti-Oja rule serves a balancing purpose, reducing connection strength to prevent excessive weight growth and thereby enhancing neural network stability [11]. The Bienenstock-Cooper-Munro (BCM) learning rule is another synaptic plasticity rule derived from the Hebbian rule. This rule states that the enhancement or reduction of synaptic weights depends on the firing frequency of the presynaptic neuron [12]. Typically, high-frequency firing leads to synaptic weight enhancement, while low-frequency firing may result in synaptic weight reduction. In the BCM rule, synaptic strength can both increase and decrease. In addition to the five rules mentioned earlier, there are also the Spike Timing-Dependent Plasticity learning rule [13], among others. Similarly, the degree to which synaptic plasticity regulations impact the efficacy of the reservoir within ESNs is not well understood and warrants further investigation.

The motivation behind this study is to delve into the effects of various rules on the performance of the reservoir within ESN. This paper reconstructs the topology of ESNs by applying, respectively, Hebbian, anti-Hebbian, Oja, anti-Oja, and BCM learning rules. As the topology of the reservoir evolves, the performance of ESN changes accordingly. We monitor the changes in reservoir parameters and network performance under various plasticity rules, including the sparsity, spectral radius, internal weights and prediction performance. Then, we explore the relationship between plasticity rules, reservoir topology, and network performance through comparative analysis. The experimental results indicate that the anti-Oja rule demonstrates superior predictive performance in the application of the NARMA system, whereas, in the predictive tasks of the Mackey-Glass system, the Oja rule performs better.

Here is the arrangement of other contents in this article. Section 2 is about some basic knowledge, including ESN and five synaptic plasticity learning rules. Section 3 is about the experimental protocol, including datasets and experimental settings. Section 4 is the experimental analysis. Section 5 is the summary of this article.

2 Preliminaries

2.1 Echo State Network

A canonical ESN is composed of three principal layers: the input, reservoir, and output layers [1], as depicted in Fig. 1. Assuming that the number of neurons in these three components are K, M, and L, in order, Fig. 1 shows the basic structure of the ESN.

If $u(t)$ indicates the input of ESN at time t, and $x(t)$ indicates the reservoir state, which is calculated by the following formula:

$$x(t) = f(W^{in}u(t) + W^r x(t-1)) \tag{1}$$

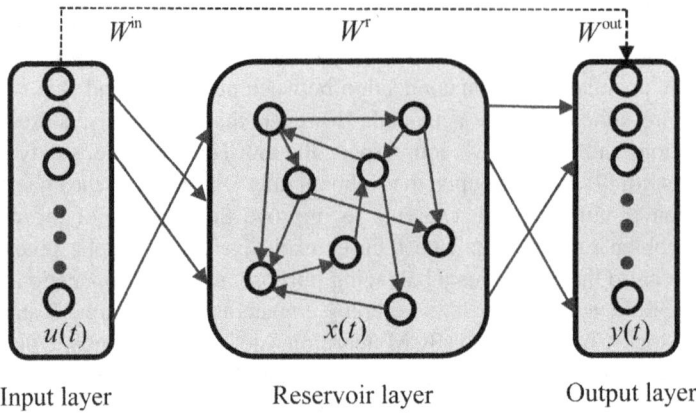

Fig. 1. Traditional ESN model.

where f indicates the activation function and it usually is the tanh function; The input weights are denoted by matrix $W^{in} \in R^{M \times K}$, while the internal reservoir connections are represented by matrix $W^r \in R^{M \times M}$, both of which are initialized randomly within the interval [−1, 1]. During the training phase of network, matrices W^{in} and W^r are kept constant. The spectral radius of the reservoir is the absolute value of the maximum eigenvalue of the internal connection weight matrix within the reservoir. The spectral radius of reservoir, which is the absolute value of maximum eigenvalue of the internal connection weight matrix, must be less than 1 to maintain the echo state property [14].

Subsequently, the output weight matrix $W^{out} \in R^{L \times M}$ is calculated using the least squares method, which is given by the formula:

$$W^{out} = (X^T X)^{-1} XY \qquad (2)$$

where X is the matrix formed by the input weights and the reservoir states, Y is the matrix of target output. The network output at time t is represented as $y(t)$ and it is calculated using the following formula:

$$y(t) = W^{out}[u(t); x(t)] \qquad (3)$$

In ESN, the selection and setting of network parameters play a crucial role in determining the network performance. Below is the introduction to the impacts of the main parameters in ESN. The sparsity of the reservoir is defined as the ratio of connected neurons to the total neuron population within the reservoir. Generally, a lower sparsity will lead to a simpler network structure. On the contrary, a higher sparsity may lead to an overfitting problem. The input scaling factor can adjust the intensity of the input signal and adjust it to a range suitable for processing by the neuron activation function.

2.2 Synaptic Plasticity Learning Rules

In our research, we examine the effects of five distinct synaptic plasticity rules on ESN performance: Hebbian, anti-Hebbian, Oja, anti-Oja, and BCM rules. First, the Hebbian rule can be represented by the following formula:

$$W_{ji}^r(t+1) = W_{ji}^r(t) + \eta x_i(t)x_j(t) \tag{4}$$

where t is the time step; i is the presynaptic neuron and j is the postsynaptic neuron; W_{ji} is the synaptic strength between the presynaptic neuron and the postsynaptic neuron; η is learning rate; $x(t)$ is the neuronal responses. This rule leads to unrestricted growth of synaptic connections.

The following formula is used to calculate the anti-Hebbian rule:

$$W_{ji}^r(t+1) = W_{ji}^r(t) - \eta x_i(t)x_j(t) \tag{5}$$

We can see that the difference between these two rules is the positive or negative weight update.

The following formula is used to calculate the Oja rule:

$$W_{ji}^r(t+1) = W_{ji}^r(t) + \eta x_j(t)\left[x_i(t) - x_j(t)W_{ji}^r(t)\right] \tag{6}$$

We can see that the Oja rule is about synaptic potentiation.
Similarly, the anti-Oja rule can be represented as follows:

$$W_{ji}^r(t+1) = W_{ji}^r(t) - \eta x_j(t)\left[x_i(t) - x_j(t)W_{ji}^r(t)\right] \tag{7}$$

The BCM rule is a learning mechanism that simulates synaptic plasticity in the brain, enabling selective responses to specific stimuli by adjusting the synaptic weights of neurons. The following equation is used to calculate the BCM rule:

$$W_{ji}^r(t+1) = W_{ji}^r(t) + \eta\left[x_j(t)(x_j(t) - \theta_{LTP})x_i(t) - \varepsilon W_{ji}^r(t)\right] \tag{8}$$

where θ_{LTP} represents the sliding threshold, which means it's not fixed, but dynamically adjusted based on the history of neural activity. ε is a constant and it is set to one here. The following equation is used to calculate θ_{LTP}:

$$\theta_{LTP} = \frac{\sum_{t'=t-h}^{t} x_j^2(t')e^{(h)}}{\sum_{t'=t-h}^{t} e^{(h)}} \tag{9}$$

where h indicates the time interval. This "sliding threshold" mechanism enables neurons to adapt to different input patterns and adjust their synaptic weights on different time scales.

3 Experimental Scheme

3.1 Datasets

The empirical evaluation is conducted using the NARMA system of order ten and the Mackey-Glass system [15], both of which serve as standard benchmarks for assessing ESN capabilities.

The NARMA dataset is a dataset generated by a nonlinear autoregressive moving average model. If the output of the system at time t is denoted by $m(t)$, while the input at time t is represented by $s(t)$. The formula of the tenth-order NARMA dataset can be given as follows:

$$m(t+1) = 0.3m(t) + 0.05m(t)\sum_{i=0}^{9} m(t-i) + 1.5s(t-9)s(t) + 0.1 \quad (10)$$

where the initial value of $m(t)$ is set to 0; and $s(t)$ is drawn from a uniform distribution from 0 to 0.5. In this dataset, we generated a total of 5200 samples. The first 2800 data are used for training, and the last 2400 are used for testing. Among them, we use the first 1000 samples as wash samples.

The Mackey-Glass system is governed by a differential equation with time delay, showcasing periodic and chaotic behaviors, making it a standard for time series forecasting challenges. The following expression is the Mackey-Glass system:

$$\frac{dx(t)}{dt} = \frac{ax(t-\tau)}{1+x^c(t-\tau)} - bx(t) \quad (11)$$

where $a = 0.2$, $b = 0.1$, $c = 10$ and $\tau = 17$. In this dataset, we generated a total of 4900 samples with zero initial state. The first 2700 data are used for training, and then 2200 are used for testing. Again, we use the first 1000 samples as wash samples.

3.2 Evaluation

To assess the forecasting accuracy of the network, we employ the Normalized Root Mean Square Error (NRMSE), detailed by the subsequent formula:

$$NRMSE = \sqrt{\frac{\sum_{t=1}^{T}(\hat{y}(t) - y^{true}(t))^2}{T\sigma^2}} \quad (12)$$

where T is the length of the time series; $\hat{y}(t)$ indicates the desired output of the network; $y^{true}(t)$ is the network output; and σ^2 represents the variance of the desired output.

3.3 Experimental Settings

To explore the correlation between the five synaptic plasticity algorithms and the topology of the reservoir layer along with network performance, we have individually applied these rules to an identical reservoir setup. In the relevant reservoir parameters, we output the internal connection weight, spectral radius and sparsity. Under the same rule, we randomly selected five different weights, while under different rules, we selected weights at the same position. In addition, when we select the values, since the internal connection weights of the reservoir range from -1 to 1, the five weights we select need to contain positive, negative and zero values at the same time. We have investigated the effects of various algorithms on the predictive accuracy of the network across two different systems. The prediction results of the network are evaluated by calculating the average and standard deviation of 30 NRMSE outcomes. Table 1 presents the common parameter values involved.

Table 1. Parameter setting for ESN-based models.

Parameters	Values
Reservoir size	50
Learning rate η	0.0001
Spectral radius	0.86
Sparsity	0.05

4 Experimental Results

4.1 NARMA System

The reservoir enters a state of complete connectivity, that is, a sparsity of 1, following the initial iteration.

Figure 2 sequentially illustrates the variations of five weight values under Hebbian rule, anti-Hebbian rule, BCM rule, Oja rule, and anti-Oja rule. As observed from Fig. 2, the initial weight values are within the range of $[-1,1]$, and after iterations, the weight values remain within this range. Under Hebbian and Oja rules, the internal connectivity weights within the reservoir steadily increase. Conversely, under anti-Hebbian and anti-Oja rules, the weights continuously decrease. The impact of BCM rule on weights is consistent with its underlying principle, leading to both increases and decreases in weights. Under the BCM rule, the weights both increase and decrease. The reason for this change is closely related to the principle of rules.

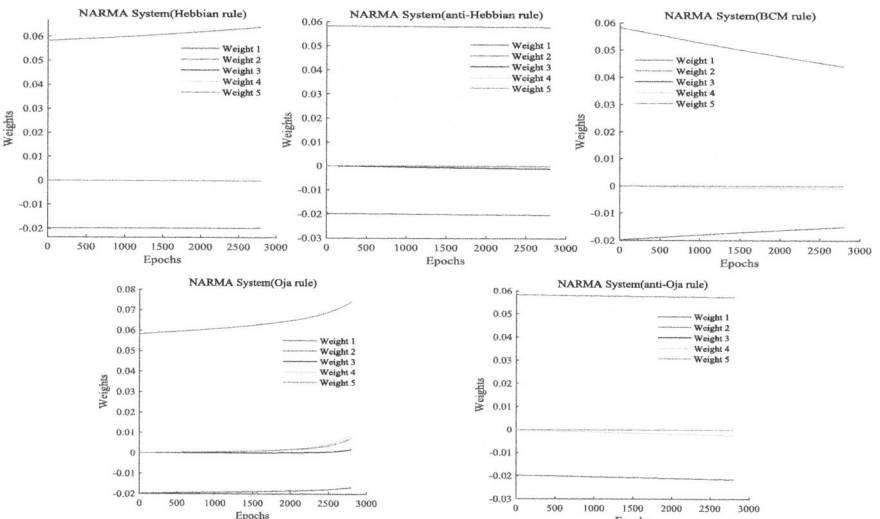

Fig. 2. The change in internal connection weights under the NARMA system.

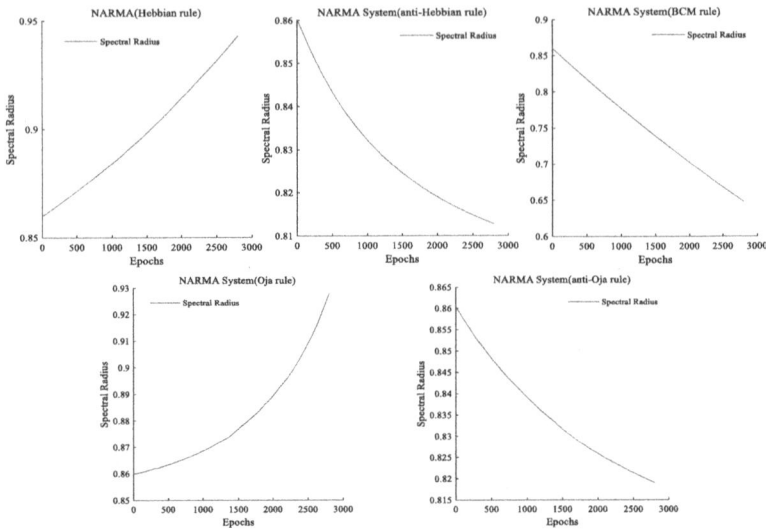

Fig. 3. The change in spectral radius under the NARMA system.

Figure 3 sequentially represents the variations in spectral radius under Hebbian, anti-Hebbian, BCM, Oja, and anti-Oja rules. As depicted in Fig. 3, the spectral radius exhibits an ascending trend under Hebbian and Oja rules. In contrast, under anti-Hebbian, anti-Oja rules and BCM rule, the spectral radius decreases. In general, the increase of spectral radius may lead to the instability of the network. Combined with the characteristics of the rules themselves, we can see that Hebbian rules are unstable. According to our results, the network stability is relatively the worst under the Hebbian rule. Under the action of the five rules, the sparsity of the reservoir is 1, and the weight and spectral radius change differently due to the characteristics of the rules themselves.

Table 1 displays the average and standard deviation of errors for the application of each of the five algorithms within the NARMA system. In the table, we can observe that the Hebbian rule and BCM rule cannot enhance the predictive performance of the network. Among them, the anti-Oja rule demonstrates a relatively better effect compared to the other rules. The reason may be that in terms of weight changes, the anti-Oja rule exhibits greater stability. Additionally, the BCM rule does not outperform the anti-Oja rule in prediction, possibly because the BCM rule is more suitable for tasks sensitive to frequency changes.

Therefore, in the NARMA dataset, we will use the anti-Oja rule first (Table 2).

Table 2. The error results under five rules.

Synaptic plasticity learning rules	Training NRMSE mean	Testing NRMSE mean	Testing NRMSE std
Hebbian rule	0.7044	0.6582	0.0039
anti-Hebbian rule	0.7040	0.6557	0.0026
Oja rule	0.7017	0.6533	0.0037
anti-Oja rule	**0.6997**	**0.6525**	**0.0012**
BCM	0.7087	0.6590	0.0034
ESN	0.7025	0.6567	0.0045

4.2 Mackey-Glass System

Analogous to observations in the NARMA system, the sparsity converges to 1 after the first iteration.

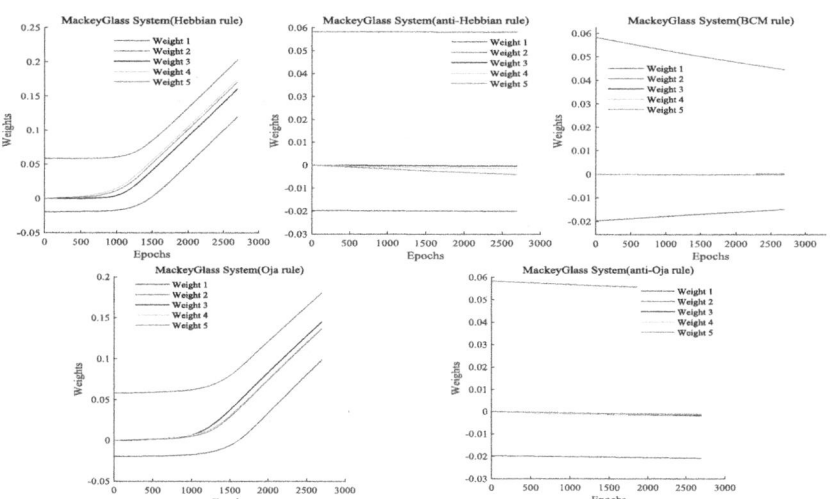

Fig. 4. The change in internal connection weights under the Mackey-Glass system.

In the Mackey-Glass system, Fig. 4 portrays the internal weight adjustments within the reservoir. Analyzing this dataset leads to a similar conclusion: under the BCM rule, the weights exhibit both increasing and decreasing behaviors. Under the Hebbian and Oja rules, the weights increase, whereas under the anti-Hebbian and anti-Oja rules, the weights decrease. Across different datasets, the influence of a rule on the weights is solely determined by the inherent characteristics of the rule itself.

Figure 5 demonstrates the alterations in the spectral radius. In this figure, the spectral radius under Hebbian rule and Oja rule is in an increasing state, and the spectral radius

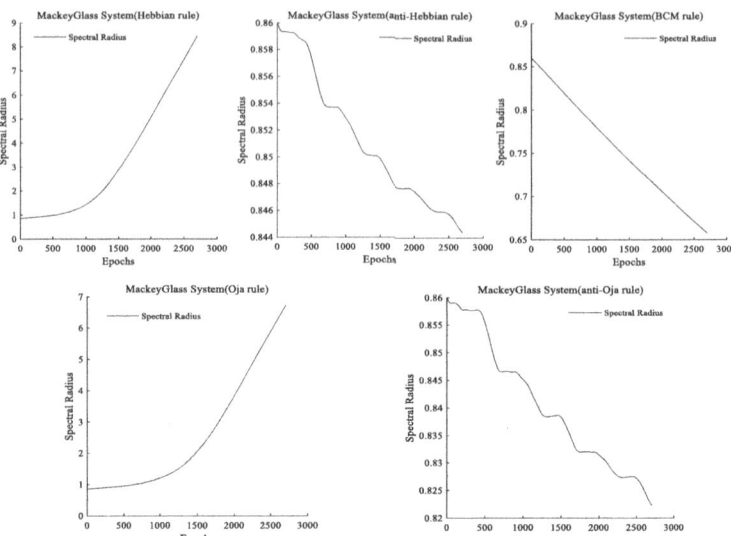

Fig. 5. The change in spectral radius under the Mackey-Glass system.

under Hebbian rule increases by more than 1. On the contrary, the spectral radius showed a decreasing trend under the other three rules. In this system, the spectral radius under Hebbian rule exceeds the original range, which is probably caused by the instability of Hebbian rule itself.

The variation of parameters in Mackey-Glass system is different from that in NARMA system. The same is the overall trend of increase and decrease, the difference is the specific change situation.

Table 3 compiles the outcomes derived from the implementation of diverse rules as well as the pristine ESN, encompassing the mean and standard deviation of errors. The data indicate that the integration of these rules has enhanced the forecasting capabilities of the network to different extents. Among the five rules, ESNs under Oja rule show better prediction performance. There is evidence for this phenomenon. Weight normalization is added to Oja rule to increase its stability, and this rule is more applicable than BCM rule in practical application.

Therefore, when considering the aforementioned parameter analysis, the Oja rule emerges as the superior choice for prediction performance within the Mackey-Glass system.

Table 3. The error results under five rules.

Synaptic plasticity learning rules	Training NRMSE mean	Testing NRMSE mean	Testing NRMSE std
Hebbian rule	3.3796	3.2100	0.5939
anti-Hebbian rule	3.4062	3.0964	0.2056
Oja rule	3.4108	**3.0086**	**0.0506**
anti-Oja rule	3.3819	3.0372	0.0514
BCM	3.3696	3.1433	0.0729
ESN	3.5348	3.3397	1.1242

5 Conclusion

This paper investigates the impact of five synaptic plasticity rules on the topological structure of the ESN reservoir. We applied these five rules to ESNs, altering the topology of reservoir, and subsequently studied the relationship between these changes and network performance. The experimental results show that in both systems, these five rules affect the topology of the reservoir according to their own characteristics. Through the analysis of parameters and error results, the anti-Oja rule outperforms the other four synaptic plasticity learning rules, demonstrating better performance in NARMA system. Oja rule shows better performance in Mackey-Glass system.

Acknowledgments. This work was supported by the Natural Science Foundation of China under Grant 62073153.

Disclosure of interest. The authors have no competing interests to declare that are relevant to the content of this article.

References

1. Xianshuang, Y., Yao, W., Di, M., Shengxian, C., Qingchuan, M.: Fractional-integer-order echo state network for time series prediction. Appl. Soft Comput. **153**, 111289 (2024)
2. Cesar, H.V., Marley, M.B.R.V., Karla, F.: Echo state networks: novel reservoir selection and hyperparameter optimization model for time series forecasting. Neurocomputing **545**, 126317 (2023)
3. Yu, X., Qi, Z., Adam, S.: Automatic topology optimization of echo state network based on particle swarm optimization. Eng. Appl. Artif. Intell. **117**, 105574 (2023)
4. Chen, H., Xie, L., Wang, Y., Zhang, H.: Postsynaptic potential energy as determinant of synaptic plasticity. Front. Comput. Neurosci. **16** (2022)
5. Citri, A., Malenka, R.C.: Synaptic plasticity: multiple forms, functions, and mechanisms. Neuropsychopharmacology **33**, 18–41 (2008)
6. Mozzachiodi, R., Byrne, J.H.: More than synaptic plasticity: Role of nonsynaptic plasticity in learning and memory. Trends Neurosci. **33**, 17–26 (2010)

7. Wang, X., Jin, Y., Du, W., Wang, J.: Evolving dual-threshold bienenstock-cooper-munro learning rules in echo state networks. IEEE Transa. Cybern. **52**, 11254–11266 (2022)
8. Jaeger, H.: Adaptive nonlinear system identification with echo state networks. In: Advances in Neural Information Processing Systems. pp. 281–287. Cambridge, MA, USA: MIT Press (1996)
9. Carlson, A.: Anti-hebbian learning in a non-linear neural network. Biol. Cybern. **64**, 171–176 (1990)
10. Oja, E.: A simplified neuron model as a principal component analyzer. J. Math. Biol. **15**, 267–273 (1982)
11. Babinec, Š, Pospíchal, J.: Improving the prediction accuracy of echo state neural networks by anti-Oja's learning. Lecture Notes Comput. Sci. **4668**, 19–28 (2007)
12. Bienenstock, E.L., Cooper, L.N., Munro, P.W.: Theory for the development of neuron selectivity: orientation specificity and binocular interaction in visual cortex. J. Neurosci. **2**, 32–48 (1982)
13. Caporale, N., Dan, Y.: Spike timing-dependent plasticity: a Hebbian learning rule. Annual Rev. Neurosci. **31**, 25–46 (2008)
14. Ozturk, M.C., Xu, D., Principe, J.C.: Analysis and design of echo state networks. Neural Comput. **19**, 111–138 (2007)
15. Jaeger, H.: The 'echo state' approach to analysing and training recurrent neural networks-with an erratum note. GMD-148 German Nat. Res. Center Inf. Technol., Bonn, Germany, Rep (2001)

LEHR: LLM-Driven Evolutionary Hybrid Rewards for Multi-agent Reinforcement Learning

Yuan Wei[✉], Xiaohan Shan, and Jianmin Li

Qiyuan Lab, Beijing, China
{weiyuan,shanxiaohan,lijianmin}@qiyuanlab.com

Abstract. Reward design is pivotal in reinforcement learning (RL), particularly in environments with sparse rewards, where the quality of the reward function has a direct impact on the algorithm's performance. This challenge becomes even more pronounced in multi-agent reinforcement learning (MARL), where balancing individual agent rewards with global rewards demands careful consideration in reward function design. To address these complexities, we propose LEHR: LLM-Driven Evolutionary Hybrid Rewards, a framework leveraging the reasoning and generation capabilities of large language models (LLMs) to develop multi-layered reward functions tailored for MARL scenarios. LLMs generate hybrid rewards that integrate global, local, and adaptive components, providing precise feedback to individual agents while promoting effective team coordination. The incorporation of a Selector module enables the LLM to generate new solutions based on selected reward functions, emulating evolutionary algorithms through selection, crossover, and mutation processes. By limiting LLMs to the reward generation phase and excluding them from the RL training process, we effectively reduce inference costs while ensuring robust reward design. Experiments in the Multi-Agent Particle Environment (MPE) reveal significant improvements in agent adaptability and collaboration, demonstrating the effectiveness of LEHR in optimizing MARL.

Keywords: Multi-Agent Reinforcement Learning · Hybrid Rewards · LLM · Evolutionary Algorithms

1 Introduction

The design of reward functions plays a pivotal role in reinforcement learning (RL), as it shapes agent behavior and drives strategic decision-making. Effective rewards provide meaningful feedback, guiding agents toward desired outcomes throughout the learning process. However, in multi-agent reinforcement learning (MARL), reward design is more complex than in single-agent settings. Relying solely on global rewards fails to reflect the impact of individual agent actions, while focusing only on selfish rewards overlooks their influence on the overall system. Balancing individual incentives with global objectives is essential but challenging.

Designing reward functions for RL traditionally relies on human intuition and trial-and-error experimentation, resulting in high training costs and inefficiencies. Despite these efforts, environments with sparse rewards often pose significant challenges, making it difficult to achieve optimal outcomes. Recent advancements in LLMs have demonstrated their potential to comprehend complex scenarios and execute logical reasoning, offering significant benefits in streamlining reward design. However, current research predominantly focuses on single-agent RL, lacking solutions that address the intricate dynamics of multi-agent systems, such as coordination, credit assignment, and conflicting goals. Additionally, many existing approaches embed LLMs directly into RL training and inference, leading to significant resource consumption and slower inference speeds.

To address these challenges, we propose a MARL framework for reward design driven by LLMs. The framework integrates internal layered guidance and external evolutionary iteration to enhance the effectiveness and adaptability of reward functions. Internally, LLMs generate multi-level rewards, including global, local, and hybrid rewards, guiding agents to align individual actions with overall system objectives. Externally, an evolutionary process optimizes reward functions through iterative selection, crossover, and mutation, ensuring the design evolves dynamically based on agent performance and feedback from the environment. In this context, LLMs act as algorithmic generators that facilitate the crossover and mutation processes, enabling the creation of superior reward functions. By leveraging the capabilities of LLMs, the framework can explore a diverse reward space, ultimately leading to more effective and adaptive reward designs.

Our main contributions are as follows: We propose a novel LLM-driven framework for multi-agent reward design, utilizing the reasoning capabilities of LLMs to generate multi-level rewards, including global, local, and hybrid rewards, ensuring a balance between individual agent objectives and overall system goals. We introduce evolutionary optimization, where reward functions are iteratively refined based on agent performance and feedback from the environment, allowing for adaptation to complex multi-agent scenarios. Additionally, we reduce inference costs by employing LLMs solely during the reward generation stage, avoiding integration into RL training and inference processes. Finally, we validate our framework through experiments in the Multi-Agent Particle Environment (MPE), comparing its performance with existing multi-agent RL algorithms to demonstrate its effectiveness in enhancing coordination and overall performance.

The remainder of the paper is organized as follows: Sect. 2 reviews the related work in the field, highlighting key advancements and existing challenges. Section 3 presents the methodology, detailing the framework design and algorithmic workflow. Section 4 describes the experimental setup and evaluation metrics, followed by a discussion of the experimental results. Finally, Sect. 5 concludes the paper, offering insights and potential directions for future research.

2 Related Work

Designing reward functions is a critical yet challenging aspect of RL. Effective rewards guide agents toward optimal behaviors, but sparse rewards and delayed feedback often hinder learning efficiency. The groundwork for reward shaping was laid by Ng et al. [18], who introduced potential-based shaping rewards. Their framework ensures that adding shaping rewards does not alter the optimal policy, providing a theoretical foundation for using auxiliary rewards. However, their approach struggles in highly sparse environments, where feedback is insufficient to guide agent behavior effectively. Building on this, Wiewiora et al. [23] extended the theory to Q-learning frameworks, enabling more robust reward shaping that accelerates learning. Harutyunyan et al. [7] further explored potential-based shaping in partially observable environments, demonstrating that shaping rewards can boost learning efficiency by providing intermediate feedback even when state information is incomplete. Recent studies have focused on making reward shaping more adaptive and responsive to complex environments. Ibarz et al. [9] introduced a method for learning from human preferences, combining demonstrations and feedback to improve agent learning in challenging tasks. Devidze et al. [4] proposed Exploration-Guided Reward Shaping (ExploRS), which combines intrinsic rewards with exploration bonuses to improve performance in environments with sparse rewards. Another recent contribution by Ma et al. [15] introduces a Self-Adaptive Success Rate (SASR)-based reward shaping mechanism. This approach uses evolving Beta distributions to model state success rates, allowing rewards to adapt dynamically over time. Designing rewards in MARL presents significant challenges, including credit assignment, ensuring coordination among agents, and dealing with non-stationarity in learning environments. Lowe et al. [13] addressed these issues through MADDPG, introducing centralized critics for stable training with decentralized execution. Rashid et al. [19] advanced this work with QMIX, which decomposes global Q-values into individual components, ensuring that agents align with collective goals while maintaining efficiency. Recently, some works have focused on how to effectively combine individual and global rewards for agents, such as Matsunami et al. [17], who integrated individual behavior rewards with penalties reflecting negative contributions to social welfare, leading to increased social utility among agents and improved cooperative policy formation. Wang et al. [21] employed the Individual Reward Assisted Team Policy Learning (IRAT) method, which combines individual and team policies with discrepancy constraints to enhance cooperation in MARL scenarios.

With the powerful understanding and reasoning capabilities of large models, an increasing number of researchers are leveraging these models for RL reward function design. A significant portion of this work has focused on single-agent reinforcement learning algorithms [2, 10, 16, 22, 24].

However, in the realm of MARL, the application of LLMs has primarily been directed towards facilitating cooperation among agents [8, 11, 12, 20].

3 Method

In this section, we introduce the LEHR (LLM-Driven Evolutionary Hybrid Rewards) framework (Fig. 1). LEHR integrates large language models with evolutionary algorithm principles.

Evolutionary algorithms are optimization techniques inspired by the process of natural selection. They typically involve three key operations: selection, crossover, and mutation. In the selection phase, the best-performing candidates from the current population are chosen based on predefined fitness criteria. Crossover then combines elements from selected candidates to produce new offspring. Mutation introduces random changes to these offspring to explore new areas of the solution space, enhancing diversity and preventing premature convergence.

In our framework, the large language model (LLM) takes on the role of evolutionary operators such as crossover and mutation. Specifically, the LLM generates new hybrid reward functions by combining components of high-performing reward functions (crossover) and introducing novel variations or adjustments to their structure (mutation). These operations are guided by the feedback provided by the Selector module, which evaluates the performance of reward functions by training in the multi-agent reinforcement learning environment.

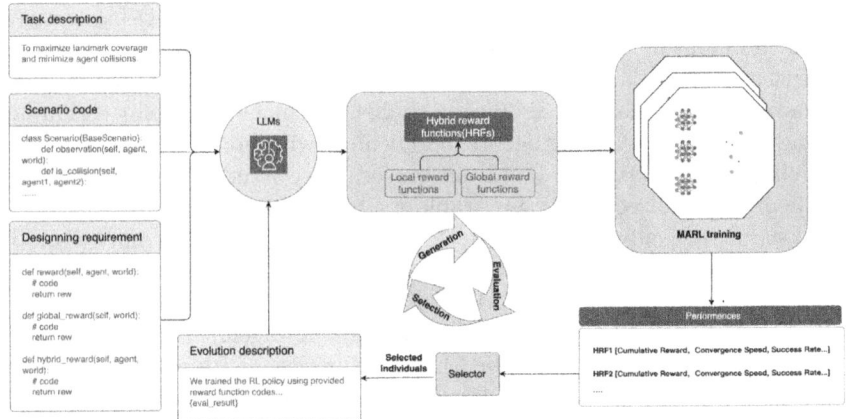

Fig. 1. The LEHR framework follows an iterative process where hybrid reward functions are generated by large language models based on task descriptions, environment code, and evolution descriptions. In each iteration, a selector module evaluates the performance of these HRFs, allowing for the selection of the most effective rewards for MARL training, ultimately enhancing agent adaptability and cooperation.

3.1 Reward Function Generation

The overall architecture of LEHR consists of several key components, as illustrated in Fig. 1. The process begins with inputting task descriptions and environment code into the LLMs. This input guides the generation of appropriate reward functions tailored to specific tasks. The LLMs produce hybrid reward functions, which are composite functions that integrate both individual and global rewards. The specific method for this combination is also determined by the large model, allowing for adaptive adjustments of weights and structures tailored to the specific scenario at hand.

The local reward function provides feedback based on the behavior of individual agents, while the global reward function evaluates the overall performance of the team. We can express the local reward function as $R_{local}(o_i, S, a_i)$, and the global reward function as $R_{global}(S, a)$, where S represents the current state, o_i denotes the observation of agent i, a_i is the action taken by agent i, and $a = \{a_1, \ldots, a_n\}$ signifies the set of actions performed by all agents. To maintain clarity, we have omitted the time step t from our notation when it does not introduce any ambiguity.

During each iteration, parallel inference is performed using the large model to produce multiple reward functions. These functions are then employed for training in the MARL environment. Importantly, the inference from the large model does not directly participate in the RL training process, which reduces inference costs and allows for a focused generation of reward functions.

3.2 MARL Training

In the MARL training phase, the framework utilizes the various reward functions generated by the LLMs in parallel. Each set of agents is trained with its corresponding hybrid reward function, serving as an evaluator for the effectiveness of the generated rewards. To enhance the efficiency of this evaluation process, we reduce the number of training iterations, allowing us to obtain a preliminary assessment of each reward function's performance. This stage-wise evaluation provides valuable feedback, which serves as input for the Selector module.

The LEHR framework is designed to be flexible, accommodating different MARL algorithms that can be selected based on the specific task or environment. This adaptability ensures that the framework can effectively integrate with various approaches while still focusing on optimizing the reward design process.

3.3 Selector Module

Our framework includes a Selector module that simulates the selection process found in evolutionary algorithms. This module evaluates the performance of different reward functions and selects the most effective ones, providing feedback to the LLMs to guide the generation of subsequent reward functions.

In the Selector module of LEHR, the selection criteria for hybrid reward functions can be customized to suit specific needs and objectives. One straightforward approach is to select the top k performing reward functions based on predefined metrics such as cumulative reward, convergence speed, or success rate.

This method ensures that the most effective rewards are continuously utilized for training.

To enhance exploration during the selection process, randomness can be introduced. For example, the performance of the selected reward functions can be quantified by the following formula: $P_{selected} = \frac{R_i}{\sum_{j=1}^{n} R_j}$, where R_i represents the performance metric of the *i-th* reward function, n is the number of candidate reward functions, and $P_{selected}$ indicates the probability of selecting that reward function. By incorporating a stochastic element, the framework can diversify the selection of reward functions, allowing for a broader exploration of the reward space and potentially uncovering more effective solutions. This approach is particularly beneficial in dynamic environments where agent behavior may vary significantly over time.

Additionally, other selection strategies from evolutionary algorithms can be referenced. For instance, techniques such as tournament selection and rank-based selection can be adapted to choose reward functions based on their relative performance, which fosters a competitive environment among the reward candidates. These strategies have been discussed in various studies, including those by Goldberg [6] on genetic algorithms, Deb [3] on multi-objective optimization, and Bck et al. [1] regarding function optimization.

By allowing for customizable selection rules, the LEHR framework can effectively adapt to different scenarios, improving the overall performance and collaboration of agents in MARL tasks.

3.4 Feedback Loop

The entire process is iterated through a feedback loop, which utilizes the performance feedback provided by the selector to refine the design of the reward functions. This mechanism simulates the selection, crossover, and mutation processes inherent to evolutionary algorithms, enabling the LLMs to generate new reward functions in each iteration. The newly generated reward functions are based on an analysis of historical performance and feedback from the selector, aiming to enhance overall cooperation through more precise evaluations of agent behaviors.

By integrating these steps, the LEHR framework effectively combines the generative capabilities of LLMs with simulated evolutionary algorithm principles, offering an innovative solution to the challenges of reward design in multi-agent environments.

4 Experiments

To validate the effectiveness of our framework, we conducted experiments in various environments.

4.1 Experiment Setup

Environments. We select the Cooperative Navigation task from the Multi-Agent Particle Environment (MPE) for validation. In this task, N agents must interact with M landmarks within the environment. A threshold is established to determine when an agent successfully detects its designated landmark. Specifically, if an agent is within this threshold distance from its landmark, it is considered to have successfully monitored that landmark. The primary objective of this task is to maximize the number of landmark observations made by all agents within a fixed time frame.

It is important to note that each landmark can only yield rewards once at any given moment. This means that if multiple agents converge on the same landmark, only one observation will be counted, rendering any redundant monitoring ineffective. Therefore, agents must strategize to locate as many distinct landmarks as possible within the allotted time.

Given the experimental setup, the rewards are spars, agents receive feedback only when they successfully observe a landmark or when collisions occur. This underscores the necessity of effective reward design, as the limited feedback encourages agents to optimize their exploration strategies to efficiently detect and monitor the landmarks.

Baselines. In this study, we used the algorithms MAPPO [25], COMA [5] and MAA2C as a foundation for our experiments. To comprehensively evaluate our LEHR framework, we designed multiple comparative experiments. To validate the effectiveness of our LEHR framework, we compared it with the original environment-provided reward function as the primary baseline. For ablation studies, we performed two sets of experiments. First, we compared the performance between local-only rewards and global-only rewards against our hybrid reward design, demonstrating the advantages of combining both reward types. Second, we implemented a non-evolutionary version of our framework (termed LHR) that generates fixed reward functions without evolution, to validate the benefits of our evolutionary mechanism.

LLM. In our experiments, we deployed the Starchat2-15B [14] model on a local server equipped with four NVIDIA A100 GPUs, accessing it through an API interface. We selected Starchat2 because of its strong performance as a helpful coding assistant among open source language models. The small scale of the model ensures fast inference speed and small deployment memory. In order to enhance speed, we integrated the VLLM acceleration framework, which significantly reduced the time required to generate each reward function. More experimental parameters are shown in Table 1.

Table 1. Experimental Configuration

Environment Parameters	
Number of agents (N)	3
Number of landmarks (M)	3
Landmark coverage radius	0.3
Collision radius	0.1
LLM Configuration	
Temperature	0.9
Top-p	1.0
Max tokens	512
Training Settings	
Training steps	50k
Evaluation episodes	100
Generated rewards per round	5
Evolution rounds	5
Population size	2

4.2 Experimental Results

In analyzing the results of our experiments, we compared the performance of the LEHR framework against the baseline configurations within the original environment, as is shown in Fig. 2. After training with three multi-agent algorithms—MAPPO, COMA, and MAA2C—our LEHR framework demonstrated a significant advantage in terms of coverage rate metrics.

For the MAPPO algorithm, the coverage rate achieved with the LEHR framework was 0.70, a notable improvement of **11%** over the baseline illustrates the effectiveness of our enhanced reward functions in encouraging agents to explore more landmarks. When examining the collision rates, MAPPO recorded the no collision rate of the LEHR framework is 0.9928, which is slightly higher than the 0.9887 achieved with baseline. This indicates that while the LEHR framework improved agents' abilities to maximize landmark observations, it did not compromise their navigation strategies, allowing them to avoid collisions effectively.

Similarly, in COMA and MAA2C algorithms, the LEHR also achieved better result than the baseline, which illustrate that the LEHR framework consistently enhances agents performance across different algorithms.

In order to further analyze the performance of our LEHR framework, we choose the MAPPO algorithm as an example and evaluate the model saved during the LEHR and baseline training process. In Fig. 3, LEHR quickly adapts the policy to improve the coverage rate over 60%, and curve of our LEHR framework consistently surpass the baseline. These clearly illustrates that the LEHR is more efficient and effective than the baseline in this multi-agent reinforcement learning task.

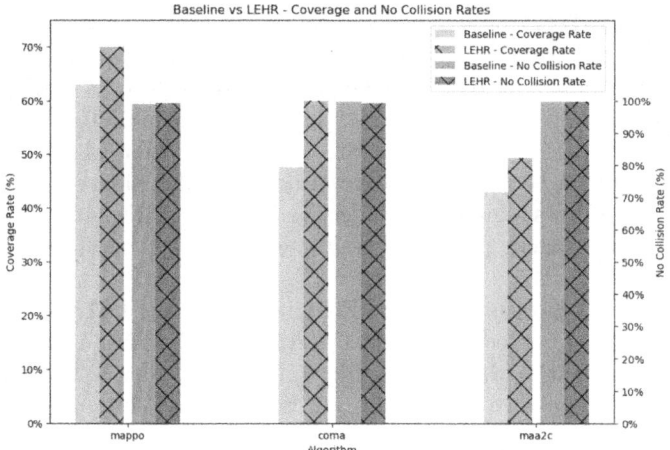

Fig. 2. LEHR vs baseline

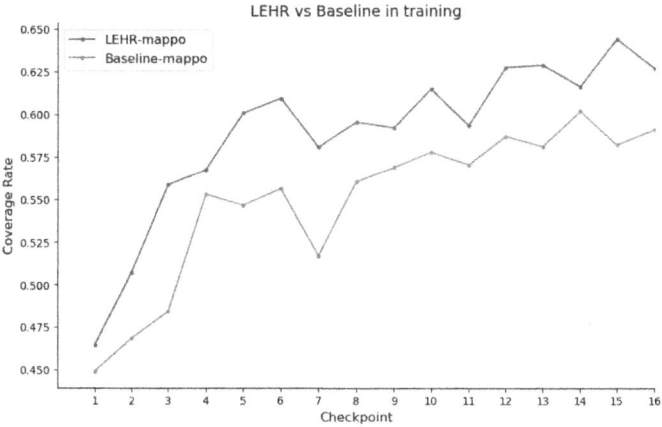

Fig. 3. LEHR vs baseline in training

4.3 Ablation Experiments

In the ablation study, we systematically compare our LEHR framework with alternative reward configurations to assess its effectiveness in enhancing multi-agent collaboration. Specifically, we evaluate LEHR against the experiment results use only local/global rewards within the original environment to understand the effectiveness of LEHR's hybrid reward design in enhancing agent performance, as is shown in Fig. 4. Additionally, we compare LEHR to a non-evolutionary version of the framework, termed LHR, with the purpose of independently analyzing the effectiveness of the evolutionary component in the framework, as is shown in Fig. 5.

LEHR vs Local/Global Reward. In the MAPPO algorithm, the LEHR framework demonstrates significant advantages over configurations using only local or only global rewards. Specifically, LEHR achieves a **45.8%** improvement in coverage rate compared to only local rewards and a **6.1%** increase over only global rewards. Additionally, LEHR maintains a no collision rate that is **0.5%** higher compared to only local rewards and **0.2%** higher than only global rewards. These results highlight the effectiveness of LEHR's hybrid reward structure in enhancing agent performance in terms of both coverage and collision avoidance.

In the COMA algorithm, the LEHR framework shows a **47.6%** improvement in coverage rate compared to only local rewards and a **25%** increase over only global rewards. For the MAA2C algorithm, LEHR achieves a **20.3%** increase in coverage rate compared to only local rewards and a **2.8%** improvement over only global rewards. The no collision rate with LEHR is similar with only local/global rewards in both algorithms.

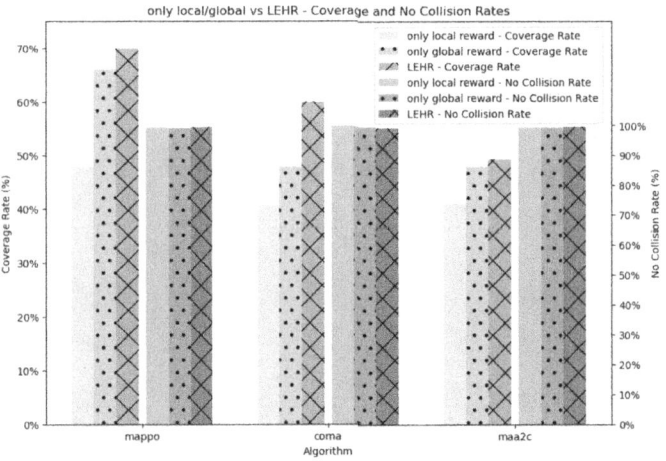

Fig. 4. LEHR vs local/global reward

LEHR vs LHR. For the MAPPO algorithm, LEHR achieves a coverage rate of 0.70, which is a substantial improvement over LHR's coverage rate of 0.6367, representing a **10.3%** increase. Additionally, LEHR maintains a no collision rate of 0.9928, slightly higher than LHR's rate of 0.9887. In the COMA algorithm, LEHR also outperforms LHR, achieving a coverage rate of 0.60 compared to LHR's 0.5433, which is a **10.4%** improvement. The no collision rate for LEHR is 0.9947, surpassing LHR's 0.9931. For the MAA2C algorithm, LEHR shows a coverage rate of 0.4933, significantly higher than LHR's 0.42, resulting in a **17.4%** improvement. The no collision rate for LEHR is 0.996, again outperforming LHR's rate of 0.9959.

Overall, the results indicate that both of the hybrid reward structure and the evolutionary component in the LEHR framework leads to improved coverage rates and better no collision performance across all three algorithms, highlighting the benefits of our LEHR in multi-agent reinforcement learning scenarios.

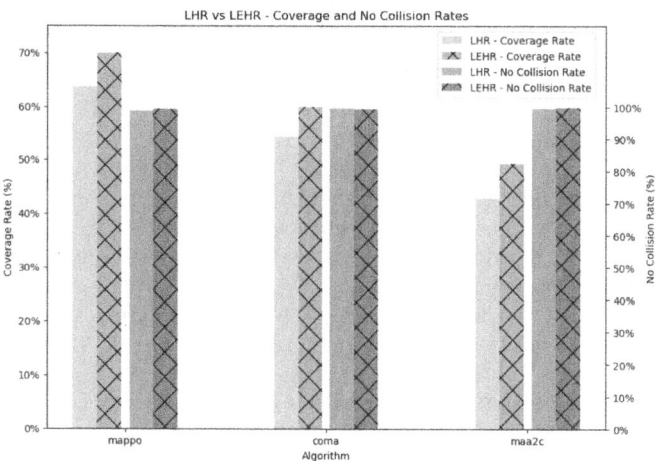

Fig. 5. LEHR vs LHR

5 Conclusion and Future Work

In this paper, we proposed the LEHR (LLM-Driven Evolutionary Hybrid Rewards) framework to address the challenges of reward design in multi-agent reinforcement learning. Our approach leverages the reasoning and generation capabilities of large language models to create multi-layered reward functions that enhance agent adaptability and cooperation. By integrating a Selector module, we mimicked principles from evolutionary algorithms to optimize the generated rewards, ensuring that our framework is flexible and effective in various scenarios.

Through rigorous experimentation in the Multi-Agent Particle Environment, we demonstrated that LEHR significantly improves agent performance compared to traditional baseline reward functions. The results showed notable advancements in agent adaptability and collaboration, highlighting the framework's ability to facilitate effective learning in complex multi-agent environments.

The findings of this study not only underscore the potential of using LLMs for reward design in MARL but also pave the way for future research in this area. Future work could explore the application of LEHR in more complex environments and investigate further enhancements to reward generation processes. Additionally, incorporating more sophisticated communication strategies among agents could lead to even better cooperative behaviors.

References

1. Bäck, T., et al.: Evolutionary algorithms for parameter optimization–thirty years later. Evol. Comput. **31**(2), 81–122 (2023)
2. Cao, Y., et al.: Survey on large language model-enhanced reinforcement learning: concept, taxonomy, and methods. arXiv preprint arXiv:2404.00282 (2024)
3. Deb, K.: Multiobjective Optimization Using Evolutionary Algorithms. Wiley, Hoboken (2001). michigan State University
4. Devidze, R., Kamalaruban, P., Singla, A.: Exploration-guided reward shaping for reinforcement learning under sparse rewards. In: Advances in Neural Information Processing Systems (NeurIPS) (2022)
5. Foerster, J.N., Farquhar, G., Afouras, T., Nardelli, N., Whiteson, S.: Counterfactual multi-agent policy gradients. In: AAAI Conference on Artificial Intelligence (2017). https://api.semanticscholar.org/CorpusID:19141434
6. Goldberg, D.E.: Genetic algorithms in search optimization and machine learning (1988). https://api.semanticscholar.org/CorpusID:38613589
7. Harutyunyan, A., Devlin, S., Vrancx, P., Nowé, A.: Expressing arbitrary reward functions as potential-based advice. In: AAAI Conference on Artificial Intelligence (2015)
8. Hong, S., et al.: Metagpt: meta programming for a multi-agent collaborative framework. arXiv preprint arXiv:2308.00352 (2023)
9. Ibarz, B., Leike, J., Pohlen, T., Irving, G., Legg, S., Amodei, D.: Reward learning from human preferences and demonstrations. In: Advances in Neural Information Processing Systems (NeurIPS) (2018)
10. Kim, C., et al.: Guide your agent with adaptive multimodal rewards. In: Advances in Neural Information Processing Systems (NeurIPS) (2023)
11. Li, H., et al.: Theory of mind for multi-agent collaboration via large language models. In: Proceedings of the 2023 Conference on Empirical Methods in Natural Language Processing (EMNLP), pp. 180–192. Singapore (2023)
12. Liu, Z., Zhang, Y., Li, P., Liu, Y., Yang, D.: Dynamic LLM-agent network: an LLM-agent collaboration framework with agent team optimization. arXiv preprint arXiv:2310.02170 (2023)
13. Lowe, R., Wu, Y., Tamar, A., Harb, J., Abbeel, P., Mordatch, I.: Multi-agent actor-critic for mixed cooperative-competitive environments. In: Advances in Neural Information Processing Systems, vol. 30 (NIPS 2017) (2017)
14. Lozhkov, A., et al.: Starcoder 2 and the stack v2: the next generation (2024)
15. Ma, H., Luo, Z., Vo, T.V., Sima, K., Leong, T.Y.: Highly efficient self-adaptive reward shaping. arXiv preprint (2024)
16. Ma, Y.J., et al.: Eureka: human-level reward design via coding large language models. In: Proceedings of the International Conference on Learning Representations (ICLR) (2024)
17. Matsunami, N., Okuhara, S., Ito, T.: Reward design for multi-agent reinforcement learning with a penalty based on the payment mechanism. Trans. Jpn. Soc. Artif. Intell. **36**(5), AG21-H_1-11 (2021)
18. Ng, A., Harada, D., Russell, S.J.: Policy invariance under reward transformations: theory and application to reward shaping. In: Proceedings of the Sixteenth International Conference on Machine Learning (ICML), pp. 278–287 (1999)
19. Rashid, T., Samvelyan, M., Schroeder, C., Farquhar, G., Foerster, J., Whiteson, S.: QMIX: monotonic value function factorisation for deep multi-agent reinforcement

learning. In: Proceedings of the 35th International Conference on Machine Learning. Proceedings of Machine Learning Research (PMLR), vol. 80, pp. 4295–4304 (2018)
20. Sun, C., Huang, S., Pompili, D.: LLM-based multi-agent reinforcement learning: current and future directions. arXiv preprint arXiv:2405.11106 (2024)
21. Wang, L., Zhang, Y., et al.: Individual reward assisted multi-agent reinforcement learning. In: Proceedings of the 39th International Conference on Machine Learning. Proceedings of Machine Learning Research (PMLR), vol. 162, pp. 23417–23432 (2022)
22. Wang, Y., et al.: RL-VLM-F: reinforcement learning from vision language foundation model feedback. In: Proceedings of the 41st International Conference on Machine Learning. Proceedings of Machine Learning Research (PMLR), vol. 235, pp. 51484–51501 (2024)
23. Wiewiora, E., Cottrell, G.W., Elkan, C.: Principled methods for advising reinforcement learning agents. In: Proceedings of the 20th International Conference on Machine Learning (ICML), pp. 792–799 (2003)
24. Xie, T., et al.: Text2reward: reward shaping with language models for reinforcement learning. In: Proceedings of the International Conference on Learning Representations (ICLR) (2024)
25. Yu, C., Velu, A., Vinitsky, E., Wang, Y., Bayen, A.M., Wu, Y.: The surprising effectiveness of PPO in cooperative multi-agent games. In: Neural Information Processing Systems (2021). https://api.semanticscholar.org/CorpusID:232092445

Magnetic Core Loss Prediction: A Data-Driven NGO-GRU Model

Ningning Hu[1], Yongqiang Mao[2(✉)], Lanmei Cong[1], Zhaohui Zhang[1], and Ziyue Han[1]

[1] Linyi University School of Automation and Electrical Engineering, Linyi 276000, Shandong, China
[2] State Grid Linyi Power Supply Company, Linyi 276000, Shandong, China
413665318@qq.com
http://www.lyu.edu.cn

Abstract. The loss in magnetic core components is closely linked to their microstructure and operating conditions. Accurate prediction of this loss is crucial for improving the efficiency and power density of power electronics. At present, there are few universally applicable and highly accurate models, making it difficult for the industry to accurately assess magnetic core losses, which in turn affects the evaluation of power converter efficiency. To enhance the accuracy of magnetic core loss prediction models, this paper introduces a new model that combines the Northern Goshawk Optimization (NGO) algorithm with a Gated Recurrent Unit (GRU) network. Initially, historical data related to magnetic core loss are preprocessed, addressing missing values using interpolation and handling outliers using the Ryder's Criterion (3σ rule). During the optimization process, we introduce adaptive weighting factors to enhance search efficiency in the exploration phase of the Northern Goshawk algorithm and incorporate nonlinear convergence factors to balance the capabilities of global search and local development in the development phase. By constructing a GRU model and optimizing key hyperparameters of the network using the improved NGO algorithm, including the number of hidden layer units, learning rate, and training epochs. Finally, comparative analysis of the models shows that the NGO-GRU model outperforms others in terms of prediction accuracy and adaptability, highlighting its strong potential for application in magnetic core loss prediction.

Keywords: Northern Goshawk Optimization(NGO) · Gated Recurrent Unit(GRU) · Ryder's Criterion (3σ rule) · Data prediction · Core Loss Prediction

1 Introduction

In the study of magnetic core losses, there are numerous traditional models, such as the core loss separation model. While these models have shown practical value under certain conditions, they perform poorly under non-sinusoidal waveforms.

Although the IGSE model has improved for non-sinusoidal conditions, it still exhibits significant errors at high frequencies and various waveform excitations.

Currently, magnetic core loss models are mainly divided into two categories: loss separation models and empirical calculation models. the industry lacks a magnetic core loss model that is both broadly applicable and capable of providing highly accurate predictions. This limitation hinders precise evaluation of losses in the design of magnetic components, which, in turn, affects the efficiency estimates of power converters. (Stenglein et al., 2021) proposed a Mn-Zn ferrite core loss model covering a wide frequency range. It spans a broad frequency spectrum but has an accuracy of only 15%, and relies on quasi-static energy loss data at low frequencies. (Barg et al., 2021) proposed a model for calculating iron losses with symmetric trapezoidal flux density waveforms. The model is simple and accurate; however, it exhibits significant errors at low duty cycles and requires additional parameters. (Matsumori et al., 2020) proposed an iron loss calculation method based on Brockmeyer's theory, which simplifies experimental conditions. However, the validation data depend on measurements from specific equipment. From the literature, it is concluded that core losses in magnetic components are influenced by factors such as operating frequency, flux density, excitation waveform, operating temperature, and core material. These factors directly impact energy consumption and operating costs. Therefore, accurate prediction and effective control of core losses are critical.

In recent years, Many scholars have used data-driven methods to develop core loss prediction models, significantly improving the accuracy and robustness of predictions compared to traditional approaches. For example, (Li et al., 2022) introduced MagNet, an open-source database designed for data-driven modeling of magnetic core losses, which facilitates the advancement of data-driven research methodologies. (Dogariu et al., 2021) demonstrate how to retrain a neural network, originally trained for specific magnetic materials and excitation conditions, with a small amount of data using the newly developed core loss dataset, MagNet. This approach simulates the core losses of other materials under similar excitations, thereby reducing data requirements and enhancing the model's multi-tasking capabilities. (Li et al., 2021) Presented MagNet, a computational framework designed for modeling losses in magnetic cores using machine learning techniques. This framework is designed to provide a sophisticated data-driven approach for modeling the characteristics of power magnetic materials. However, its performance on new excitations or material types not included in the training set has not yet been validated. (Serrano et al., 2023) further emphasize the importance of data-driven models for the characteristics of power magnetic materials, analyzing the core losses and hysteresis loops of manganese-zinc ferrites. This analysis quantifies the complexity of modeling core losses and provides guidance for using data-driven approaches to model power magnetic materials. However, the universality and adaptability of the models remain a challenge, especially for new materials or conditions not covered previously.

In summary, developing a magnetic core loss prediction model that can span different material types and operating conditions has become crucial. This paper introduces a magnetic core loss prediction model leveraging the NGO-GRU method. The key contributions are outlined as follows:

- A data-driven magnetic core loss prediction model, NGO-GRU, is proposed, which effectively improves the accuracy of magnetic core loss predictions.
- By introducing adaptive weighting factors and nonlinear convergence factors, the search efficiency of the Northern Goshawk Optimization algorithm is improved during the exploration phase. A GRU model is constructed, and several key hyperparameters of the network are optimized using the enhanced NGO algorithm.
- This study designed multiple comparative experiments, the results of which effectively validate the predictive accuracy of the proposed model, providing strong empirical support for research on magnetic core loss prediction models.

2 Data Sources and Pre-processing

Although traditional magnetic core loss models have value under certain conditions, they generally suffer from accuracy limitations and narrow applicability. Given this situation, the industry is increasingly focused on developing more accurate, data-driven predictive models. Research has shown that core loss is closely related to factors such as material, frequency, waveform, and flux density. Therefore, we collected 12,404 data samples from different materials for training and prediction. Since external factors or human error may cause data loss or anomalies, which would affect data quality, preprocessing is required. The detailed data preprocessing steps are shown in Fig. 1.

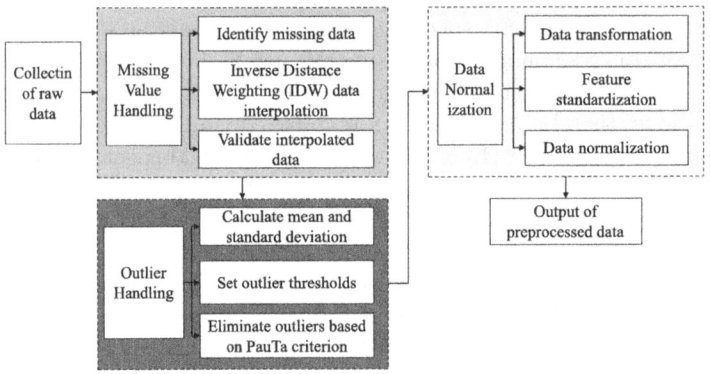

Fig. 1. Flowchart of Data Preprocessing Steps

a. For missing values, suppose each sample data has several pieces of information, of which several are missing, then there exists

$$P_m = \frac{n}{N} \times 100\% \tag{1}$$

In this equation, P_m represents the missing data rate of the sample. If the missing rate is too high, the sample becomes meaningless and can be discarded. Otherwise, missing values can be restored using interpolation. In our research, we utilized the Inverse Distance Weighting (IDW) technique to address missing data and maintain the integrity of the dataset. The details are as follows:

The IDW method operates on the principle that the value at an unsampled location is determined by the weighted average of the known values from surrounding points within a specific area. Where closer points are assigned greater weights. Let $Z^*(x_0)$ be the estimated value of the variable Z at the unsampled point x_0 based on the surrounding data $Z(x_i)$:

$$Z^*(x_0) = \sum_{i=1}^{n} \omega_i Z(x_i) \tag{2}$$

Here, ω_i is the weight for each $Z(x_i)$, and n is the number of neighboring points used for estimation. The formula for inverse distance weighting is:

$$\omega_i = \frac{1/d_i^p}{\sum_{i=1}^{n} 1/d_i^p} \tag{3}$$

Here, d_i is the distance between the estimation point and the sampling points, and p is the exponent parameter.

b. For outlier values, The Ryder's Criterion (3σ rule) is used to eliminate outliers. This criterion is utilized to exclude data points that exceed 3 times the standard deviation from the mean. By applying this criterion, there isa probability of 0.27% of making an error by discarding the true data. as formula (4).

$$v_b = |x_b - \bar{x}| > 3\sigma, 1 \leq b \leq n \tag{4}$$

as formula (5).

$$\hat{\sigma} = s = \sqrt{\frac{1}{n-1} \sum_{i=1}^{n} (x_i - \bar{x})^2} \tag{5}$$

By applying the Ryder's criterion, outliers are removed; however, this criterion cannot be applied in practice when data $n < 10$. The algorithm identifies outliers in the data and handles them appropriately, effectively ensuring data integrity and effectiveness, which strongly supports the development of subsequent classification models.

c. To eliminate the effects of different units of measurement, the processed data is normalized using the following formula:

$$X' = \frac{X - \mu}{\sigma} \tag{6}$$

In the formula: X represents the original data; X' represents the normalized data; μ and σ represent the mean and standard deviation of the original sample data, respectively.

Through the aforementioned processes of handling missing values, detecting outliers, and data normalization, we obtained a complete and consistent dataset, ensuring the quality and applicability of the data. This preprocessed data forms a solid foundation for building subsequent models and effectively predicting magnetic core losses. Some of the data is shown in Table 1.

Table 1. Table of Partial Data After Preprocessing.

Number	Temperature	Frequency	Core Material	Excitation Waveform	Core Loss Value
1	50	199500	Material One	Sine Wave	452.1
2	25	158750	Material One	Trapezoidal Wave	1382332.1
3	90	112180	Material One	Trapezoidal Wave	12338.7
4	70	158510	Material Two	Sine Wave	17979.5
5	25	396820	Material Two	Triangular Wave	86672.7
6	25	50020	Material Three	Sine Wave	6183.1
7	90	125890	Material Three	Triangular Wave	1457906.8
8	25	70780	Material Four	Triangular Wave	32950.2
9	50	89200	Material Four	Triangular Wave	26988.0
10	50	89210	Material Four	Trapezoidal Wave	3382.3

3 NGO-GRU Model Prediction

3.1 Northern Goshawk Optimization (NGO)

In 2022, inspired by the natural hunting strategies of the Northern Goshawk, Mohammad and colleagues introduced the Northern Goshawk Optimization algorithm (NGO). This algorithm models two primary behaviors observed during the goshawk's hunting process: identification and attack, alongside pursuit and evasion.

a. Population initialization:

$$x = lb + rand \times (ub - lb) \tag{7}$$

Here, ub and lb respectively represent the upper and lower position boundaries of the Northern Goshawk.

b. Identification and attack:

During the identify and attack phase, the algorithm's exploratory power is enhanced by randomly selecting prey within the search space, aiming to conduct a global search to locate the best area. Introduce the adaptive weighting factor into the current position of the Northern Goshawk to better update the direction of the optimal position. The mathematical representation of the adaptive weighting factor (γ) is

$$\gamma = \sin(\frac{\pi \cdot t}{2 \cdot T} + \pi) + 1 \tag{8}$$

$$x_{i,j}^{new,P1} = \begin{cases} \gamma x_{i,j} + r(p_{i,j} - Ix_{i,j}), & Fp_i < F_i \\ \gamma x_{i,j} + r(x_{i,j} - p_{i,j}), & Fp_i \geq F_i \end{cases} \quad (9)$$

Here, Represents the latest state of the Northern Goshawk in the dimension. r and I are random numbers used in the identification and attack phase; next, the update of position is determined based on whether the fitness function decreases:

$$X_i = \begin{cases} X_i^{new,P1}, & F_i^{new,P1} < F_i \\ X_i, & F_i^{new,P1} \geq F_i \end{cases} \quad (10)$$

c. Pursuit and evasion:

After the Northern Goshawk initiates an attack, the prey attempts to escape. Therefore, in a series of chases and evasions, the Northern Goshawk continues to pursue the prey. With the Northern Goshawks high-speed maneuverability, it can chase prey in virtually any situation and eventually capture them. This behavioral simulation boosts the algorithms capacity to utilize local searches within the search space, closely associated with the attack position within a radius of . Changing the value of the coefficient can determine the search method of the algorithm.

$$x_{i,j}^{new,P2} = x_{i,j} + R(2r - 1)x_{i,j} \quad (11)$$

$$R = 0.02(1 - \sin(\eta \cdot \frac{t}{T}\pi + \varphi)) \quad (12)$$

$$X_i = \begin{cases} X_i^{new,P2}, & F_i^{new,P2} < F_i \\ X_i, & F_i^{new,P2} \geq F_i \end{cases} \quad (13)$$

The current iteration is represented by t; the maximum number of iterations is represented by T; the new state of the i Northern Goshawk in the second hunting phase is represented by $X_i^{new,P2}$; the new position of the i Northern Goshawk in the j dimension during the second hunting phase is represented by $X_{i,j}^{(t+1)}$; the fitness value in the new state is represented by $f(X_i^{(t+1)})$; the radius r gradually decreases with the increase of iteration t, causing the algorithm to iteratively converge. γ and φ are parameters related to the expression.

3.2 Gated Recurrent Unit (GRU)

The GRU is a type of recurrent neural network that effectively captures the dynamic variations in time series. The fundamental concept involves using a gate mechanism to manage information flow, thereby enhancing the memory of historical information and forecasting future states. The GRU primarily consists of update and reset gates; The update gate determines how much past information is carried into the current state, while the reset gate regulates the amount of past information preserved in the memory of the current state.

The update gate adjusts the frequency of information updates by deciding the ratio of maintaining old hidden states to incorporating new candidate states, thus facilitating a balance between retention and updating in the model and effectively capturing long-term dependencies. The formula for calculating the update gate is as follows:

$$Z_t = \sigma(W^{(z)}x_t + U^{(z)}h_{t-1}) \tag{14}$$

The reset gate controls the impact of the previous hidden state, dynamically discarding irrelevant information to alleviate the vanishing gradient issue and improve processing capabilities for varying sequence lengths. The formula for calculating the reset gate is as follows:

$$r_t = \sigma(W^r x_t + U^r h_{t-1}) \tag{15}$$

Current memory content:

$$h'_t = \tanh(Wx_t + r_t \odot Uh_{t-1}) \tag{16}$$

Final memory of the current time fabric:

$$h_t = z_t \odot h_{t-1} + (1 - z_t) \odot h'_t \tag{17}$$

3.3 NGO-GRU Prediction Model

We present a magnetic core loss prediction model based on the NGO-GRU approach. Initially, historical data undergo preprocessing, including handling missing data via interpolation, outlier removal using the Grubbs' test (3σ), and data standardization. Next, a GRU model is constructed, and an improved NGO optimization algorithm is used for dynamic hyperparameter tuning, optimizing hyperparameters such as the number of hidden units, learning rate, training

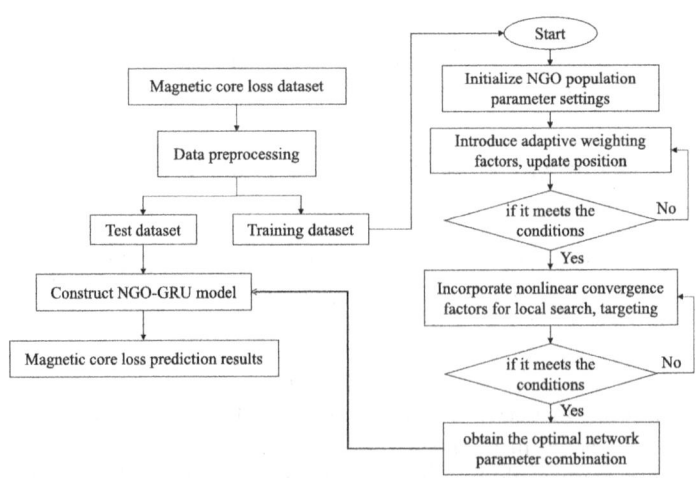

Fig. 2. Flowchart of the NGO-GRU-based Magnetic Core Loss Prediction Model

epochs, and the L2 regularization coefficient. The performance of the NGO-GRU model in predicting magnetic core loss is then analyzed through comparative experiments against other models. Ultimately, the NGO-GRU model outperforms traditional models across various performance metrics, demonstrating higher prediction accuracy. The flowchart of the model is shown in Fig. 2.

4 Experiment

4.1 Parameter Settings

During the training process, the NGO optimization algorithm iteratively optimizes the hyperparameters of the GRU network. NGO, simulating the hunting behavior of hawks, dynamically adjusts the parameters of the GRU model in an attempt to find the global optimum, thereby minimizing the loss function. This paper utilizes the NGO to optimize parameters in the GRU model, such as the number of hidden units, training epochs, L2 Regularization, and learning rate, through network search methods for hyperparameter optimization. The selected hyperparameters used in the experiments are shown in Table 2.

Table 2. NGO-GRU Hyperparameter Settings

Number of Hidden Units	Training Epochs	L2Regularization	Learning Rate
128	200	0.001	0.005

4.2 Experiment 1: Comparison of Optimization Algorithm Performance

To demonstrate the effectiveness of the optimization algorithms discussed in this paper, we conducted comparative experiments to assess fitness among the following optimization algorithms. Each optimizer was configured with the following parameters: search $agents = 30$, $maxiterations = 50$. The experiments compared their performance on the challenging F5 function from the CEC2005 function test set.

From Fig. 3, it can be seen that NGO and PSO converge faster compared to SA and GA. NGO is capable of finding the optimal solution within fewer iterations, making it suitable for optimization problems with high efficiency requirements. It can also approximate theoretical values more closely.

4.3 Experiment 2: Prediction Accuracy Comparison

In this section, we compare the proposed model with commonly used core loss prediction models. All models utilize the same dataset and feature inputs, maintaining consistency in the optimizer, learning rate, batch size, and evaluation

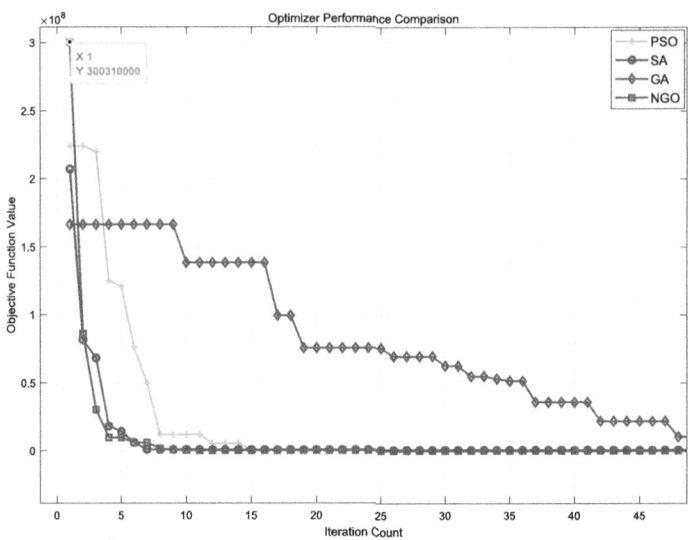

Fig. 3. Precision comparison of different models

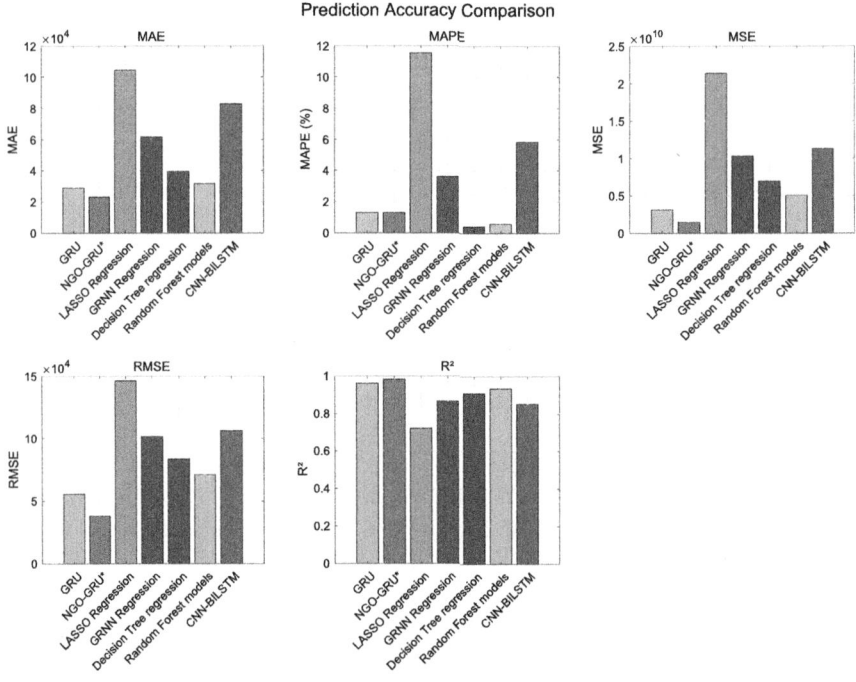

Fig. 4. Precision comparison of different models

metrics. Table 3 illustrates the accuracy comparison among the GRU, LASSO regression, GRNN regression, Decision Tree regression, and Random Forest models (Fig. 4).

Three commonly used regression metrics, root mean square error (RMSE), mean Absolute Error (MAE), and the coefficient of determination (R^2), are used as evaluation criteria. RMSE and MAE are error metrics, with lower values indicating better performance. R^2 measures the predictive power of the statistical model, with values closer to 1 indicating stronger explanatory power of the model for data variance.

Table 3. Comparative experiment of model superiority.

	MAE	RMSE	R^2
NGO-GRU	**23003.3809**	**38058.8906**	**0.98099**
GRU	28715.9863	55721.4727	0.95925
LASSO regression	104423.143	146147.7014	0.71966
GRNN regression	61662.2191	101666.982	0.86434
Decision Tree regression	39381.0428	83358.9224	0.9088
Random Forest	31384.1707	71051.0301	0.93374
CNN-BILSTM	82746.2422	106273.6641	0.85176

As shown in Table 3, The NGO-GRU model excels across multiple important indicators, particularly achieving a coefficient of determination R^2 of 0.98099, indicating its exceptionally strong capability in explaining data variability, far surpassing other models. Compared to the GRU model, NGO-GRU is closer to an R^2 of 1, with significant improvements in error metrics, demonstrating its predictive performance superiority. Compared to LASSO regression and CNN-BILSTM, NGO-GRU exhibits significantly lower errors, while these two models show relatively higher errors, especially in terms of MAE and RMSE. For applications requiring precise prediction and high data interpretability, the NGO-GRU model is a superior choice. To visually observe the predictive effects of different models, the prediction curves of various models are illustrated in Fig. 5 below.

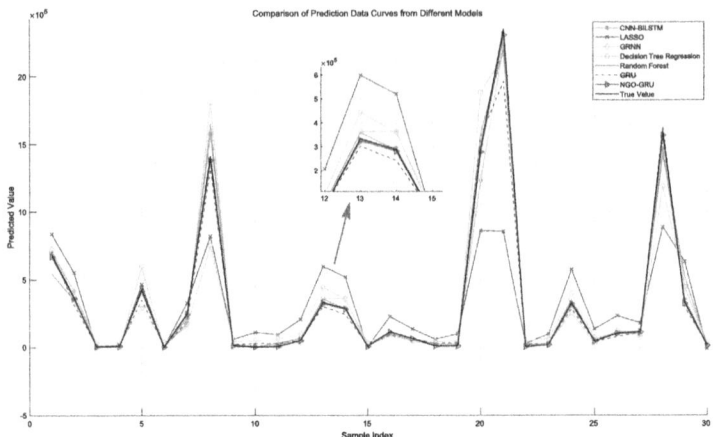

Fig. 5. Comparison of Prediction Data Curves from Different Models

5 Conclusion

Through comparative analysis, the NGO-GRU model demonstrates exceptional performance in predicting core loss. It utilizes the Northern Goshawk Optimization algorithm to enhance the traditional GRU model, making it more effective in handling complex and long-term dependent sequence data. The NGO-GRU model surpasses other comparison models in all performance evaluation metrics, particularly showing efficient predictive capability in accuracy and model fitting. The successful application of this model underscores the importance of integrating advanced optimization algorithms with deep learning technologies, providing an effective tool for precise prediction and management of core losses in power electronics and other technical fields. Furthermore, the development of the NGO-GRU model opens new avenues for the design optimization of magnetic components, helping to improve the overall performance and energy efficiency of devices.

References

Stenglein, E., Dürbaum, T.: Core loss model for arbitrary excitations with DC bias covering a wide frequency range. IEEE Trans. Magn. **57**(6), 1–10 (2021)

Barg, S., Bertilsson, K.: Core loss calculation of symmetric trapezoidal magnetic flux density waveform. IEEE Open J. Power Electron. **2**, 627–635 (2021)

Matsumori, H., Shimizu, T., Kosaka, T., Matsui, N.: Core loss calculation for power electronics converter excitation from a sinusoidal excited core loss data. AIP Adv. **10**(4), 045001 (2020)

Li, H., et al.: MagNet: an open-source database for data-driven magnetic core loss modeling. In: 2022 IEEE Applied Power Electronics Conference and Exposition (APEC), pp. 588–595 (2022)

Dogariu, E., Li, H., Serrano López, D., Wang, S., Luo, M., Chen, M.: Transfer learning methods for magnetic core loss modeling. In: 2021 IEEE 22nd Workshop on Control and Modelling of Power Electronics (COMPEL), pp. 1–6 (2021)

Li, H., Lee, S.R., Luo, M., Sullivan, C.R., Chen, Y., Chen, M.: MagNet: a machine learning framework for magnetic core loss modeling. In: 2020 IEEE 21st Workshop on Control and Modeling for Power Electronics (COMPEL), pp. 1–8 (2020)

Serrano, D., et al.: Why MagNet: quantifying the complexity of modeling power magnetic material characteristics. IEEE Trans. Power Electron. **38**(11), 14292–14316 (2023)

Rodriguez-Sotelo, D., Rodriguez-Licea, M.A., Soriano-Sanchez, A.G., Espinosa-Calderon, A., Perez-Pinal, F.J.: Advanced ferromagnetic materials in power electronic converters: a state of the art. IEEE Access **8**, 56238–56252 (2020)

Steinmetz, C.P.: On the law of hysteresis. Trans. Am. Inst. Electr. Eng. **IX**(1), 1–64 (1892)

Venkatachalam, K., Sullivan, C.R., Abdallah, T., Tacca, H.: Accurate prediction of ferrite core loss with nonsinusoidal waveforms using only Steinmetz parameters. In: 2002 IEEE Workshop on Computers in Power Electronics, 2002. Proceedings, pp. 36–41 (2002)

Deng, J., Wang, W., Ning, Z., Venugopal, P., Popovic, J., Rietveld, G.: High-frequency core loss modeling based on knowledge-aware artificial neural network. IEEE Trans. Power Electron. **39**(2), 1968–1973 (2024)

Shen, X., Martinez, W.: Machine learning model for high-frequency magnetic loss predictions based on loss map by a measurement kit. In: 2023 25th European Conference on Power Electronics and Applications (EPE 2023 ECCE Europe), pp. 1–8 (2023)

Wang, X., Han, Q., Li, J., Jin, Y.: Research on prediction model of epileptic EEG signal based on GRU. In: 2021 International Conference on Electronic Information Engineering and Computer Science (EIECS), pp. 9–12 (2021)

Research on Method of Control Surface Jamming Fault Injection in Fly Test of Fly-by-Wire Aircraft Based on Multiple Control

Lei Ming[✉], Xie Qingping, and Li Yajing

Chinese Flight Test Establishment, Xi'an 710089, China
`leiming061012@163.com`

Abstract. In view of the fact that the Fly-By-Wire (FBW) aircraft needs to evaluate whether the aircraft can continue to fly and land safely after the control surface is jammed through flight tests. This paper proposes a fault injection method of control surface jamming based on negative feedback control and control instruction pre-locking, so as to realize the control surface jamming based on multiple control modes. First, negative feedback method based on control surface instruction is studied. Control surface jamming instruction and negative feedback instruction are injected into flight control system simultaneously by instruction signal generation system to realize control surface jamming in flight test. Secondly, research is conducted on the method of control surface jamming faults injection based on pre-locking control instruction. During flight, the control instruction locking mode is switched, and an external control surface jamming signal is injected into the flight control system to achieve the flight test control surface jamming. Through modeling simulation, flight control iron bird platform testing, and ground test verification, the results show that the second control surface jamming control mode based on control instruction pre-locking is better and more versatile, and the final flight test adopts the second method to realize efficient and safe control surface jamming flight test.

Keywords: fly-by-wire flight control system · external instruction control · negative feedback control · control instruction locking · control surface jamming flight test

1 Introduction

Article 25.671 (c) of the General Provisions of Part 25 of the Civil Aviation Regulations of China stipulates that "Analysis, testing, or a combination of both must be used to demonstrate that, in the event of any of the following faults or jamming in the flight control system and control surfaces (including trimming, lift, resistance, and sensation systems) within the normal flight envelope, the aircraft can continue to fly and land safely without special driving skills or physical exertion. Possible functional abnormalities must have only a minor impact on the operation of the control system and must be easily countered by the pilot: Any jamming in the control positions used normally during takeoff, climb, cruise, normal turning, descent, and landing. Unless this type of jamming is shown to have a very low probability or can be alleviated. If the probability of the flight control device slipping to an unfavorable position and subsequent jamming is not extremely low, then this slip and jamming must be considered" [1]. The loss, malfunction of flight attitude control, flutter suppression, and other functions, or the non-directive operation of the flight control system can have a serious impact on the safety of the aircraft, crew, and passengers, leading to a deterioration of the aircraft's handling characteristics [2] and even catastrophic failure consequences [3, 4]. The development and evolution of flight control systems have always been aimed at improving aircraft flight safety and performance [5]. With the increasing demand for flight safety in FBW aircraft, control surface jamming flight tests have gradually become a necessary flight test subject for FBW aircraft, especially FBW transport aircraft, to confirm that the aircraft has sufficient safety.

The control surface jamming flight test is a high-risk flight test subject, with high risks and technical difficulties. There are many control surface jamming tests conducted on mature flight simulators abroad, and only Bombardier Aerospace Group's CRJ-700 aircraft has undergone airworthiness verification flight tests for control surface jamming. The publicly available information is only a list of test subjects, without specific implementation plans or technical solutions for control surface jamming. Wang Ting et al. [6] proposed three schemes to achieve primary control surface jamming, including fixing control surface scheme, control device disengagement scheme, and jamming fault simulation scheme. The fixing control surface scheme can only be implemented before takeoff, and the jamming angle of the control surface cannot be arbitrarily changed during flight; The detachment scheme of control devices results in poor accuracy of control surface jamming and difficulty for pilots to operate; The scheme of jamming faults simulation is affected by flight state fluctuations and the accuracy of jamming is difficult to guarantee. Moreover, this method is designed for a specific flight control system and lacks universality.

This article proposes a control surface jamming fault injection method based on negative feedback control and pre-locking control instruction for FBW aircraft flight tests, to solve the problems of difficulty in changing the control surface jamming angle during flight, poor control surface jamming accuracy, and high pilot control difficulty faced by fixing control surface schemes, control device disengagement schemes, and jamming fault simulation schemes. The method of control surface jamming faults injection of the FBW aircraft flight test based on multiple control modes includes the control mode method based on negative feedback of control instruction and the control mode method based on pre-locking control instruction. The latter control mode method introduces jamming control mode on the basis of normal flight control mode to achieve pre-locking of control instruction. Through the modeling and simulation, flight control iron bird platform testing, and on-board ground testing, the results show that the second control surface jamming fault injection control mode based on pre-locking control instructions is more effective. Finally, the second method was adopted in the flight test to achieve the control surface jamming flight test.

2 Introduction to Aircraft FBW Control System and Control Surface Jamming Flight Test

2.1 The FBW Control System

The most typical feature of a FBW control system is the fly by wire operation system. In order to meet the increasingly sophisticated overall design requirements of aircraft and the flying needs of pilots [7], the aircraft control system is constantly updated and improved, gradually shifting from mechanized control to electronic control, known as the Fly-By-Wire control system. The FBW control system uses sensors to convert the pilot's control signals and displacement into electrical signals, which are then processed by a computer and transmitted to the servo to control the aircraft's attitude. The FBW control system is an important milestone in the development history of aircraft. It can reduce the workload of pilots, improve the comfort of flying and sitting, and optimize the design of aircraft. The FBW control system mainly consists of various sensor components, input devices, flight control computers, servos, and electrical transmission lines that collect signals [8, 9], and is interconnected with important systems such as display systems, alarm systems, flight parameters systems, electromechanical systems, and engine control systems.

The first fighter aircraft to use FBW control system was F-111, which began flying in 1964 with a simulated system with triple redundancy and mechanical backup. In 1971, the YF-16 light fighter jet from the United States successfully made its maiden flight, becoming the first aircraft to adopt the FBW control system without any mechanical backup. This also marked the beginning of the development period of FBW control system flight without any mechanical backup. Since the 1980s, FBW control systems have undergone significant development, and many newly developed military and civilian aircraft have adopted FBW control systems. The civilian aircraft A320, which was put into operation in 1986, adopted a digital FBW control system with mechanical backup. The civilian aircraft B777, which was put into operation in the mid-1990s, also adopted a digital FBW control system. This aircraft was the first Boeing aircraft in the United States to use the FBW flight control system [10]. The flight control system of B777 mainly consists of three parts: the FBW control system (main flight control system), the automatic flight control system, and the automatic throttle system. The following mainly introduces the primary flight control system closely related to the control surface jamming fault injection method proposed in this article. The main components of the B777 aircraft's primary flight control system include [5]: Primary Flight Computer (PFC), Actuator Control Electronic (ACE), Power Control Unit (PCU), and displacement sensors. The main components of the primary flight control system are shown in Fig. 1.

During manual operation, the movement of the control panel, pedals, and speed brake handle controlled by the pilot is sensed by corresponding displacement sensors and converted into analog electronic signals. These signals are sent to ACE and converted into digital signals to be sent to PFC by ACE. PFC receives signals from atmospheric data inertial reference units and other systems, calculates and generates corresponding control instructions based on the designed control law, and sends them to ACE. ACE sends these instructions to systems such as power control components. The power components

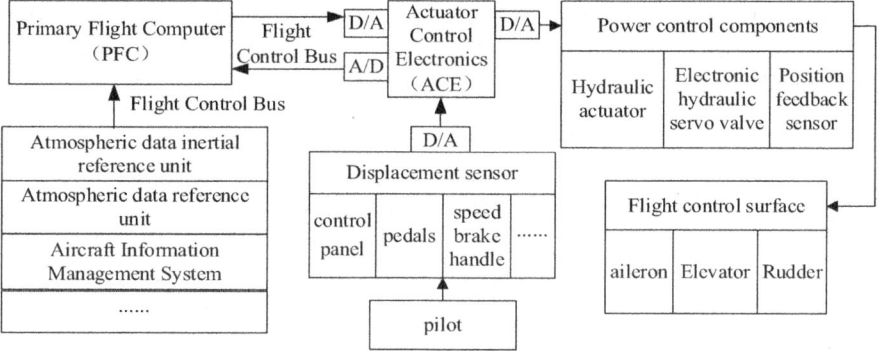

Fig. 1. Main components of the primary flight control system

control the aircraft control surface, including ailerons, elevators, rudders, and spoilers, ultimately achieving flight control.

2.2 Control Surface Jamming Flight Test

Article 25.671 (c) of the General Provisions of Part 25 of the Civil Aviation Regulations of China clearly stipulates that control surface failure and jamming tests must be conducted to complete flight tests related to aircraft maneuverability and stability, meeting airworthiness requirements. The key issue of the control surface jamming flight test is how to keep the control surface jamming at the required position and quickly restoring it to a normal and controllable state if necessary.

The Fig. 1 shows that ACE not only receives input from the pilot, but also from the primary flight control system. The generation of this input is more complex, not only related to the aircraft's speed, but also to parameters such as the aircraft's three-dimensional attitude angle, three-dimensional angular velocity, and overload, and has higher authority. When conducting the control surface jamming test, the aircraft not only needs to perform stable level flight, but also needs to perform longitudinal pushing/pulling rod step control, left /right turn at different slopes, and quickly establish a fixed slope on the other side while flying at a fixed slope on one side. At this time, the input instructions generated by the primary flight computer cannot be ignored.

3 Control Mode Method of Negative Feedback Control Surface Jamming Fault Injection Based on Control Surface Instruction

3.1 Working Principle of the Method

The negative feedback control surface jamming fault injection method based on control surface instructions uses the analog acquisition function of the instruction signal generation system to collect the original control instruction signal A of the control surface. The instruction signal generation system performs inverse sign operation on the original control instruction signal A to obtain the -A signal, and then adds the signal to the target signal B to obtain the control surface jamming fault injection instruction signal C (C = B-A). Utilize the signal superposition port of ACE, as shown in Fig. 2, to output the C signal into ACE.

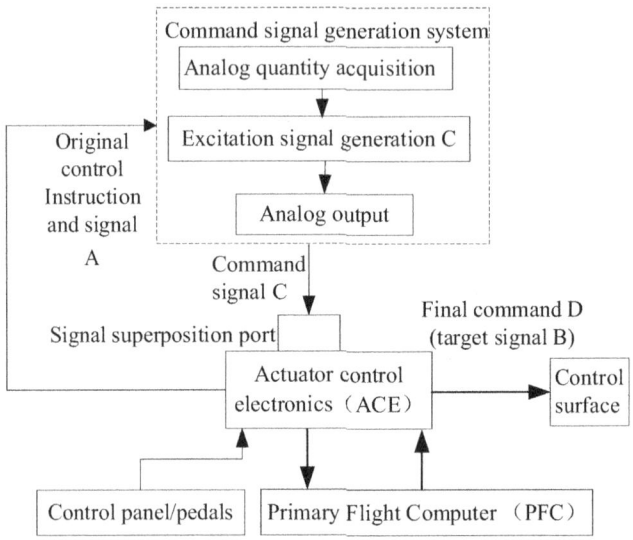

Fig. 2. Principle of negative feedback control surface jamming fault injection method based on control instruction of the control surface

ACE adds the original control instruction signal A of the control surface and the injection instruction signal C of the control surface jamming fault to obtain the final control surface jamming instruction D(D = A + C). From the above, it can be seen that D = A + C = A + B − A = B which means that the control surface moves according to the target signal B. The working principle of this method is shown in Fig. 2.

3.2 Simulation Verification

Taking the aircraft aileron control surface as an example, verify the feasibility of the above method. Firstly, identify the circuit model of the aileron servo control input to the control surface position response using test data. The control surface circuit models are all shown in Eq. (1).

$$H_1 = \frac{a_n s^{n-1} + a_{n-1} s^{n-2} + \cdots\cdots + a_2 s + a_1}{b_n s^{n-1} + b_{n-1} s^{n-2} + \cdots\cdots + b_2 s + b_1} \quad (1)$$

The final identified control surface circuit model is:

$$H_a = \frac{10^{-3}(0.4s^6 + 1.8s^5 - 6.8s^4 + 9.5s^3 - 6.3s^2 + 1.9s - 0.2)}{s^7 - 5.56s^6 + 13.74s^5 - 19.64s^4 + 17.64s^3 - 10s^2 + 3.33s - 0.5} \quad (2)$$

The comparison between the simulation results and test results of the aileron identification model is shown in Fig. 3.

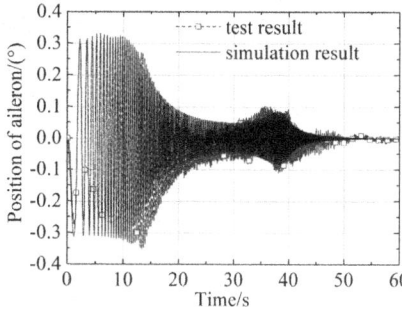 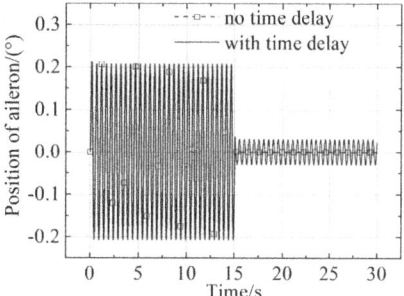

Fig. 3. Comparison between simulation results and test results of aileron

Fig. 4. Simulation results of aileron jamming

Based on the control surface circuit model, establish a negative feedback control surface jamming fault injection simulation model based on control surface control instruction, and carry out control surface jamming simulation. Figure 4 shows the entire process from the analog acquisition of the instruction signal generation system to the analog output. The figure shows the aileron jamming states of no time delay and with time delay when the input of the aileron control is a sine signal. Figure 4 shows that when there is no time delay, the control surface can be completely locked. When there is a time delay, the control surface will exhibit a certain amplitude of oscillatory response.

3.3 Flight Control Iron Bird Platform Verification

During the test of control surface jamming of the flight control iron bird platform, control instruction is added to the aircraft through the control panel of the flight control iron bird platform to observe the response of the aircraft control surface. When performing aileron jamming, control the aircraft ailerons through the control panel. Figure 5 shows the application of a 6° jamming signal to the aileron, while Fig. 6 shows the application of a –6° jamming signal to the aileron. It can be seen that when there is no aileron control instruction, the aileron control surface position can be jammed at the required value; When there is an aileron control instruction, the position of the aileron control surface can generally be jammed at the required value, with slight fluctuations at the jammed position. The position fluctuation during aileron jamming is shown in Fig. 7, with values ranging from –0.15° to 0.15°. The aerodynamic impact of this fluctuation on the aircraft can be almost ignored. The schematic diagram in Fig. 8 shows the time delay between the final control instruction signal of the control surface and the aileron control instruction of the operation panel when the aileron is jammed at 6°.

After the final control surface jamming instruction is added, when the control panel does not apply the aileron control instruction, the aileron can stabilize at a fixed deflection; When the steering control panel disturbance input, the ailerons exhibit a finite fluctuation response with the disturbance input. There is a time-delay between the negative feedback in the final control surface jamming instruction and the aileron instruction on the control panel, resulting in the jamming signal being unable to completely cancel out the control instruction on the control panel, causing the aileron to exhibit disturbance

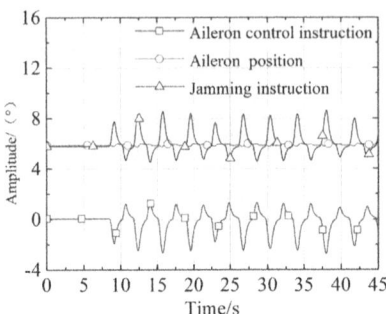

Fig. 5. Iron Bird test 6° aileron jamming

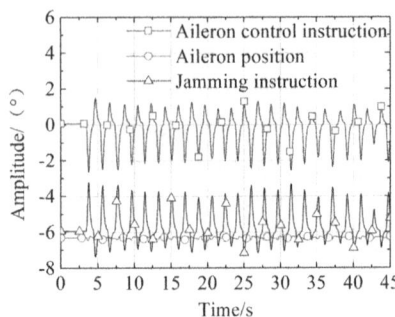

Fig. 6. Iron Bird test –6° aileron jamming

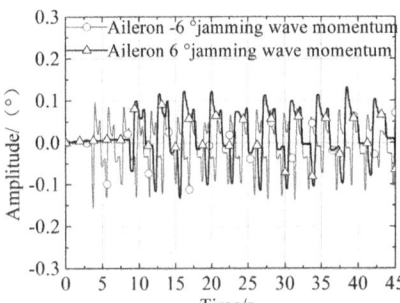

Fig. 7. Position fluctuation of aileron jamming

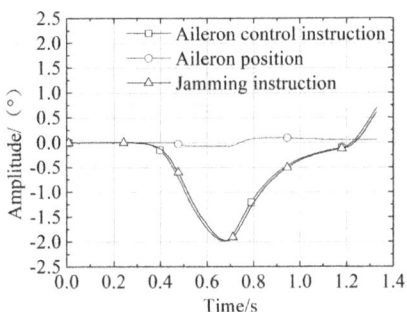

Fig. 8. Time delay of 6° right aileron step excitation

response. When the time delay is fixed, the slope of the signal curve, i.e. the rate at which the pilot manipulates the control surface, determines the magnitude of the disturbance amplitude. From Fig. 7, it can be seen that the fluctuation is relatively small and does not affect the effect of the control surface jamming. On the other hand, during the control surface jamming flight test, the aircraft's stable level flight, left and right roll, and left and right rudder control movements are generally slow movements, without significant maneuvering or aggressive control. Therefore, this method can be used to achieve control surface jamming flight tests, and compared to small FBW aircraft, large FBW aircraft have slower control speeds, so the effectiveness of this method will be better.

4 Control Mode Method of Control Surface Jamming Faults Injection Based on Pre-locked Control Instruction

4.1 Working Principle of the Method

This article proposes a second control mode method, that is the control surface jamming fault injection method based on pre-locked control instruction. On the basis of the original FBW control mode, an FBW control surface jamming control law is designed to lock the original control instruction of the control surface at a fixed value, without being affected by the control panel/pedals and feedback instruction from the flight control system. On this basis, the instruction signal generation system injects jamming instruction signals through the ACE signal superposition port, causing additional deflection of the control surface and achieving jamming of the control surface in flight test. During the flight test, the FBW control mode switch is used to switch the control mode. Based on the above, the principle of the control surface jamming fault injection method based on control law switching is formed as shown in Fig. 9. During the flight process, the FBW control mode switch is first used to switch the control law from normal control law to control surface jamming control law, achieving control surface instruction locking. When the jamming signal is added to the control surface by instruction signal generation system, the control surface will not be affected by the feedback instructions from the control panel/pedals and the flight control system, thus achieving control surface jamming.

In order to enable the aircraft to enter and exit the jamming state of the control surface and have as little transient response as possible, the addition and exit of jamming instructions should be a gradual process, and cannot be a completely step adding and exiting form. To this end, the form of adding and exiting the jamming instruction is designed as a first-order link $h(t) = 1 - e^{-\frac{1}{T}t}$, as shown in Fig. 10, and the time constant should be proportional to the jamming angle of the control surface.

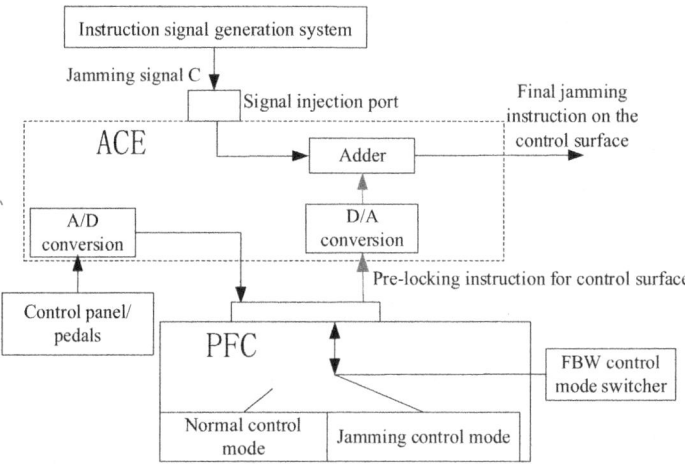

Fig. 9. Principle of control surface jamming fault injection method based on control mode switching

4.2 Flight Control Iron Bird Platform Verification and On-Board Ground Test

The control surface jamming test of flight control Iron Bird platform and on-board ground test consist of two parts, including the control surface jamming test without control panel/pedal input and the control surface jamming test with control panel/pedal input, with different jamming angles for each test. Jamming tests on control surfaces such as ailerons, elevators, and rudders was conducted. Figure 11 shows the test results

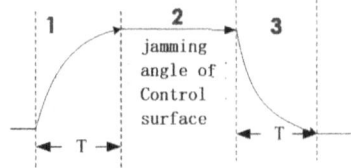

Fig. 10. Form of adding and exiting jamming signals

of the –3.5° rudder jamming on flight control iron bird platform. The figure shows that when the pedals are not operated, the rudder can be jammed at the required deviation. When there is pedal input, the rudder can still be jammed at a fixed deviation, and the first-order jamming instruction adding and exiting are achieved. Figure 12 shows the result of –7° elevator jamming. The figure shows that when there is no control panel input, the elevator can be jammed at the required deflection, and when there is control wheel input, the elevator can still be jammed at a fixed deflection.

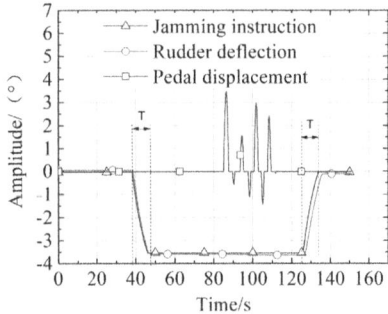
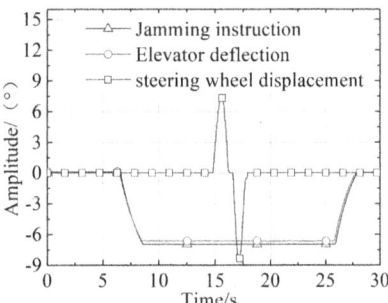

Fig. 11. Result of –3.5° Rudder jamming test (flight control iron bird platform)

Fig. 12. Result of –7° Elevator jamming test (ground test)

4.3 Control Surface Jamming Flight Test

Flight test was conducted to verify the control mode method of control surface jamming faults injection based on pre-locking control instruction. The result of the –8° elevator jamming flight test is shown in Fig. 13, and the result of 10° aileron jamming flight test is shown in Fig. 14. The figures show that the elevator and aileron can stably jam at the required position in the test, without being affected by the control instructions of the control panel.

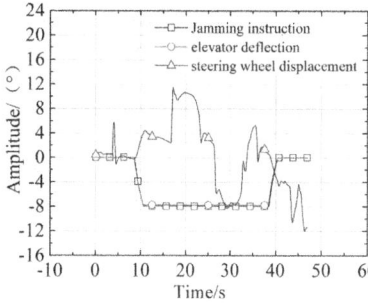
Fig. 13. Flight test results of –8° elevator jamming

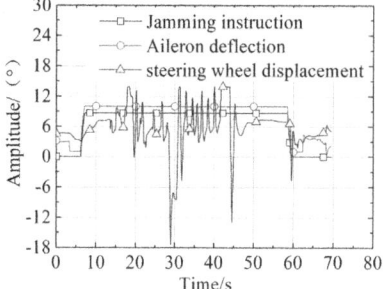
Fig. 14. Flight test results for 10° aileron jamming

5 Conclusion

On the basis of external control instruction, this article studies the method of control surface jamming fault injection based on negative feedback control and pre-locking control instruction for FBW aircraft flight tests, realizes control surface jamming based on multiple control modes, and completes verification of modeling and simulation, flight control iron bird platform test, on-board ground test, and final flight test verification. The following conclusions are drawn:

1. On the basis of external control instruction, the control mode method of control surface jamming faults injection in FBW aircraft flight tests based on negative feedback control can achieve control surface jamming during flight tests. Due to the time delay between the negative feedback control instruction in the jamming signal and the original control instruction of the control surface, the control surface cannot be stabilized at a constant value. However, the fluctuation is small and will not affect the stability characteristics of the aircraft, nor will it affect the jamming flight test effect of the aircraft control surface. Moreover, compared with small FBW civil aircraft, the control speed of large FBW civil aircraft is slower, so this method will work more effective.
2. On the basis of external control instruction, the control mode method for the control surface jamming fault injection of the FBW aircraft flight test based on pre-locking control instruction can achieve control surface jamming of the flight test. This method ultimately was verified by the flight test and was able to stabilize the control surface at a fixed position. The method has strong universality and better performance, and can be applied to the control surface jamming flight test of large and small FBW aircraft.

References

1. Civil-Aviation-Administration-of-China. CCAR-25-R4Chinae Civil-aviation-regulations: Part 25-Airworthiness-standards: Transport-category-airplanes. Beijing: Civil-Aviation-Administration-Of-China (2011). (in-Chinese)

2. Zhong, J.J., Wang, L.X., Le, T.: The HQRM method applicable to FBW military transport aircraft. In: 8th Youth Science and Technology Forum Proceedings of the Chinese Aeronautical Society. Beijing, pp. 232–238. (2018). (in Chinese)
3. Luo, L., Lu, Z., Song, H., et al.: Safety analysis for fly-by-wire system based on fault injection model. Acta Aeronautica et Astronautica Sinica, **44**(09), 277–290 (2023). (in Chinese)
4. Xue, Y.: The depression methodology against loss of airfoil control and airfoil oscillation for Fly-By-Wire flight control system in civil aircrafts. Shanghai Jiao Tong University, pp. 2–5 (2020). (in Chinese)
5. Wang, H.: Analysis on security influence factors in architecture design of fly-by-wire flight control system. In: 17th China Aerospace Measurement and Control Technology Annual Conference Proceedings, pp. 255–259. Measurement & Control Technology, Xi'an (2020). (in Chinese)
6. Wang, T., Zou, Q., Liu, Y.: Flight test method study on civil aircraft primary flight control surfaces jamming. Adv. Aeronaut. Sci. Eng. **5**(03), 343–349 (2014). (in Chinese)
7. Wu, S.T., Fei, Y.H.: Flight Control System, pp. 78–82 Beihang University Press, Beijing, (2009). (in Chinese)
8. Lian, L.: Research on fly by wire control system. In: Aviation Test Technology Academic Proceedings in 2017, pp. 83–85. Measurement & Control Technology, Wuhan (2017). (in Chinese)
9. Li, L.: Analysis of fly by wire flight control system and flight test. In: 16th China Aerospace Measurement and Control Technology Annual Conference Proceedings, pp. 89–92. Measurement & Control Technology, Beijing (2019)
10. Yu, T.: Boeing 777 Flight Control System. Civil Aircraft Design and Research, vol. 02, pp. 4–6 (1998). (in Chinese)

Tactile-Based Manipulation for Wire Following

Jiazhen Cai[1], Jing Cui[1(✉)], and Zhongyi Chu[2]

[1] College of Mechanical and Energy Engineering, Beijing University of Technology, Beijing 101100, China
cuijing@bjut.edu.cn
[2] School of Instrumental Science and Opto-Electronics Engineering, Beihang University, Beijing 100191, China

Abstract. Flexible wires are widely used in mechanical and electrical products. As a kind of deformable linear object (DLO), the posture of the wire in the gripper changes dynamically especially during the following process in a limited operating space. For the following manipulation of flexible wires, we design a compact parallel gripper based on a flexible hinge and a tactile sensor is installed on the gripper. And we adopt a suitable algorithm for wire posture estimation. In addition, we build a planar pulling model of the wire-gripper and construct the relationship between the end position adjustment and the wire posture to keep the wire at the center of the sensor. Experiments are carried out to verify the accuracy of posture estimation algorithm and the effect of following wire.

Keywords: deformable linear object · flexible hinge · tactile sensor · posture estimation

1 Introduction

Deformable linear object (DLOs) are targets such as wires and cables that one dimension is especially bigger than other dimensions [1]. The operation of DLOs is common in both industrial and non-industrial tasks, especially in the automotive and aerospace sectors, and involves a variety of applications such as wire following [2, 3].

Wire following manipulation is a complex task because the wire is narrow and the operating space is limited, and the wire's posture changes dynamically while the gripper is following its profile [4]. Many researchers have made out many classic work on DLO operation. Zhu et al. [5] wrap the wire around a hook using an end-effector attached to the end of the wire and another fixed actuator that passively allows the cable to slide in order to pull out longer wire. The system allows to slide but loses the ability of sensing and controlling the state of the wire while sliding. Jiang et al. [6] use a gripper with rollers in the jaws to track cables in a wire harness. This gripper uses a spring to passively adjust the gripping force to accommodate wires of different sizes. It can sense and control the force perpendicular to the translational motion along the cable in order to follow the wire. However, it is only suitable for wires with a diameter of 10 mm or more, and it is different to feel the gripping force for wires with small diameters. In able to obtain the status of the clamped wire, She et al. [7] and Wilson et al. [4] adopt parallel gripper with a GelSight sensor which is an optical tactile sensor to perform the wire operations. They extract the contacting region by thresholding the depth image and get the principal axis of the wire on the sensor by Principal Component Analysis (PCA) to get the orientation and posture. Vision-based tactile sensors convert haptics into vision by visualizing the deformation of the contact surface. Due to its high spatial resolution, this type of sensor shows unique advantages and has been successfully used for different robotic manipulation tasks such as contour tracking and in-hand manipulation. But visual perception of DLO state estimation is very difficult in workspaces where the space of objects changes dynamically and is often occluded. In addition, force and tactile sensors are used in many wire manipulation articles. Abegg et al. [8] use force-torque sensor to detect the changes in the state of contact, such as the movement of a rope from free space to contacting a rigid object. Yue and Henrich [9] use a force-torque sensor to detect the vibration frequency of the wire before counteracting the vibration. But it can't obtain the posture information. Accordingly, some posture estimation ways used in dexterous manipulation of electrical wires for assembly field are instructive to us [1, 10–12].

The grippers used in previous work either lack sensing capability or were only suitable for large diameter wires, and the sensors used for wire posture estimation are easily affected by the environment. To address the above challenges, we design a compact gripper for thin wire and only rely on haptic information for posture measurement. Our work is summarized as follows. First, we design a parallel gripper based on parallelogram mechanism, which is based on flexure hinge. And we equip the gripper with a tactile sensor to endow it with the ability to feel. Second, we propose a recognition strategy based only on the use of tactile data to estimate the wire's posture in gripper. Third, we build the plane pulling model of wire-gripper and determine the relationship between the gripper position adjustment and the system state of the wire-gripper.

2 Hardware of the Parallel Gripper

2.1 The Gripper's Mechanism Design

In terms of the gripper design, the work of She [7] et al. has a great inspiration for us: parallelogram mechanisms are used to generate lateral displacement and crank-slider mechanisms are used to drive the parallelogram mechanisms. In this work, we use the combination of them to promote parallel gripping. To make a compact actuation mechanism, we adopt a linear actuator for actuation.

As shown in Fig. 1: includes a linear actuator, slider, two cranks, two parallelogram mechanisms, two fingers with the tactile sensor. One end of the crank rod is connected to the slider and the other is coupled with the rocker of the parallelogram mechanism. And the motor's pushrod is connected with the other end of the slider by screw thread as a whole. The sensor is embedded inside the finger.

2.2 The Gripper's Mechanism Dimensions

To design suitable dimensions for our gripper, the guidelines we extract are as follows: (1) the max opening distance of the gripper is set at 18 mm which corresponds to

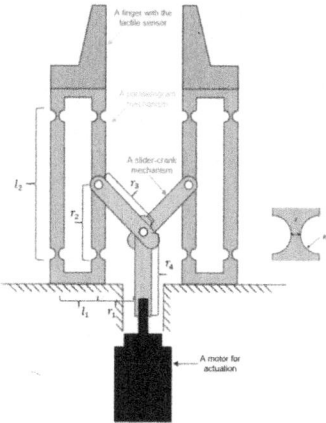

Fig. 1. Mechanism structure design

a displacement of 9 mm for each finger. (2) the parallelogram mechanism should be designed to accommodate the sensor. (3) reduce the gripper's overall size and weight as much as possible. We design the gripper with the dimensions in Table 1.

Table 1. Dimension detail

Parameters	Value	Parameters	Value
r_1	9 mm	l_1	16 mm
r_2	24 mm	l_2	48 mm
r_3	12 mm	T	1 mm
r_4	11 mm	R	2 mm

2.3 The Gripper's Flexible Hinge Design

We use flexible hinge to replace the rigid parallelogram mechanism to an equivalent complaint parallelogram mechanism. In this way, it can greatly reduce the number of parts to simplify the assembly process and provide the same kinematics functionality. So the design of the flexible hinge is especially important. In order to achieve high motion accuracy and appropriate rotation range, we adopt circular flexible hinge. From the formula in (1), we can know the rotational stiffness of the flexible hinge contains four critical parameters: E(the material's Elastic modulus), b(the flexible hinge's thickness), R and t(described in Fig. 1).

$$K_a = \frac{Eb}{\int_0^\pi \frac{12R \sin \alpha}{(2R+t-2R \sin \alpha)^3} d\alpha} \quad (1)$$

Common materials used to make flexible hinges include 7075-T6AL, 60Si2Mn, Polypropylene (PP) and PLA for 3D printing. According to the formula in (1) and combined with the length of l_2, we can get the required force F for different materials to reach the desired displacement(maximum displacement on each side: d = 9 mm)

$$F = \frac{K_\alpha \arcsin \frac{d}{l_2}}{r_2} \qquad (2)$$

Next, we will simulate four materials and combinations of different R, t values in Abaqus with a force that achieves the desired displacements to observe the maximum stress values at the deformation. From the above simulation results (shown in Fig. 2), it can be seen: when the hinges rotate a sufficient angle, the maximum stress of the hinges made of 7075-T6AL and PP is higher than the permissible stress. The hinges made of 60Si2Mn need extremely big force. Hinges made of PLA satisfy the stress requirement and the required force is less than others (Table 2).

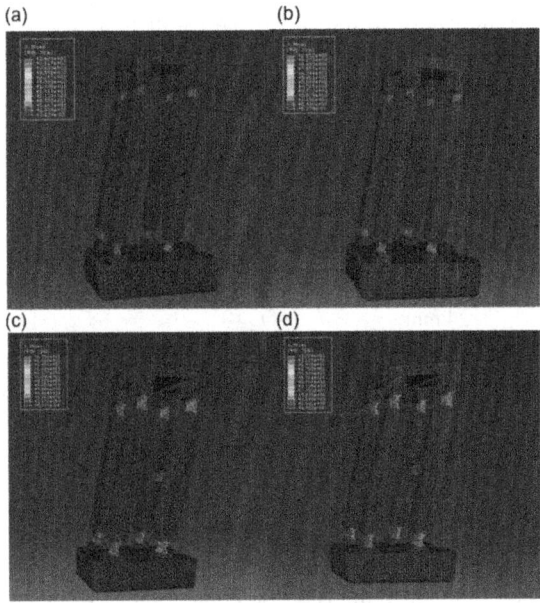

Fig. 2. Simulation results: (a) 7075-T6AL (b) 60Si2Mn (c) PP (d) PLA

2.4 The Sensor's Composition and Detection Principle

In the previous work at our laboratory, a small and ultra-thin capacitive tactile sensor arrays with classical 3-D measurement structures (C3DMS) has been developed [13, 14]. As a example, a C3DMS is shown in Fig. 3(a). Assuming that the bottom of the sensor is fixed and the surface is subjected to 3-D force, the dielectric layer will deform and drive the common electrode to move we use a tactile sensor with 64 cells to detect the

Table 2. Simulation parameter

Material	E(GPA)	R, t(mm)	Force(N)	Max Stress	Permissible Stress
7075-T6AL	71 GPA	1.75,0.5	27.3 N	340 MPA	328 MPA
60Si2Mn	206 GPA	1.75,0.5	79.3 N	1012 MPA	1375 MPA
PP	2.5 GPA	2,1	5.24 N	17.7 MPA	12 MPA
PLA	3.3 GPA	2,1	3.92 N	23.3 MPA	85 MPA

contact position. According to the principle of parallel-plate capacitors, tangential force changes the plate area of the C3DMS, generating differential capacitance output but not affecting the plate distance, while normal force changes the plate distance, altering the sum of the four capacitors, but almost not affecting the differential output. The principle of 3-D measurement based on parallel-plate capacitors can be described as

$$\begin{cases} \Delta C_n = \varepsilon_c A/(d_0 - \Delta d) - \varepsilon_c A/d_0 \\ \Delta C_s = \varepsilon_c \Delta A/d_0 = \varepsilon_c \Delta x L/d_0 \end{cases} \quad (3)$$

where ΔC_n and ΔC_s are the changes in capacitance when the capacitor is subjected to normal force F_n and shear force F_s, respectively. ε_c is the dielectric constant of the dielectric layer, $A = L \times L$ is the facing area of the capacitor (L is the side length of the square electrode), d_0 is the initial thickness of the dielectric layer, Δd is the change in plate area (the pole plate is moved horizontally by the shear force F_s). According to Hooke's law

$$\begin{cases} \Delta d/d_0 = F_n/(A_Z E) \\ \Delta x/d_0 = F_s/(A_Z G) \end{cases} \quad (4)$$

where A_Z is the area of the dielectric layer, E is the elastic modulus, G is the shear modulus, and $\Delta d \ll d$. Equation (3) can be written as

$$\begin{cases} \Delta C_n = \varepsilon_c L F_n/(d_0 A_Z E) = K_n F_n \\ \Delta C_s = \varepsilon_c L F_s/(d_0 A_Z G) = K_s F_s \end{cases} \quad (5)$$

where K_n and K_s are the linear coefficients. This equation explains that within a certain range, the normal deformation of the ultrathin dielectric layer will be much less than its original thickness. Therefore, in principle, a parallel-plate capacitor can achieve linear 3-D measurement. According to the measurement principle of C3DMS, the force output can be measured as

$$\begin{cases} F_x = k_x(-\Delta C_1 + \Delta C_2 - \Delta C_3 + \Delta C_4) \\ F_y = k_y(-\Delta C_1 - \Delta C_2 + \Delta C_3 + \Delta C_4) \\ F_z = k_z(\Delta C_1 + \Delta C_2 - \Delta C_3 + \Delta C_4) \end{cases} \quad (6)$$

where F_x, F_y, and F_z are the triaxial components of the load in the sensitive region; and k_x, k_y, and k_z reflect the linear relationship between the capacitance and the force under the applied load. As illustrated in Fig. 2(b), the sensing unit is expanded horizontally and vertically to achieve the 4×4 cm^{-2} 3-D tactile array output. In particular, the sensor area of the array sensor is designed to be within 1 cm^2. The internal structure of the tactile sensor is shown in Fig. 2(c). The gray part is the PCB substrate, embedded with orange sensing electrodes, and the green part is the designed interpenetrating topology dielectric layer that enhances the sensitivity of the sensor with lower equivalent elastic modulus and equivalent area(where the purple color is the overlapping part of the dielectric layer and the yellow color is the nonoverlapping part). As shown in Fig. 2(d), the tactile array further enhances the electrode density to 8×8 cm^{-2}. This provides us with the assurance of sensing the wire profile. So in this work, we use the tactile sensor to achieve accurate measurement of the wire position.

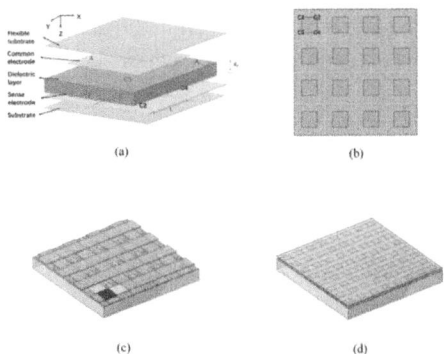

Fig. 3. (a) C3DMS (b) Tactile sensor array with C3DMS (c) Internal structure of the tactile sensor array (d) 8*8 tactile sensor array

3 Estimation and Adjustment of the Grasped Wire's Posture

3.1 Wire Posture Estimation Algorithm

We are well inspired by the approach of Pirozzi et al. [10] to perform posture estimation, on which we use more cells for estimation and adopt a more appropriate data fitting method. We define a physical reference frame (O, X, Y) at the tactile sensor's surface (shown in Fig. 4). For each cell, we use a couple of coordinates (x_i, y_i) for positioning. According to the physical distances of the each cell from the sensor center, the x-coordinates of the columns are –4.375 mm, –3.125 mm, –1.875 mm, –0.625 mm, 0.625 mm, 1.875 mm, 3.125 mm, 4.375 mm, from left to right, and the y-coordinates of the rows are 4.375 mm, 3.125 mm, 1.875 mm, 0.625 mm, –0.625 mm, –1.875 mm, –3.125 mm, –4.375 mm, from top to bottom. Hence, the 64 cells are organized as a matrix, where each cell can be identified by the coordinates of its row and column indices, as shown in Fig. 4.

Due to the regular shape of the wires and their inherent hardness (as shown in Fig. 5 (a) not (b)), so estimating the grasped wire's posture means to estimate the k and b

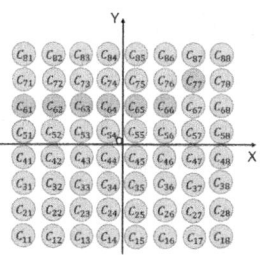

Fig. 4. The sensor flame

parameters characterizing longitudinal axis of the wire. In our work, to estimate the grasped wire's posture, we consider the wire as a straight line and model it in the frame with equation (Fig. 7)

$$y = kx + b \qquad (7)$$

where k and b are the two key parameters to be estimated according to the tactile data.

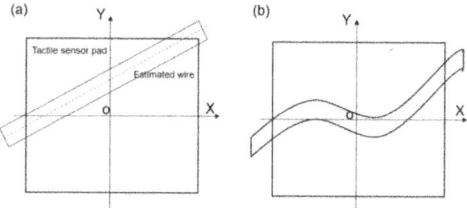

Fig. 5. Different wire's postures

The process for the wire posture estimation is constituted by two steps. In the first step, the centroid coordinates for each column are calculated. In the second step, we calculate the model parameters k and b by applying the Random Sample Consensus (RANSAC) method to the data set constituted by the coordinates of the column centroids.

In practical terms:

Step 1:

the y coordinates y_j^c of the column centroids are calculated by tactile datas as

$$y_j^c = \frac{\sum_{i=1}^{8} y_i \Delta C_{ij}}{\sum_{i=1}^{8} \Delta C_{ij}} \quad j = 1, ..., 8 \qquad (8)$$

where y_i is the mechanical y coordinate of the i-th row, y_j^c is the centroid of the j-th column, ΔC_{ij} is the measured capacitance variation corresponding to the cell in coordinate (x_i, y_j). The x_j^c is the mechanical x coordinate of the j-th column centroid.

Step 2:

After the first step, we can gain a data set constituted by the coordinates (x_j, y_j^c) of the 8 column centroids. Then the Random Sample Consensus (RANSAC) method is adopted to estimate the model parameters k, b.

3.2 Wire Posture Adjustment Relationship

The actual model of the wire-following operation is shown in Fig. 6(left). We focus on the posture of the wire in the gripper. We build this problem as a planar pulling model. As shown in Fig. 6(right), the area where the wire is in contact with the tactile sensor is represented as a planar rigid sliding object. We describe the position and orientation of the wire relative to the x-axis of the sensor frame with b and θ. In addition, we ulteriorly define the angle α which is between the center of the wire in the gripper and the fixed end. The whole system is determined by three parameters [b, θ, α]. We also define ϕ as the required pulling angel according to the state of the system. θ and b can be obtained from the previous sensing information, and α is calculated by (9).

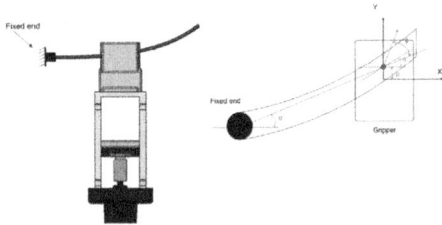

Fig. 6. (Left) Wire following diagram (Right) Planar wire-pulling model

$$\alpha = \arctan \frac{y_m + b}{x_m} \tag{9}$$

where x_m and y_m are the distances that the gripper moves along the x-axis and y-axis, respectively. We creat the linear model as (10).

$$\dot{x} = Ax + Bu \tag{10}$$

where x = [b, θ, α], u = [ϕ], A and B are the coefficients of the model. We use the A and B matrices from the model to build an LQR controller. Further, we can get the optimal feedback gain K, which provides us with the optimal control quantity in reverse, u = −k*x. The LQR controller parameters we use are Q = [5,1,0.1] and R = [0.1], since it is more important to regulate y and u to ensure that the wire is not crossed out. We convert the ϕ to the gripper moving distance along x and y axes in the world frame with the following kinematic relation:

$$\begin{cases} d_x = d(\cos(\alpha + \phi)) \\ d_y = d(\sin(\alpha + \phi)) \end{cases} \tag{11}$$

where d is the predefined magnitude of the moving distance of the gripper.

4 Experiment

4.1 Wire Posture Estimation Experiment

To verify the effectiveness of the above method, the experiment device is set up as shown in Fig. 8. The tactile sensor is attached to the bottom plate which is ensured to be coaxial with the top plate through 4 dowel pins. A transparent PMMA plate is adhered to the center circular area of the top plate to facilitate the photographing and observation of the measured wire and we use a standard weight of 300 g for pressing. The tactile sensor datas are transmitted to the PC through serial communication and dynamically visualized through Python.

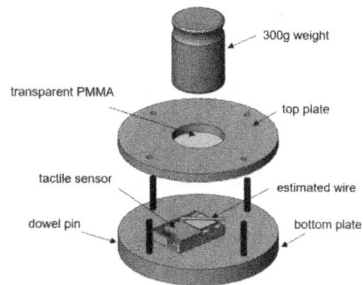

Fig. 7. Posture estimation experiment setup

The experiment procedure is to place the wire above the tactile sensor at different tilt angles and offset distances, and take pictures to record through the transparent PMMA plate, then press a 300 g standard weight above the PMMA plate and pick it up for 5 s. The above loading process is repeated four times and the pressing time is about 30 s respectively. In the above process, we record the capacitance change value of each cell. Due to the excessive number of cell, we only extract the ones with significant changes to show in the Fig. 8. And the values of k and b computed by the above algorithm the data are shown in the Fig. 9.

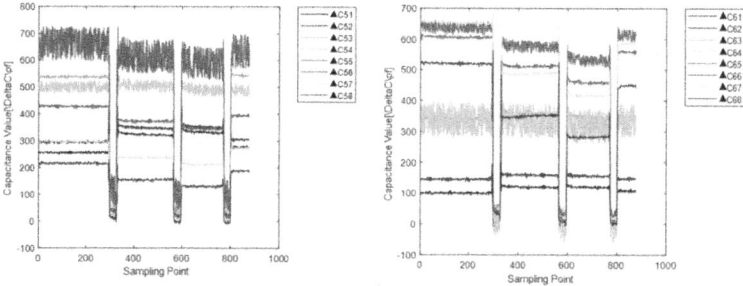

Fig. 8. Capacitance change values for special taxel

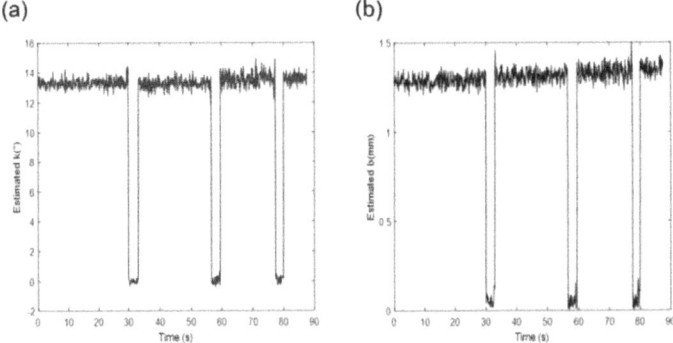

Fig. 9. Estimated k and b in the experiment

The calculation results of the algorithm in the other two experiments are shown in Fig. 10 and Fig. 11. We compare the posture information obtained from the image processing of the pro-taken photos as standard values with the values obtained from the above algorithm to verify its validity. The truth number and algorithm results are shown in Table 3. In the first experiment, the maximum difference between the measured value of k and the standard value is 1.23°, and the maximum difference of b is 0.1 mm. In the second and third experiments, the maximum difference of k is 1.37°and 2.18°, respectively, and the maximum difference of b is 0.25 mm and 0.22 mm, respectively. From the above data, it can be seen that the calculation results of the algorithm are comparable to those of the image processing method, and the maximum difference generally occurs at the movement of the weights.

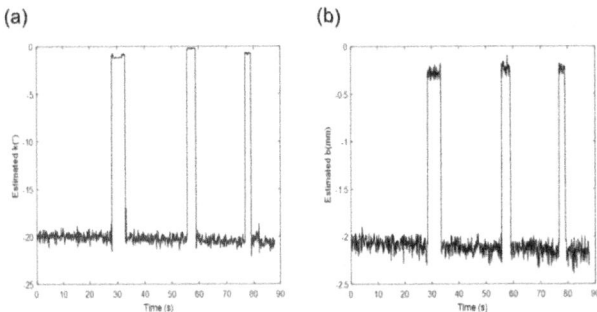

Fig. 10. Estimated k and b in the second experiment

4.2 Wire Following Experiment

We integrated the gripper at the end of the Franka robotic arm and performed wire-following experiments. The diagram of the experimental setup and two kinds of USB wires are shown in Fig. 12. The white one is wire1 with 3.5 mm diameter and the black

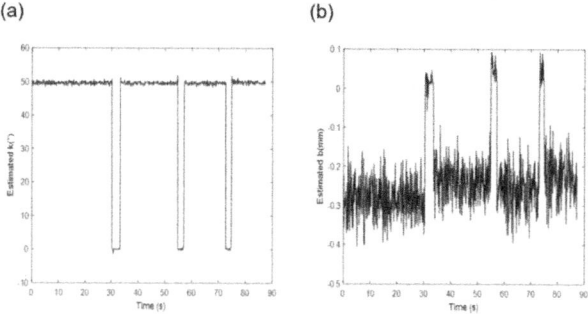

Fig. 11. Estimated k and b in the third experiment

Table 3. Comparison of the estimation method

Num	Truth	Estimation
1	k: 14° b: 1.31 mm	k: 12.77°— 14.12° b: 1.25 mm—1.41 mm
2	k: −20° b: −2.14 mm	k: −21.5°— −18.63° b: −2.39 mm—−2.03 mm
3	k: 49° b: −0.31 mm	k: 48.65°— 51.18° b: −0.40 mm—−0.09 mm

one is wire2 with 2.5 mm diameter. We secured one end of the wire and use the gripper to grasp the other end to move. During the moving process, the wire will be randomly shifted towards different directions due to its own properties. Simultaneously, we record the trajectory of the end-effector (position along the axes in Fig. 6) and wire's posture of θ and b.

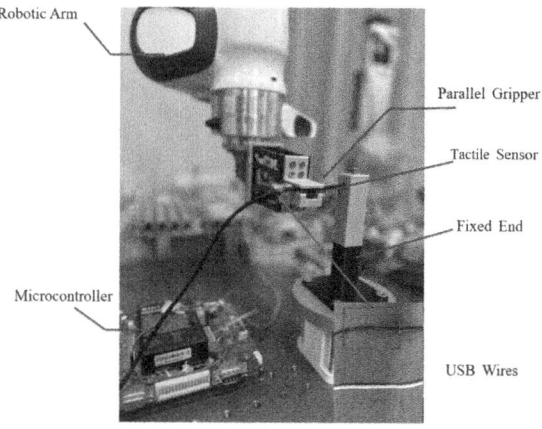

Fig. 12. The diagram of the experimental setup

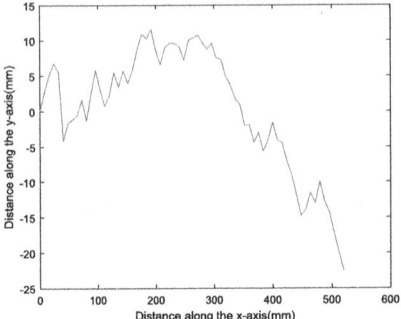

Fig. 13. The trajectory of wire1 following

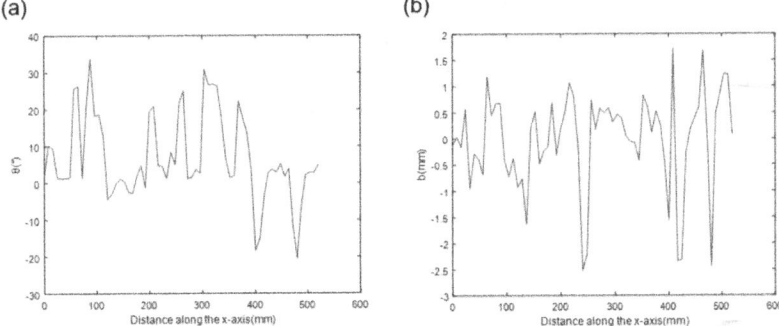

Fig. 14. The parameters variation of wire1 following

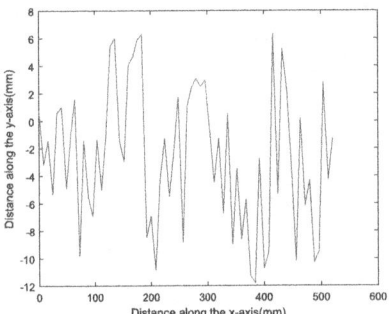

Fig. 15. The trajectory of wire2 following

The length of following experiment is about 52 cm. The datas of two different diameters wires during the following experiments are shown in Fig. 13, 14, 15 and 16. From the data, it can be seen that when θ and b change significantly, the robotic arm will make corresponding changes in position. Both θ and b can be roughly adjusted to around 0 (the center of the sensor) when they are offset. And larger diameter wire is slightly more adjustable than smaller diameter wire. This may be due to the fact that

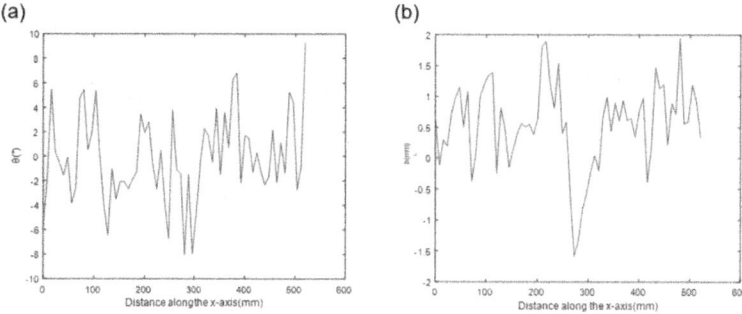

Fig. 16. The parameters variation of wire2 following

larger diameter cables have more internal padding and are relatively harder, which makes them less prone to deformation and therefore easier to slide on the surface of the finger. The following process is described as Fig. 17. It can be seen that the wire is relatively smooth at the beginning, as the length increases, the wire occurs irregular deflection due to its own softness. Overall, the posture adjustment during wire following is good.

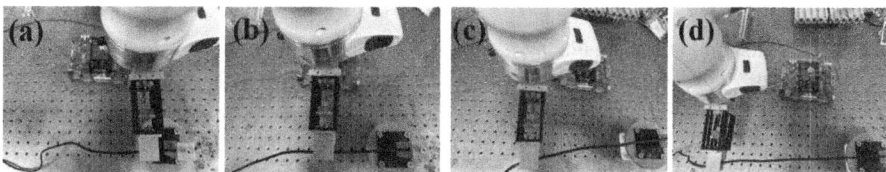

Fig. 17. The following process

5 Conclusions

In this paper, to solve the requirement of the wire following manipulation in limited space, we design a compact and easy-to-assemble parallel gripper to grasp different wires. We propose an algorithm that relies purely on tactile information for posture estimation, and we establish the regulation strategy to keep the wires from falling. Finally, the effectiveness of the posture estimation algorithm and the regulation strategy is experimentally verified.

Acknowledgments. This research is supported by National Natural Science Foundation of China (Grant 52375006) and the Ministry of Science and Technology of China (Grant 2018AAA0102900, the New Generation of Artificial Intelligence Technology Innovation 2030 Major Project).

References

1. Monguzzi, A., Pelosi, M., Zanchettin, A.M., Rocco, P.: Tactile based robotic skills for cable routing operations. In: 2023 IEEE International Conference on Robotics and Automation (ICRA), London England, pp. 3793–3799 (2023)
2. Sanchez, J., Corrales, J.-A., Bouzgarrou, B.-C., Mezouar, Y.: Robotic manipulation and sensing of sensing of deformable objects in domestic and industrial application: a survey. Int. J. Robot. Res. **37**, 688–716 (2018)
3. Yin, H., Varava, A., Kragic, D.: Modeling, learning, perception, and control methods for deformable object manipulation. Sci. Robot. **6**, eabd8803 (2021)
4. Wilson, A., Jiang, H.L., Lian, W.Z., Yuan, W.Z.: Cable routing and assembly using tactile-driven motion primitives. In: 2023 IEEE International Conference on Robotics and Automation(ICRA), London England, pp.10408–10414 (2023)
5. Zhu, J.H., Navarro, B., Passama, R., Fraisse, P., Crosnier, A., Cherubini, A.: Robotic manipulation planning for shaping deformable linear objects with environmental contacts. IEEE Robot. Autom. Lett. **5**, 16–23 (2020)
6. Jiang, X., Nagaoka, Y., Ishii, K., Abiko, S., Tsujita, T., Uchiyama, M.: Robotic recognition of a wire harnedd utilizing tracing operation. Robot. Comput. Integr. Manuf. **34**, 52–61 (2015)
7. She, Y., Wang, S.X., Dong, S.Y., Sunil, N., Rodriguez, A., Adelson, E.: Cable manipulation with a tactile-reactive gripper. Int. J. Robot. Res. **40**, 1385–1401 (2021)
8. Abegg, F., Remde, A., Henrich, D.: Force and vision based detection of contact state transitions. Adv. Manuf.111–134 (2000)
9. Yue, S., Henrich, D.: Manipulating deformable linear objects: Sensor-based fast manipulation during vibration. In: proceedings 2002 IEEE International Conference on Robotics and Automation (ICRA), pp.2467–2472 (2002)
10. Pirozzi, S., Natale, C.: Tactile-based manipulation of wires for switchgear assembly. IEEE-ASME Trans. Mechatron. **23**, 2650–2661 (2018)
11. Palli, G., Pirozzi, S.: A tactile-based wire manipulation system for manufacturing applications. Robotics **8**, 46 (2019)
12. Cirillo, A., Laudante, G., Pirozzi, S.: Tactile sensor data interpretation for estimation of wire features. Electronics **10**, 1458 (2021)
13. Hu, Z.K., Chu, Z.Y., Wang, Y.J., Cui, J.: Two-stage supe-resolution for classical 3-D tactile sensor arrays with DT-driven enhancement. IEEE Trans. Instrum. Meas. **73**, 1–12 (2024)
14. Hu, Z.K., Wang, Y.J., Feng, K.M., Chu, Z.Y., Cui, J., Sun, F.C.: A viscoelastic compensator for force sensors with soft materials. IEEE Trans. Instrum. Meas. **72**, 1–10 (2023)

Author Index

A
Aiju, Li I-3

B
Bi, Ying II-268

C
Cai, Jiazhen I-319
Cai, Shipei II-292
Cao, Mingwei I-254
Chen, Hang II-170
Chen, Ken I-186
Chen, Qili I-243, I-265
Chen, Xiangyong II-55, II-161
Chen, Zhuxiang I-75
Cheng, Shaoqi I-272
Chu, Zhongyi I-319
Cong, Lanmei I-296
Cui, Huixia I-152
Cui, Jing I-319

D
Deng, Fang II-222
Du, Lipeng I-243
Du, Pengcheng II-37
Du, Yanwei I-214
Du, Yu II-37

F
Feng, Jiayue I-272
Feng, Xiaoliang II-89
Fu, Deqian II-182
Fu, Yijun II-268
Fu, Zunwei II-20, II-47

G
Gao, Ji Xian I-144
Gao, Jixian I-231
Gong, Ping II-215
Gong, Shuli II-235

Guo, Ming II-55
Guo, Weifeng II-268
Guo, Zijie II-255

H
Hai, Zhao II-280
Han, Weixin I-131
Han, Ziyue I-296
Hao, Li II-280
He, Liping II-134, II-198
He, Yuesheng I-63, II-149
Hu, Ningning I-296
Huo, Qingyun I-27

J
Ji, Yang I-88
Jia, Chun II-246
Jia, Senping I-152
Jiang, Du I-75, II-74
Junhong, Zhang II-280

K
Kang, Yuntong I-254

L
Lei, Chen II-134, II-198
Lei, Yidi II-170
Letian, Zhao II-280
Li, Boao II-74
Li, Fanjun I-272
Li, Jianmin I-283
Li, Meng I-231
Li, Rui II-120
Li, Wenjie II-111
Li, Yuxuan I-27
Li, Zhangliang II-182
Li, Zhaoxing I-178
Liang, Jing II-268
Liang, Jinling II-292
Liang, Tian I-117

Lin, Wenshuai II-255
Liu, Jianhua II-161
Liu, Longjing II-198
Liu, Mingtao I-27
Liu, Shufen II-161
Liu, Ying II-74
Liu, Yunzhuo II-99
Liu, Ziqi II-182
Long, Liu II-280
Lyu, Yi I-186

M

Ma, Bing I-15
Ma, Jianliang II-222
Ma, Xiaolei II-170
Ma, Zhaowei II-99
Mao, Yongqiang I-296
Ming, Lei I-308

N

Niu, Yifeng II-99

P

Pan, Guangyuan I-54, I-243
Pan, Qiang I-15
Peng, Jinzhu I-231

Q

Qi, Mingdong I-214
Qiao, Xunyu I-96
Qiao, Zhongli II-182
Qin, Yaoyao II-198
Qingping, Xie I-308
Qiu, Jianlong II-55, II-182
Qiu, Sen I-254

S

Shan, Liqiu I-39
Shan, Wu I-117
Shan, Xiaohan I-283
Shen, Huang I-88
Shen, Xinyu I-272
Shi, Shaoguang II-66
Sun, Junfeng II-304
Sun, Wenqi I-54

T

Tang, Xinyang I-165
Tao, Bo I-75
Tian, Yunlong I-39

W

Wang, Bing I-165
Wang, Chunxiang I-165
Wang, Dawei I-186
Wang, Fei I-63, II-149
Wang, Haozhu II-74
Wang, Jinjiang I-198
Wang, Keqing II-182
Wang, Mengyun II-99
Wang, Shuo II-89
Wang, Tong I-243
Wang, Xing I-39
Wang, Xixin II-66
Wang, Yaonan I-231
Wang, Yifang II-27
Wang, Yizhen II-47
Wang, Youchuan II-304
Wang, Yunyu I-27
Wang, Zhelong I-254
Wang, Zheng II-20
Wang, Zhujun II-235
Wei, Yuan I-283
Wen, Qinglin II-246
Wu, Chengcheng I-186
Wu, Mingkuo I-231

X

Xue, Jinqiu II-3

Y

Yajing, Li I-308
Yang, Chenxi I-165
Yang, Haichuan II-304
Yang, Haobin I-96
Yang, Jian II-222
Yang, Meng I-27
Yang, Ming I-63, I-165, II-149
Yang, Yicong I-265
Yang, Yuqi II-99
Yao, Haoran I-254
Yao, Shuanglong I-39
Yao, Ximing I-15
Yao, Yuan I-15
Yining, Du I-3

Author Index

Yu, Hongliang II-55
Yu, Junyao II-235
Yuan, Wei I-165
Yue, Zhengyuan II-182
Yun, Juntong I-75, II-74
Yusen, Zhao I-117

Z

Zhang, Ancai I-54, II-235, II-304
Zhang, Congcong II-47, II-66
Zhang, Fang Fang I-144
Zhang, Guoqiang I-198
Zhang, Haiying I-96
Zhang, Hangwei I-15
Zhang, Heng I-214
Zhang, Jiachang II-246
Zhang, Jiage II-134
Zhang, Jiyuan II-292
Zhang, Lele II-222
Zhang, Rui II-120
Zhang, Wenyin II-3
Zhang, Xin I-27
Zhang, Xingzhao I-178
Zhang, Yiman I-186
Zhang, Yingxin I-152
Zhang, Yuan I-131
Zhang, Zhaohui I-296
Zhang, Zhengqiang II-111
Zhang, Zilong II-255
Zhao, Feng II-161
Zhao, Letian II-170
Zhao, Lin I-243
Zhao, Nan I-231
Zheng, Wenbo I-54, II-304
Zheng, Xiaohong II-255
Zhou, Kun II-55
Zhou, Qi I-198
Zhou, Xinhao I-243
Zhou, Yangyang II-170
Zhu, Yunlong II-20
Zhu, Zhen Yu I-144
Zong, Zihan I-63, II-149

Made in the USA
Monee, IL
03 May 2026